Conservation of Plant Genetic Resources

Saikat Gantait • Paweł Chmielarz
Editors

Conservation of Plant Genetic Resources

Strategies, Progress and Prospect

Springer

Editors
Saikat Gantait
Crop Research Unit (Genetics & Plant Breeding)
Bidhan Chandra Krishi Viswavidyalaya
Mohanpur, West Bengal, India

Paweł Chmielarz
Institute of Dendrology
Polish Academy of Sciences
Kórnik, Poland

ISBN 978-981-96-8330-7 ISBN 978-981-96-8331-4 (eBook)
https://doi.org/10.1007/978-981-96-8331-4

This Springer imprint is published by the registered company Springer Nature Singapore Pte Ltd.
The registered company address is: 152 Beach Road, #21-01/04 Gateway East, Singapore 189721, Singapore

Preface

The beneficial plants are now threatened, and they are being collected more extensively ex situ in gene banks rather than in situ. However, the management or preparation of plant materials is a barrier to easy ex situ conservation. Under this context, this book discusses the most verified in vitro biotechnological treatments that have resulted in a shift in the system of conserving and using plant genetic resources. Collection, characterization, in vitro culture setup, multiplication, short- to long-term preservation, and distribution are the basic know-how that contribute to an in vitro conservation system, and they are thoroughly covered in each chapter. Commencing with in vitro strategies for seed germination of rare and endangered plant species and conservation of them using slow-growth approaches, to the use of cell suspension cultures, protoplasts, pollen banking, embryos and embryogenic axes, somatic embryos, and embryogenic callus/cultures as the plant materials for cryopreservation have been comprehensively described. Furthermore, dedicated chapters provide deeper insights into prolonged conservation methods that were successful for ornamentals, trees (gymnosperms and angiosperms), fruit, medicinal plants, etc. Finally, cutting-edge techniques like AI and multi-omics, successfully utilized in conservation, have been covered meticulously. In due course, featuring 10 diverse chapters with a comprehensive bibliography contributed by subject experts and scientists across 12 different countries (Belgium, Brazil, China, Finland, India, Iran, New Zealand, Poland, Spain, Türkiye, UK, USA), this book provides immediate access to scientific data on the in vitro biotechnological interventions for the conservation and utilization of plant genetic resources worldwide. Aimed at scientists, researchers, ecologists, dendrobiologists, botanists, and students, it offers invaluable insights into the recent achievements and future avenues toward biotechnological approaches for ex situ conservation of plant genetic resources.

We are highly obliged to Springer Nature Singapore Pte. Ltd. for bringing out this project with successful accomplishment.

Mohanpur, West Bengal, India — Saikat Gantait
Kórnik, Poland — Paweł Chmielarz

Contents

Editors and Contributors

About the Editors

Saikat Gantait is an Assistant Professor in Genetics and Plant Breeding at Crop Research Unit, Bidhan Chandra Krishi Viswavidyalaya. He gained advanced research experience while working as a Research Associate in Govt. of India-funded project, followed by working as a Post-doctoral Researcher at Universiti Putra Malaysia. He guided/co-guided multiple MSc and PhD students. His 18-year research works include molecular markers, plant tissue culture, polyploidy induction, synthetic seed, and cryopreservation of medicinal and ornamental plant germplasms. He is acting as an Editorial Board Member in multiple journals of Springer Nature and Elsevier. He reviewed 450+ manuscripts as an ad hoc reviewer in 75+ eminent journals of Springer, Elsevier, Taylor and Francis, PLoS, Wiley, etc. His name has been included in the list of Top 2% Scientists Worldwide (2023 and 2024) by Stanford University. To date, he has published 175+ articles/chapters in peer-reviewed journals/books (cumulative 375+ SCI Impact Factor, 5200+ Google Scholar citations) and has edited 4 books.

Paweł Chmielarz is a Professor in Natural Sciences at the Institute of Dendrology Polish Academy of Sciences. His research focuses on seed biology, in vitro culture, cryopreservation, germplasm ex situ conservation, and epigenetics. He served as an expert on cryopreservation in the Scientific Board of the Kostrzyca Forest Gene Bank. He was a recipient of the Award of the Minister of the Environment, Warsaw. He examines seeds and seedlings of forest tree species in changing climates. Working on the plant in vitro culture, he cloned 800-year-old monumental oaks. He serves as a reviewer for international journals. He is a member of European Cost Action 21157 COPYTREE (2022–2026) and Global Conservation Consortia's Cryopreservation Working Group and a Principal Investigator of national and international scientific grants. He has guided MSc and PhD students. He has published 90+ articles/chapters in peer-reviewed journals/books (with 1450+ Google Scholar citations with 22 H-index) on ex situ conservation of plant germplasms.

Contributors

Tuija Aronen Natural Resources Institute Finland (LUKE), Savonlinna, Finland

Daniel Ballesteros Department of Botany and Geology, University of Valencia, Burjassot, Spain

Botanical Garden of the University of Valencia, Valencia, Spain

Trait Diversity and Function, Royal Botanic Gardens, Kew, Wakehurst, UK

Jose L. Canón-Prieto Department of Botany and Geology, University of Valencia, Burjassot, Spain

Subhash Chander ICAR-National Bureau of Plant Genetic Resources (NBPGR), Pusa Campus, New Delhi, India

Hongying Chen Kunming Institute of Botany, Chinese Academy of Sciences, Kunming, China

Yelda Özden Çiftçi Faculty of Science, Department of Molecular Biology and Genetics, Gebze Technical University, Gebze, Kocaeli, Türkiye

Josimare Q. Conceição Universidade Federal do Recôncavo da Bahia, Cruz das Almas, BA, Brazil

Diego F. M. Cortes Universidade Federal do Recôncavo da Bahia, Cruz das Almas, BA, Brazil

Eva M. R. Costa Universidade Federal do Recôncavo da Bahia, Cruz das Almas, BA, Brazil

Flavio Jose da Silva Dantas Laboratorio de Radio e Fotobiologia, Universidade do Estado do Rio de Janeiro, Rio de Janeiro, RJ, Brazil

Jamine de Almeida Pettinelli Núcleo de Biotecnologia Vegetal, Universidade do Estado do Rio de Janeiro, Rio de Janeiro, RJ, Brazil

Bianka de Oliveira Soares Instituto Federal de Educação, Ciência e Tecnologia do Rio de Janeiro, Rio de Janeiro, RJ, Brazil

Nicolas Desoignies Phytopathology, Microbial and Molecular Farming Lab, Center D'Etudes et Recherche, Appliquée-Haute Ecole Provinciale du Hainaut Condorcet, Ath, Belgium

Everton Hilo de Souza Universidade Federal do Recôncavo da Bahia, Cruz das Almas, BA, Brazil

Lloyd Donaldson Forest Genetics and Biotechnology, Scion, Rotorua, New Zealand

Morgan Fays Phytopathology, Microbial and Molecular Farming Lab, Center D'Etudes et Recherche, Appliquée-Haute Ecole Provinciale du Hainaut Condorcet, Ath, Belgium

Raquel Folgado The Huntington Library, Art Museum, and Botanical Gardens, San Marino, CA, USA

Jeremy Foster Chicago Botanic Garden, Glencoe, IL, USA

Rachel Fatima Gagliardi Núcleo de Biotecnologia Vegetal, Universidade do Estado do Rio de Janeiro, Rio de Janeiro, RJ, Brazil

Ravi Gowthami ICAR-National Bureau of Plant Genetic Resources (NBPGR), Pusa Campus, New Delhi, India

Teresa Hazubska-Przybył Institute of Dendrology, Polish Academy of Sciences, Kórnik, Poland

Onildo N. Jesus Embrapa Mandioca e Fruticultura, Rua Embrapa, s/n, Chapadinha, Cruz das Almas, BA, Brazil

Fundação Getúlio Vargas, FGV Conhecimento, Rua Jornalista Orlando Dantas, Botafogo, Rio de Janeiro, RJ, Brazil

Amir Khoddamzadeh Department of Earth and Environment, Institute of Environment, Florida International University, Miami, FL, USA

Paulina Kosek Kostrzyca Forest Gene Bank, Miłków, Poland

Jana Krajnakova Forest Genetics and Biotechnology, Scion, Rotorua, New Zealand

Dariusz Kulus Laboratory of Horticulture and Landscape Architecture, Department of Biotechnology, Faculty of Agriculture and Biotechnology, Bydgoszcz University of Science and Technology, Bydgoszcz, Poland

Carlos A. S. Ledo Embrapa Mandioca e Fruticultura, Rua Embrapa, s/n, Chapadinha, Cruz das Almas, BA, Brazil

Fundação Getúlio Vargas, FGV Conhecimento, Rua Jornalista Orlando Dantas, Botafogo, Rio de Janeiro, RJ, Brazil

P. Magudeeswari Vanavarayar Institute of Agriculture, Pollachi, Coimbatore, Tamil Nadu, India

Anju Mahendru-Singh ICAR-National Bureau of Plant Genetic Resources (NBPGR), Pusa Campus, New Delhi, India

Julien Miseur Phytopathology, Microbial and Molecular Farming Lab, Center D'Etudes et Recherche, Appliquée-Haute Ecole Provinciale du Hainaut Condorcet, Ath, Belgium

Hellen C. P. Moura Universidade Federal do Recôncavo da Bahia, Cruz das Almas, BA, Brazil

Ayoub Molaahmad Nalousi Department of Horticultural Sciences, Faculty of Agriculture, University of Kurdistan, Sanandaj, Iran

Research Center of Medicinal Plants Breeding and Development, University of Kurdistan, Sanandaj, Iran

Anna Nebot Botanical Garden, University of Valencia, Valencia, Spain

Małgorzata Pałucka Kostrzyca Forest Gene Bank, Miłków, Poland

Valerie Pence Center for Conservation and Research of Endangered Wildlife, Cincinnati Zoo and Botanical Garden, Cincinnati, OH, USA

Ana María López Peralta Empresa de Transformación Agraria, S.A. (TRAGSA), Guadalajara, Spain

Juliana L. S. Pereira Universidade Federal do Recôncavo da Bahia, Cruz das Almas, BA, Brazil

Megan Philpott Center for Conservation and Research of Endangered Wildlife, Cincinnati Zoo and Botanical Garden, Cincinnati, OH, USA

Laura Prieto Centro Nacional de Recursos Genéticos Forestales 'Alaquàs', Subdirección General de Política Forestal y Lucha Contra la Desertificación, Alaquàs, Spain

Hugh W. Pritchard Kunming Institute of Botany, Chinese Academy of Sciences, Kunming, China

P. E. Rajasekharan Formerly with ICAR-Indian Institute of Horticultural Research, Bengaluru, Karnataka, India

Ronilze L. Silva Universidade Federal do Recôncavo da Bahia, Cruz das Almas, BA, Brazil

Fernanda V. D. Souza Embrapa Mandioca e Fruticultura, Rua Embrapa, s/n, Chapadinha, Cruz das Almas, BA, Brazil

Fundação Getúlio Vargas, FGV Conhecimento, Rua Jornalista Orlando Dantas, Botafogo, Rio de Janeiro, RJ, Brazil

Veysel Süzerer Vocational School of Health Services, Department of Pharmacy Services, Bingöl University, Bingöl, Türkiye

Oksana Sytar Agronomic and Agri-food Center of the Province of Hainaut (CARAH), Ath, Belgium

Engin Tilkat Faculty of Science and Literature, Department of Biology, Batman University, Batman, Türkiye

Yu Tu Kunming Institute of Botany, Chinese Academy of Sciences, Kunming, China

Yavar Vafaee Department of Horticultural Sciences, Faculty of Agriculture, University of Kurdistan, Sanandaj, Iran

Research Center of Medicinal Plants Breeding and Development, University of Kurdistan, Sanandaj, Iran

Sakari Välimäki Natural Resources Institute Finland (LUKE), Savonlinna, Finland

Saila Varis Natural Resources Institute Finland (LUKE), Savonlinna, Finland

Lisa Wallace Department of Biological Sciences, Old Dominion University, Norfolk, VA, USA

Urszula Wasileńczyk Kostrzyca Forest Gene Bank, Miłków, Poland

Mikołaj Krzysztof Wawrzyniak Institute of Dendrology, Polish Academy of Sciences, Kórnik, Poland

Max Winkeljohn Center for Conservation and Research of Endangered Wildlife, Cincinnati Zoo and Botanical Garden, Cincinnati, OH, USA

Advancements in In Vitro Seed Germination of Rare, Endangered Plants: Optimization Methods and Environmental Factors

1

Oksana Sytar, Morgan Fays, Julien Miseur, and Nicolas Desoignies

Abstract

The germination of seeds plays a vital role in the propagation and preservation of rare and endangered plant species. This study examines in vitro culture techniques to optimize germination under controlled conditions, focusing on factors critical for rare plants. The key aspects of optimization include disinfection methods; nutrient media; and environmental conditions, namely, moisture, oxygen, temperature, and light. Disinfection techniques, such as sodium hypochlorite and ethanol, effectively reduce microbial contamination while preserving seed viability. Nutrient media such as Murashige and Skoog (MS) and Gamborg's B5 are optimized for rare plant species. Moisture availability is essential for seed imbibition, and oxygen is vital for respiration and metabolism. Innovations such as hydrogels and controlled-release water maintain optimal moisture and increase transplantation success. Plant growth regulators (PGRs), namely, gibberellins, ethylene, and abscisic acid, are key for overcoming dormancy and promoting germination in rare plants. Oxygen levels influence hormone pathways, and managing reactive oxygen species (ROS) supports seedling development while preventing premature senescence. In summary, optimizing in vitro conditions—moisture, oxygen, and hormonal balance—improves seed germination and supports the conservation of rare plant species, addressing critical biodiversity and sustainability challenges.

O. Sytar (✉)
Agronomic and Agri-food Center of the Province of Hainaut (CARAH), Ath, Belgium
e-mail: o.sytar@carah.be

M. Fays · J. Miseur · N. Desoignies
Phytopathology, Microbial and Molecular Farming Lab, Center D'Etudes et Recherche, Appliquée-Haute Ecole Provinciale du Hainaut Condorcet, Ath, Belgium

S. Gantait, P. Chmielarz (eds.), *Conservation of Plant Genetic Resources*,
https://doi.org/10.1007/978-981-96-8331-4_1

Keywords

Biodiversity conservation · Disinfection methods · Hormonal treatments · In vitro culture · Nutrient media composition · Plant propagation

1.1 Introduction

In vitro culture has emerged as a pivotal technique for the conservation of rare, endangered, and threatened plant species, playing a crucial role in safeguarding biodiversity. These plants face significant threats, including habitat destruction, climate change, and overexploitation, which can drastically reduce or even eliminate their populations in the wild (Corlett 2016; Wrobleski et al. 2023). Traditional conservation methods, including seed banking and the creation of natural reserves, often fall short in addressing the intricate challenges linked to preserving these species. Conversely, in vitro culture provides a controlled and sterile setting for plant propagation, serving as an effective ex situ conservation strategy to safeguard species survival under adverse or unpredictable natural circumstances.

In vitro techniques not only preserve genetic diversity but also enable large-scale propagation of endangered species, providing opportunities for reintroduction into their natural habitats or the establishment of new populations. Over the years, substantial progress has been made in applying in vitro culture to various rare species, such as orchids (Li et al. 2021), carnivorous plants such as *Nepenthes* (Rathore et al. 1991; Miguel et al. 2020), Hawaiian silverswords (*Argyroxiphium sandwicense*) (Bruegmann et al. 2003; Meesakul et al. 2023), baobabs (*Adansonia* spp.) (Lamine N'Doye et al. 2012; Rolli et al. 2014), and African violets (*Saintpaulia* spp.) (Silva et al. 2017). These examples highlight the versatility and efficacy of in vitro culture in preserving a diverse array of plant species.

One of the most significant barriers to the conservation of rare plants is the challenge of seed germination. Many species exhibit poor germination rates due to factors such as seed dormancy, specific environmental requirements, and low seed viability (Klupczyńska and Pawłowski 2021; Reed et al. 2022). Dormant seeds often rely on specific environmental cues to break dormancy, a strategy that evolved to ensure survival against predators and pathogens (Qaderi 2023; Hubert et al. 2024). For some species, seeds are either scarce or highly sensitive to stressors, further hindering natural regeneration efforts. Consequently, traditional germination methods often fail, necessitating alternative approaches such as in vitro germination to overcome these limitations.

The germination of seeds is a vital stage in plant growth and significantly influences species reproduction and agricultural yields (Carrera-Castaño et al. 2020; Balde et al. 2021). In vitro culture provides a controlled environment to optimize germination by manipulating variables such as light, temperature, humidity, and nutrient accessibility. These conditions not only increase seed viability and germination rates but also ensure the uniform growth of seedlings (Hesami et al. 2021; Rezaei et al. 2023). This technique has proven particularly effective in improving the germination of rare medicinal plants and other species with challenging propagation requirements.

The optimization of in vitro seed germination involves multiple critical steps, including selecting effective disinfection methods to control microbial contamination (Si et al. 2022; Gilbert et al. 2023), developing nutrient-rich growth media (Lindsey et al. 2017a; Jia et al. 2022), and applying hormonal treatments to stimulate growth (Gammoudi et al. 2022; Hasnain et al. 2022). For example, while many seeds are contaminated with surface microorganisms, heat-based disinfection methods may harm seed viability (Gilbert et al. 2023). Chemical and physical treatments have been widely investigated, highlighting their efficacy in promoting germination rates and enhancing seedling quality (Fu et al. 2024).

Recent advancements in biotechnology have further refined in vitro germination methods. The application of plant growth regulators and molecular tools such as polymerase chain reaction (PCR), molecular cloning, DNA sequencing, and protein analysis techniques (e.g., Western blotting) has significantly improved our understanding of seed biology. These innovations not only support the conservation of rare plants but also contribute to advancements in agriculture and sustainable practices (Breman et al. 2021).

With global problems such as food security and shifting climate patterns, the conservation of rare medicinal plants through in vitro germination has become increasingly important (Salgotra and Chauhan 2023; Fanzo and Miachon 2023). The growing demand for sustainable solutions to increase biodiversity and crop yields underscores the value of these methodologies (Kulak et al. 2022). This chapter explores advanced strategies to optimize in vitro seed germination and aligns with broader objectives of conserving threatened species, supporting ecological balance, and addressing global environmental challenges.

1.2 Advantages of Seed Disinfection for In Vitro Seed Germination

In vitro methods for seed germination have become an innovative approach in contemporary plant science, providing accurate regulation of germination environments. These methods, which were initially developed for research and commercial crop improvement, have expanded to play a pivotal role in biodiversity conservation, particularly for underutilized and endangered species (Radomir et al. 2023; Tarraf and De Carlo 2024). By creating an artificial yet optimal environment, in vitro techniques not only increase germination success but also allow for the study of plant reactions to various biotic and abiotic factors. The resistance of plants to both biotic and abiotic stresses can be developed by applying selection agents to culture media (Nawrot-Chorabik et al. 2022). For example, NaCl can be used to induce salt resistance, whereas PEG or mannitol can promote drought tolerance. Moreover, pathogen culture filtrates, phytotoxins, or the pathogens themselves can be utilized to increase disease resistance, strengthening the plant's capacity to endure environmental challenges (Rai et al. 2011). On the other hand, different plant taxonomic groups employ various physiological adaptive responses to cope with stress and defensive stimuli, which enable them to synthesize specific secondary metabolites (Isah 2019).

The first and most crucial step in the development of seed germination in an in vitro environment is the effective disinfection of seeds. Disinfection eliminates microbial contaminants such as fungi, bacteria, and viruses that can compromise seed viability, nutrient availability, and overall germination success (Silva et al. 2015). Thermal treatments are unsuitable for seed sterilization, as the elevated temperatures needed to minimize microbial contamination may also adversely impact seed germination. For example, Gilbert et al. (2023) reported that two chlorine-based sanitizers—diluted household bleach (0.6% sodium hypochlorite) and freshly prepared hypochlorous acid (800 ppm chlorine)—were the most efficient at achieving up to a 5-log reduction in bacterial presence while maintaining seed germination across 14 tested plant cultivars. The disinfection process is particularly important under in vitro conditions, where the absence of microbial competition creates a highly conducive environment for pathogen growth if contaminants are not thoroughly eradicated.

Disinfection protocols are highly specific and must be tailored to the species of plant being studied (Cuba-Díaz et al. 2020). The antimicrobial and phytotoxic impacts of calcium hypochlorite ($Ca(ClO)_2$), silver nitrate ($AgNO_3$), and silver nanoparticles (AgNPs) were evaluated across different concentrations and durations of exposure for both native and introduced Antarctic plant species. In *D. antarctica*, $Ca(ClO)_2$ at 100 mg/mL effectively controlled microbial contamination when applied for 20 min, whereas the same treatment was effective for *C. quitensis* populations; a 15-min application showed effectiveness. In the case of *J. bufonius*, the application of silver nitrate ($AgNO_3$) at a concentration of 10 mg/mL for 10 min inhibited microbial proliferation but induced oxidative stress. Silver nanoparticles (AgNPs), however, fail to control contamination and do not harm plant tissues (Cuba-Díaz et al. 2020).

Factors such as seed size, surface texture, and susceptibility to damage influence the choice of disinfection agent and method (Gilbert et al. 2023). For example, seeds with hard outer coatings may tolerate stronger chemical treatments, whereas delicate seeds require milder solutions to preserve their viability.

Disinfection of seeds is a critical first step in plant tissue culture, micropropagation, and other laboratory processes aimed at growing plants in a controlled and sterile environment. Proper disinfection helps eliminate microbial contaminants such as bacteria and fungi, which can compete with or kill developing plant tissue (Okoroafor 2022). Moreover, this microbial growth is supported by the medium richness. Proper disinfection techniques (using agents such as sodium hypochlorite or ethanol) are needed to avoid contamination without damaging the seed. However, improper disinfection can decrease seed viability, reduce germination rates, or stop growth (Mangena and Mokwala 2019). The test of viability is important for the chosen seed. A small batch of seeds can be tested for germination before committing a large set of seeds to disinfection, allowing the optimization of the treatment nature, time, and concentration. This enables the determination of the right balance between disinfection strength and crucial seed protection.

For the disinfection of seeds, the structure of seed is always important. Seeds with thick seed coats (e.g., legumes: soybeans (*Glycine max*), peas (*Pisum sativum*), lentils (*Lens culinaris*), beans (*Phaseolus vulgaris*); nuts: walnuts (*Juglans* spp.), almonds (*Prunus dulcis*), chestnuts (*Castanea* spp.); tree seeds: acorns (*Quercus* spp.) and horse chestnuts (*Aesculus hippocastanum*), fruit seeds of avocado (*Persea americana*) and pomegranate (*Punica granatum*)) may require and tolerate longer disinfection times or stronger solutions (Geneve 1998; Baskin and Baskin 2014), as the sterilant needs to penetrate the outer layer (Smýkal et al. 2014; Hong et al. 2023).

For seeds with delicate or soft coats, the exposure duration should be minimized to avoid harm to the embryo or internal tissues (Maiti 2013). These seeds generally germinate more easily and do not require additional treatment to break the seed coat. They can quickly absorb water, making them ideal for fast germination under the right conditions. For example, the seeds with thin or soft seed coats are tomatoes (*Solanum lycopersicum*), peppers (*Capsicum* spp.), lettuce (*Lactuca sativa*), cucumber (*Cucumis sativus*), orchids (Orchidaceae family), spinach (*Spinacia oleracea*), and rice (*Oryza sativa*).

It is also an important pretreatment process before disinfection and post-disinfection care. While not typically used for seed disinfection, autoclaving can be effective for sterilizing tools, containers, or growth media used in seed germination. However, the seeds themselves are too sensitive to high heat and pressure and are not autoclaved directly. Seed rinsing can be important before the seeds are disinfected. After disinfection with disinfection agents such as Triton X-100, 70% ethanol, and hypochloride Na or Ca, seeds must be thoroughly rinsed with sterile water to ensure that no residual sterilizing agents remain, which could otherwise continue to affect seed viability (Agogbua and Okoli 2022).

For each plant species, an optimal disinfection strategy must be designed that considers the appropriate type and concentration of sterilizing agent. The most common disinfectants used are summarized in Table 1.1. Frequently, a combination of treatments (e.g., a quick ethanol dip followed by sodium hypochlorite soak) is applied to maximize disinfection while reducing seed damage. For example, *Arabidopsis* seeds have optimized disinfection protocols using either liquid-phase bleaching or vapor-phase chlorine gas, both of which achieve high germination rates with minimal contamination (Lindsey et al. 2017a, b). Rezaei et al. (2023) explored the effects of various disinfectants and their concentrations for *Petunia* seeds, including sodium hypochlorite (NaOCl), mercuric chloride ($HgCl_2$), hydrogen peroxide (H_2O_2), calcium hypochlorite ($Ca(ClO)_2$), and carbon nanotube-based compounds, highlighting their effectiveness in seed disinfection. Sodium hypochlorite (0.6%) and freshly generated hypochlorous acid achieved up to a 5-log reduction in bacterial counts without harming seed germination (Gilbert et al. 2023).

The results of the seed disinfection studies demonstrated that different disinfection protocols are required for seeds of varying colors within a single plant species. Specifically, the percentage of brown seeds of *Suaeda aralocaspica* that germinated was 100%, whereas the percentage of black seeds was only 17%. As a result, brown

Table 1.1 Most often used surface disinfection agents for in vitro seed germination

Disinfection agents	Advantages	Disadvantage	References
Calcium hypochlorite ($Ca(ClO)_2$)	Highly effective: Calcium hypochlorite efficiently eliminates a broad range of microbial contaminants. Stable disinfectant: It remains stable longer than sodium hypochlorite, maintaining its strength over time in solution	Aggressive on seeds: Calcium hypochlorite can be too harsh on seeds with thin or soft coats, potentially damaging their surface or embryo. Residual toxicity: Residues left on seeds can reduce viability and affect germination. Complex preparation: It requires careful handling and more steps in preparation compared to liquid disinfectants	Phuong Quynh et al. (2022), Maiti (2013)
Sodium hypochlorite (NaOCl)	Highly effective against a wide range of microorganisms	Overexposure can weaken or kill the seed embryo, especially in small or delicate seeds	Debergh and Zimmerman (1991); George et al. (2008); Gamborg and Phillips (2013); Lindsey et al. (2017a, b)
Ethanol (70%)	Fast-acting and can be used in conjunction with other sterilants for stronger disinfection	Ethanol can damage seed coats if used for extended periods, so exposure time must be strictly controlled	Pierik (1997); George et al. (2008); Schreiber et al. (2008); Lindsey et al. (2017, b)
Hydrogen peroxide (H_2O_2)	Safer for seeds, can reduce microbial load while being less damaging	May be less effective against persistent contaminants	Khan et al. (2018); González-Benito et al. (2019); Siddiqui et al. (2020); Kumar et al. (2021)
Mercuric chloride ($HgCl_2$)	Effective against a wide range of pathogens	High toxicity and environmental concerns make this method less favorable in modern labs	Cheng et al. (2018); Khan et al. (2019); Gómez et al. (2020); Malik et al. (2021)

seeds were identified as more suitable for subsequent culture experiments. Among the disinfection methods evaluated, sodium hypochlorite (NaClO) was found to be more effective than mercuric chloride ($HgCl_2$) for brown seeds. The optimal disinfection procedure included rinsing with 75% ethanol for 60 s, followed by NaClO treatment for 8 min, and cultivation on pH 8.0 Murashige and Skoog (MS) (Murashige and Skoog 1962) medium for 7 days, which was successfully applied to the in vitro culture of *S. aralocaspica* (Si et al. 2022).

The impacts of five disinfecting agents (ethanol, mercuric chloride, flugal, nystatin, and sodium hypochlorite) on the germination and growth of *Achyranthes aspera* (Amaranthaceae), a significant medicinal species, were assessed. Parameters such as contamination rates, germination percentage, seed color, growth pattern, and shootlet formation were analyzed (Sen et al. 2013). Sodium hypochlorite treatment for 30 min was the most effective treatment. While higher concentrations of all disinfectants reduce microbial contamination, they also lower survival rates (Sen et al. 2013). Cultures without sterilization clearly exhibited contamination, highlighting the importance of sterilization. In contrast, for another species, *Gloriosa superba* (L.), contamination levels decreased as the concentration of sterilizing agent and duration of exposure increased. (Mosoh et al. 2023). These findings highlight that each plant species may require an adapted disinfection protocol to optimize both contamination control and survival outcomes during in vitro culture.

1.3 Nutrient Media Composition

The formulation of nutrient media plays a vital role in achieving successful in vitro seed germination, as it supplies the necessary nutrients, vitamins, and hormones required to sustain plant growth and development.

Nutrient availability is influenced by pH, with an optimal pH range of 5.5–6.2 in growing media being ideal for most plants to effectively absorb nutrients (Yang et al. 2018; Funnekotter et al. 2023). The key components, including macroelements, microelements, nitrogen sources (such as amino acids), carbon sources, and organic supplements (primarily vitamins and other additives), must be incorporated at fixed concentrations. To facilitate organogenesis, plant growth regulators—including auxins, cytokinins, gibberellins, abscisic acid, and ethylene—are incorporated alongside solidifying agents. The balance of these growth regulators varies depending on the specific tissue culture technique employed (Sudheer et al. 2022).

The selection of media can strongly influence germination rates, seedling strength, and overall plant development. Table 1.2 provides a summary of the most frequently utilized nutrient media for enhancing seed germination in certain underutilized plant species. These media formulations have been optimized on the basis of the unique physiological requirements of the seeds, which increase germination rates and overall viability (George et al. 2008; Sharma and Kaur 2018).

Kunakhonnuruk et al. (2018) investigated the effects of various nutrient media on the seed germination of the endangered Thai rheophytic orchid *Epipactis flava* Seidenf. Seeds were cultivated for 2 months in both semisolid and liquid media: modified Vacin and Went medium (VW), MS, Burgeff's medium (BM), Mitra medium (MM), and Knudson C medium (KC). The results indicated that both semisolid and liquid VW media yielded the highest germination rates, reaching 70.2% and 70.4%, respectively. These findings indicate that the VW medium is the most effective for enhancing seed germination in this endangered species. Similarly, a study comparing tropical and temperate epiphytic and terrestrial orchids across various growth media revealed that temperate terrestrial species, such as *Pterostylis banksii* and *Thelymitra nervosa*, require more nutrient-rich media, such as Norstog and BM1, for optimal seed germination and seedling development. In contrast, tropical epiphytic species,

Table 1.2 Common nutritional media and their applications

Nutrient medium	Formulation	Application for plant objects	References
Murashige and Skoog (MS) medium	*Macronutrients:* Nitrogen, phosphorus, potassium, calcium, magnesium, and sulfur *Micronutrients:* Iron, manganese, zinc, copper, boron, and molybdenum *Vitamins:* Thiamine (vitamin B1), nicotinic acid, and pyridoxine (vitamin B6) *Growth regulators (optional):* Auxins and cytokinins can be added to regulate growth and differentiation	Effective in germinating rare and endangered species such as *Euphorbia* and *orchids*	Murashige and Skoog (1962)
Gamborg's B5 medium	*Macronutrients:* Ammonium nitrate, potassium nitrate, calcium chloride, magnesium sulfate, and potassium dihydrogen phosphate *Micronutrients:* Iron, manganese, zinc, copper, boron, and molybdenum *High content of vitamins:* Thiamine (vitamin B1), nicotinic acid, and pyridoxine (vitamin B6) *Sucrose:* Serves as the primary carbon source *Additional components:* Glycine and myo-inositol, which play roles in plant cell metabolism	Rare tree species such as *Azadirachta indica* (neem tree) *Swietenia macrophylla* (big-leaf mahogany) *Tectona grandis* (teak tree) *Shorea robusta* (sal tree) *Pterocarpus santalinus* (red sandalwood)	Gamborg et al. (1968), Roja and Heble (1996), George et al. (2008), Parasharami et al. (2007), Rai et al. (2009), Pérez and Villalobos (2016)
Nitsch and Nitsch medium	*Macronutrients*: Nitrogen, phosphorus, potassium, and calcium, essential for basic cellular functions like protein synthesis, energy transfer, and maintaining ion balance *Micronutrients*: Iron, manganese, zinc, copper, and boron, which are required in small amounts for enzyme function and metabolic processes *Sucrose*: Provides the necessary carbohydrate source for energy and carbon skeletons in the absence of photosynthesis during in vitro culture *Vitamins*: Typically includes thiamine, pyridoxine, and nicotinic acid, important for cellular metabolic processes *Growth Regulators*: The medium often incorporates *auxins* and *cytokinins* (e.g., indole-3-acetic acid or kinetin) to promote cell division, differentiation, and morphogenesis	Suitable for rare orchids such as *Paphiopedilum* species, *Dendrobium* orchids, *Vanda* orchids, *Cymbidium* species	Nitsch and Nitsch (1967), Huan and Tanaka (2004), Utami and Hariyanto (2019), Maharjan et al. (2019), Tikendra et al. (2022)

(continued)

Table 1.2 (continued)

Nutrient medium	Formulation	Application for plant objects	References
Woody plant medium (WPM)	*Macronutrients*: NH_4NO_3 (ammonium nitrate), KNO_3 (potassium nitrate), KH_2PO_2 (potassium dihydrogen phosphate), $MgSO_4{\cdot}7H_2O$ (magnesium sulfate), $CaCl_2{\cdot}2H_2O$ (calcium chloride) *Micronutrients*: H_3BO_3 (boric acid), $MnSO_4{\cdot}H_2O$ (manganese sulfate), $ZnSO_4{\cdot}7H_2O$ (zinc sulfate), $Na_2MoO_4{\cdot}2H_2O$ (sodium molybdate), $CuSO_4{\cdot}5H_2O$ (copper sulfate), $CoCl_2{\cdot}6H_2O$ (cobalt chloride) *Iron source*: $Na_2EDTA{\cdot}2H_2O$ (disodium ethylenediaminetetraacetate), $FeSO_4{\cdot}7H_2O$ (iron sulfate) *Vitamins*: Myo-inositol, thiamine·HCl (Vitamin B1), nicotinic acid, pyridoxine·HCl (vitamin B6) *Carbon source*: Sucrose	Woody plants such as *Acer* spp., *Quercus* spp., *Betula* spp., *Eucalyptus* spp. etc.	McCown and Lloyd (1981), Watt et al. (2003), Bowen-O'Connor et al. (2007), Iliev et al. (2010), Fadladeen et al. (2021), Ree et al. (2024)
Hartig medium	*Macronutrients*: *Nitrates*: Ammonium nitrate (NH_4NO_2), calcium nitrate ($Ca(NO_3)_2$) *Phosphates*: Potassium dihydrogen phosphate (KH_2PO_4) Sulfur: Magnesium sulfate ($MgSO_4$) *Micronutrients*: *Iron*: Iron chelate (Fe-EDTA) *Other elements*: Manganese sulfate ($MnSO_4$), zinc sulfate ($ZnSO_4$), copper sulfate ($CuSO_4$), boric acid (H_3BO_3), sodium molybdate (Na_2MoO_4) *Vitamins*: Nicotinic acid (B3), pyridoxine (B6), thiamine (B1) *Agar*: Agar (if solid medium is required) *pH*: Adjust the pH to approximately 5.8–6.0 using hydrochloric acid (HCl) or sodium hydroxide (NaOH)	Rare species like *Taxus* (yew) and various conifers	Nhut et al. (2007)

(continued)

Table 1.2 (continued)

Nutrient medium	Formulation	Application for plant objects	References
Custom media formulations develop custom media based on preliminary studies	Formulations may include specific ratios of macronutrients, micronutrients, and growth regulators tailored to the species in question	Specific rare plants	Baskin and Baskin (1998), Mallon et al. (2014)

such as *Dendrobium strebloceras* and *Dendrobium lineale*, present relatively low nutrient requirements. Fine-tuning in vitro nutrient conditions is crucial for the successful ex situ conservation of orchid species (Diantina et al. 2020).

Numerous studies have highlighted the efficacy of specific growth media, such as MS and Gamborg's B5, in promoting the germination of underexploited and rare plant species. These media play a key role in conservation efforts by supporting the successful propagation of species that are otherwise difficult to germinate, contributing to both conservation and agricultural practices (Hussain et al. 2019; Raghavan 2005). For example, in the rare plant *Alstroemeria*, pretreatment with potassium nitrate (KNO_3) combined with modified ½ MS medium led to consistent germination and a considerably greater germination rate than the control. (Volpi et al. 2024). A different study examined the in vitro germination of seeds from 4-month-old hand-pollinated *Phalaenopsis amboinensis* orchids cultured on various concentrations of MS and VW media: MS, ½ MS, Vacin and Went (VW), and ½ VW. The highest seed germination rate, 90.7%, was observed on VW media, highlighting its effectiveness in promoting the growth of this rare orchid species (Utami and Hariyanto 2019). Such approaches are essential for the conservation and propagation of rare species, supporting biodiversity preservation.

1.4 Major Highlights of Breaking Seed Dormancy

The process of overcoming seed dormancy is an essential physiological and biochemical event involving complex interactions between environmental signals, biochemical pathways, and the internal conditions of a seed. The scheme in Fig. 1.1 describes this process.

Breaking seed dormancy is a multistep process that integrates environmental signals with internal biochemical pathways (Farooq et al. 2022). Embryonic ABA is crucial for initiating and sustaining seed dormancy while suppressing the shift from embryonic development to germinative growth (Rodríguez-Gacio et al. 2009). The key ABA signaling genes ABI3, ABI4, and ABI5 perform diverse and essential functions during seed development, including roles in degreening, desiccation tolerance, maturation, dormancy, and seed vigor, often in coordination with various cofactors (Ali et al. 2021). Water absorption triggers enzyme activation, which is

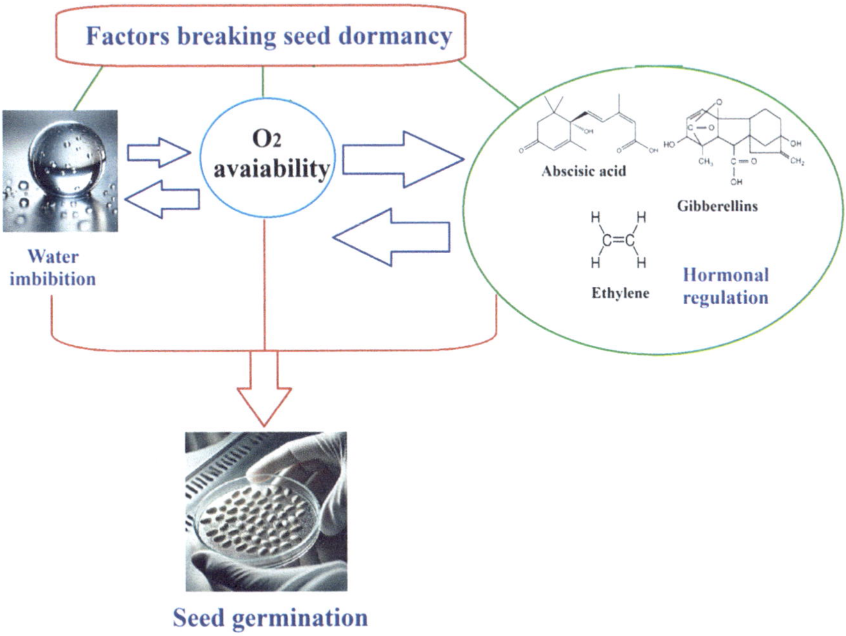

Fig. 1.1 Factors that affect the breaking of seed dormancy. (Source: Diagram prepared by authors)

mediated by hormonal changes (mainly GA, ABA, and ethylene) (Ge et al. 2020; Ali et al. 2021), whereas physical disruptions of the seed coat and the provision of oxygen further enable seedling growth. ABA inhibits ethylene activity by reducing its biosynthesis (Ahammed et al. 2020), with nitric oxide (NO) suggested as a key mediator in the crosstalk between ABA and ethylene during seed regulation (Arc et al. 2013). These processes ensure that the seed breaks dormancy and begins germination under favorable conditions.

1.4.1 Water Imbibition and Enzyme Activation

Water uptake activates various enzymes, especially hydrolases such as amylases, proteases, and lipases. During germination, the aleurone layer and scutellum contain the highest concentration of enzymes involved in the process. In addition, enzymes such as xylanases, proteases, and β-glucanases are found in the endosperm, whereas β-glucanases and lipases have been identified in the embryo (Guzmán-Ortiz et al. 2018). These enzymes breakdown stored nutrients, including starches, proteins, and lipids, into simpler compounds such as sugars, amino acids, and fatty acids, which are essential for supporting the embryo during its early growth stages, typically beginning on Day 4 of germination in most cereals. Germination triggers the hydrolysis of macromolecules and compounds, including β-glucans and phytic acid (Guzmán-Ortiz et al. 2018). Enzyme activation is influenced by the levels of gibberellins (GAs), which often increase during dormancy release. GA influences the process in two ways: first, by increasing the growth potential of the embryo, and second, by stimulating the activity of hydrolytic enzymes (Gupta and Chakrabarty 2013).

1.4.2 Hormone Regulation

Hormones play crucial roles in overcoming dormancy and triggering germination. The primary hormones involved include abscisic acid (ABA), gibberellins (GA), and ethylene. The initial loci associated with dormancy and germination identified in *Arabidopsis* mutants were genes related to the biosynthesis, perception, and signaling of abscisic acid (ABA) and gibberellins (GA) (Koornneef et al. 1982, 1984; Steber et al. 1998; Holdsworth et al. 2008). Additionally, brassinosteroids (BRs), cytokinins (CKs), auxins, and jasmonates (JAs) contribute to the regulation of seed germination (Miransari and Smith 2014; Van de Poel et al. 2015).

Gibberellins (GA) These phytohormones stimulate the production of enzymes, such as α-amylase, in the aleurone layer, which is crucial for the breakdown of stored starches into fermentable sugars. Gibberellin (GA) specifically enhances seed germination by promoting embryo growth potential and facilitating cell elongation (Ge et al. 2020). Gibberellins are often exogenously applied to seeds to break dormancy, especially in species with physiological dormancy (Lee et al. 2022). Bioactive gibberellins (GAs) are produced in the final steps catalyzed by the enzymes GA20ox and GA3ox. The deactivation of gibberellins (GAs) by GA2ox oxidase, along with transcriptional feedback mechanisms, plays a critical role in regulating GA homeostasis, which is essential for the successful progression of seed germination (Claeys et al. 2014; Vishal and Kumar 2018). An altered GA/ABA ratio plays a key role in embryo development, alleviates the mechanical restrictions of the seed coat, and facilitates radicle emergence in recalcitrant *Panax notoginseng* seeds (Ge et al. 2023).

Abscisic acid (ABA) ABA generally inhibits seed germination and maintains dormancy. Recent research has revealed multiple regulatory factors that functionally interact with key transcription factors, including ABA INSENSITIVE3 (ABI3), ABI4, and ABI5, which are essential for modulating seed ABA levels (Sano and Marion-Poll 2021). A reduction in ABA levels and/or an increase in GA concentrations is pivotal for breaking dormancy. The interaction between ABA and GA is fundamental in governing both dormancy release and seed germination (Ma et al. 2017).

Ethylene In some seeds, ethylene also helps in breaking dormancy, particularly in seeds that require mechanical disruption of the seed coat. Research involving inhibitors of ethylene biosynthesis or action, alongside studies of mutant lines with disruptions in key ethylene signaling genes (such as etr1, ein2, ain1, etr1, and erf1), highlights the role of ethylene in controlling seed germination and dormancy. Ethylene antagonizes ABA by modulating its metabolism and associated signaling pathways (Corbineau et al. 2014). Ethylene acts antagonistically to ABA but synergistically with GAs in regulating *Arabidopsis* seed germination. Its role in seed dormancy and germination regulation is driven by reciprocal interactions with both ABA and GA biosynthesis and signaling pathways (Van de Poel et al. 2015). Ethylene facilitates germination by interacting with key regulators of seed dormancy and germination, such as ABA, GAs, and ROS. However, the sensitivity of seeds to ethylene increases during the release of dormancy, whether it is induced by chilling or after-ripening during dry storage (Corbineau 2024).

1.4.3 Physical and Mechanical Disruption

The seed coat often presents a physical barrier to embryo growth. Seed coat impermeability functions as a protective strategy, delaying germination until optimal environmental conditions are achieved. The factors contributing to hard seed development are categorized as internal or external. Internal factors, such as the structural properties of the seed coat, include species-specific traits, genetic variability, and seed morphology. External factors include environmental factors such as soil nutrients, water availability, humidity, temperature, and light, all of which affect seed coat permeability and germination capacity. Moreover, seed maturity, drying procedures, and storage conditions play crucial roles in the development of hard seeds (Ermis et al. 2024). The combined effect of these factors determines a plant's propensity for producing hard seeds. To overcome dormancy induced by the impermeability of the seed coat, various techniques—physical, chemical, or mechanical—have been employed. In some species, breaking dormancy requires physical disruption, such as exposure to light, cold stratification, or scarification (mechanical abrasion of the seed coat) (Yang et al. 1999). For example, seed germination in sedges is hindered by a hard seed coat and physiologically dormant embryos. However, dormancy can be effectively overcome through a combination of bleach soaking and subsequent cold stratification (Rosbakh et al. 2019). Another study demonstrated that the germination of Saskatoon berry seeds can be effectively improved through various dormancy-breaking treatments. Stratification with KNO_3 (0.2%) or smoke water (1:100) is particularly beneficial for unstored seeds, significantly increasing germination rates up to 87% (Górnik et al. 2023). The processes of cold stratification or scarification either soften or crack the seed coat, allowing the embryo to expand as it absorbs water.

1.4.4 Oxygen Availability

Oxygen is essential for respiration and supplies the energy required for the initial stages of seed germination (Ray et al. 2016). Some researchers utilize oxygen consumption (Q) measurements to identify and select high-quality seeds (Tu et al. 2023). Oxygen (O_2) is essential for seed germination; however, its availability can be reduced by excessive water content, compaction, compression, and hard surfaces. Reduced O_2 concentrations may negatively affect both the germination success (proportion of seeds that germinate) and the speed (time to reach 50% germination, t_{50}) of seed germination (Yasin and Andreasen 2016).

Dormant seeds are typically in a low metabolic state, but once they start imbibing water, respiration rates increase significantly, requiring oxygen for oxidative phosphorylation and energy production. The dormant seeds consume a large amount of oxygen through glycolysis and the TCA cycle. Song and Zhu (2019) demonstrated that the maintenance of primary dormancy in *Pinus koraiensis* seeds is associated with attenuated biosynthetic metabolic processes, a lower utilization rate of amino acids, and greater activity of glycolysis and the tricarboxylic acid (TCA) cycle in embryos.

Seed germination is a highly energy-demanding process that relies on the breakdown of stored reserves within the seed to meet its energy requirements. Under normoxic conditions, where oxygen is readily available from the surrounding air, the aerobic metabolism of these reserves supplies the energy needed for germination. In contrast, submergence limits oxygen availability, creating abiotic stress conditions referred to as anoxia (complete absence of oxygen) or hypoxia (reduced oxygen levels, approximately 3%), which significantly hinder the germination process (Narsai et al. 2015).

1.4.5 Temperature and Light Signals

Temperature fluctuations and specific light conditions are crucial in breaking seed dormancy (Abubakar and Attanda 2022). Dormancy mechanisms and the conditions required for seed germination vary significantly between species (Bhatt et al. 2022). Some seeds require exposure to cold (cold stratification) or heat (warm stratification) to overcome dormancy and initiate germination. An effective method for growing healthy plants of the rare wild species *Mandragora autumnalis* was developed, where cold stratification and 100 mg/L gibberellic acid (GA_3) treatment increased the in vitro germination rate of *Mandragora* seeds to 96% (Al-Ahmad 2020). Plants grown in warmer climates may produce seeds with altered germination responses to temperature and/or cold stratification (Milbau et al. 2009). For example, under historical plant temperatures, a seed size–number trade-off was observed. However, under future warming scenarios, plants produce fewer and smaller seeds, with no evidence of a trade-off (Notarnicola et al. 2023).

On-plant seed storage (serotiny) and seeds eaten by fruit-eating animals are considered special types of physical dormancy because they share similar traits with hard-coated seeds. Warm stratification does not require seeds to be wet, as it is simply a physical process. Heat can help both hard and soft seeds germinate by increasing the permeability of their outer layers (Lamont and Pausas 2023).

The ability to detect optimal conditions for germination and seedling establishment, particularly through temperature fluctuations, is especially vital for small seeds (Xu and Du 2023). Although warming resulted in reductions in both seed size and quantity, germination itself remained unaffected, despite the correlation between seed size, quantity, and germination traits. Interestingly, warming increased germination rates, particularly in larger seeds; however, it ultimately caused a more than fourfold decline in overall plant fitness (Notarnicola et al. 2023).

Significant interactions between cold stratification and temperature were found to influence germination traits in *Spartina alterniflora*, a species native to the United States (Cheng et al. 2022). Germination traits exhibited linear correlations with latitude, revealing that seeds from various provenances germinated at different times and adopted diverse germination strategies. Under warmer regional temperature conditions, seeds from higher latitudes germinate earlier and at increased rates, potentially promoting the northern range expansion of this invasive species (Cheng et al. 2022).

In a warming subarctic climate, the germination and recruitment of dwarf shrubs (*Vaccinium myrtillus*, *V. vitis-idaea*, *Betula nana*), trees (*Alnus incana*, *Betula pubescens*), and grasses (*Calamagrostis lapponica*, *C. purpurea*) are expected to improve. Elevated temperatures are expected to facilitate earlier germination, enhancing the establishment, growth, and long-term persistence of these species within the region (Milbau et al. 2009).

However, the degree of change in the germination response across different species is likely influenced by their seed dormancy characteristics (Bernareggi et al. 2016). Research suggests that approximately 90% of the peatland plant species tested in the Changbai Mountains exhibit seed dormancy. These findings reveal that seed traits, including the embryo-to-seed ratio and seed size, as well as abiotic factors such as pH and temperature seasonality, significantly influence the germination patterns of these species. This highlights seed dormancy as a prevalent adaptive strategy among peatland plants in temperate mountainous regions (Wang et al. 2024b).

Light, in conjunction with temperature, may promote germination, especially under less favorable temperature conditions. Environmental factors such as temperature and light appear to play a key role in the germination of the studied *Cistus* species once physical seed dormancy is overcome (Luna et al. 2022).

Light can also trigger dormancy release in certain species, especially those with photo-dormancy, where light exposure induces the production of hormones such as gibberellins. A study investigating the effects of light on the germination of fresh and naturally aged tobacco seeds revealed contrasting outcomes. In fresh seeds, light promoted germination by increasing GA_3 signaling and suppressing abscisic acid (ABA) signaling. Conversely, in naturally aged seeds, light inhibited germination by suppressing GA signaling and stimulating ABA signaling. These findings highlight the critical role of light in regulating seed germination through its influence on the metabolism and signaling pathways of the plant hormones ABA and GA_3 (Wang et al. 2021).

The regulatory influence of blue light on dormancy release and the enhancement of germination capacity in *Arabidopsis* seeds was examined through a detailed model study. Stawska and Oracz (2019) demonstrated that the HY5 gene, encoding a transcription factor, likely governs the expression of genes involved in gibberellic acid (GA) and abscisic acid (ABA) metabolism and/or signaling pathways during blue light (BL)-mediated dormancy alleviation in *Arabidopsis* seeds. Furthermore, the functional roles of HYH and HFR1 were associated with germination regulation (Stawska and Oracz 2019).

1.4.6 Reactive Oxygen Species (ROS) Regulation

Reactive oxygen species (ROS) are naturally generated as by-products of metabolic activity, particularly during the initiation of germination processes (Huang et al. 2019; Guo et al. 2023). Although excessive ROS can be harmful, they also function as critical signaling molecules that govern the release of seed dormancy and support early seedling development (El-Maarouf-Bouteau and Bailly 2008). Proper management of ROS is critical for ensuring successful germination. Reactive oxygen species (ROS) regulate seed germination by modulating hormone levels via Ca2+ and NO signaling, as well as through the MAPK cascade pathway. ROS contributes to the germination process by affecting DNA methylation, chromatin remodeling, and protein posttranslational modifications (Wang et al. 2024a). Additionally, ROS have potential roles in remodeling the nuclear genome and influencing the function of certain epigenetic modifiers, although the specific mechanisms involved in seed germination remain largely unknown.

Reactive oxygen species (ROS) and reactive nitrogen species (RNS) may play critical roles in oxidative and nitrosative signaling pathways (Khan et al. 2023), which are essential for regulating seed germination within the optimal oxidative window (Kumar et al. 2021). Different organelles within plant cells contain the enzymes necessary for generating reactive oxygen species (ROS) and reactive nitrogen species (RNS), as well as their associated scavenging mechanisms. The spatial and temporal patterns of these enzymes are pivotal in shaping the signaling pathways they influence (Khan et al. 2023). The activity of RNS is modulated by their concentration, duration of exposure, and specific environmental or physiological context. RNS are involved in nitrating nucleic acids and fatty acids, facilitating posttranslational protein modifications (such as nitration and S-nitrosylation) and regulating ROS metabolism (Ciacka et al. 2020).

Both ROS and nitrogen species contribute to germination by sensing environmental cues such as light and temperature. They help facilitate germination under favorable conditions, but at the same time, they can act as protective agents by inhibiting germination under high-temperature stress (Jeevan Kumar et al. 2021). This dual role allows the plant to balance the need for germination with protection from adverse environmental conditions. Nitrogen species, in particular, are involved in modulating these responses by integrating both light and temperature signals, ensuring that germination occurs at the most appropriate time while mitigating the risks associated with heat stress (Puglia 2024). Furthermore, the interaction between ROS and RNS is complex, as they can either synergistically enhance seed germination or act as signaling molecules to delay it when environmental stress exceeds optimal thresholds.

1.5 Factors for the Optimization of Seed Germination

The factors influencing seed germination optimization are illustrated in Fig. 1.2. Key determinants include light, water, temperature, and phytohormones, with water and temperature identified as the primary factors. While light and phytohormones play supporting roles, they are essential for enhancing germination processes. Water availability is crucial for seed hydration and the activation of metabolic pathways, whereas optimal temperature ranges stimulate the enzymatic activities necessary for germination (Bewley et al. 2013; Thompson et al. 2020a). Additionally, light exposure can trigger specific photoreceptors that influence germination in certain species, and phytohormones such as auxins and gibberellins regulate growth and developmental responses during the germination process (Finch-Savage and Leubner-Metzger 2006; Toorop et al. 2008). Understanding these factors allows for tailored approaches to maximize germination rates in various plant species.

Concerning the primary factors influencing seed germination, water and temperature are crucial, alongside the application of presoaking or scarification for hard-coated seeds and stratification (either cold or warm) to simulate natural germination cues. These techniques play a vital role in breaking seed dormancy and facilitating the initiation of germination by replicating the environmental conditions that seeds encounter in their natural habitats (Finch-Savage and Leubner-Metzger 2006; Klupczyńska et al. 2021).

Water and temperature are essential factors influencing seed germination by driving enzyme activation and hormone regulation within the aleurone layer of the seed, which is crucial for initiating growth (Farooq et al. 2022). Water absorption hydrates the aleurone layer, activating enzymes that breakdown stored nutrients and soften the seed coat, allowing the embryo to grow and emerge (Yan et al. 2014). Temperature also plays a significant role, as it can break seed dormancy, a physiological block that prevents germination under unfavorable conditions (Finch-Savage et al. 2006). Certain plant species, particularly those adapted to colder climates, require a chilling period, known as stratification, to overcome dormancy and initiate germination (Chen et al. 2021). This mechanism ensures that seed germination occurs only when the environmental conditions are optimal for seedling establishment and survival.

Dormant seed embryos, after 4 weeks, presented increased concentrations of metabolites linked to carbohydrate metabolism, particularly those involved in the pentose phosphate pathway and the tricarboxylic acid cycle while displaying diminished amounts of intermediates from amino acid metabolism compared with

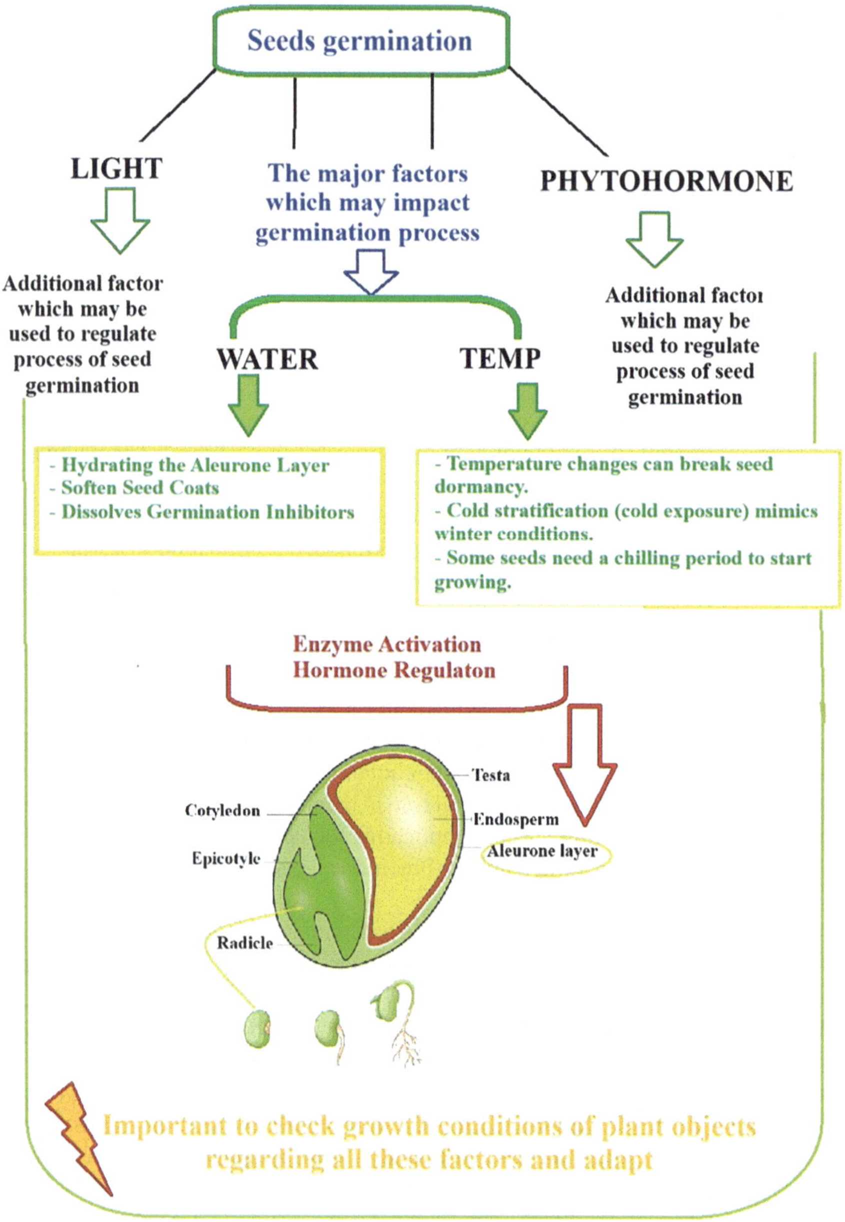

Fig. 1.2 Factors influencing the optimization of seed germination. (Source: Diagram prepared by authors)

nondormant seeds incubated for 1 week. These findings suggest that changes in carbohydrate and amino acid metabolism within dormant seed embryos may contribute to the maintenance of primary dormancy, providing insights into the mechanisms governing seed dormancy (Song et al. 2021).

In raspberry (*Rubus idaeus* L.) breeding programs, double dormancy can hinder germination. Pergolotti et al. (2023) investigated an optimized in vitro germination protocol focusing on the role of temperature. For the experimental cultivars, the raspberry seeds were subjected to transverse and longitudinal cutting techniques and were subjected to cold stratification at 4 °C for 2 weeks, followed by direct culture at 24 °C. The best results were obtained from transversely cut seeds cultured directly at 24 °C, with germination rates ranging from 73 to 78%, with initial germination occurring just 4 days posttreatment. This highlights the importance of stratification in enhancing germination success in raspberry seeds (Fig. 1.3).

Table 1.3 presents a comprehensive overview of the environmental factors that are instrumental in enhancing seed germination. These factors include light, water, and temperature, each of which play a critical role in initiating and regulating the germination process.

The seeds of some species require exposure to low light following a stratification period in darkness for successful germination (Goggin et al. 2010). During

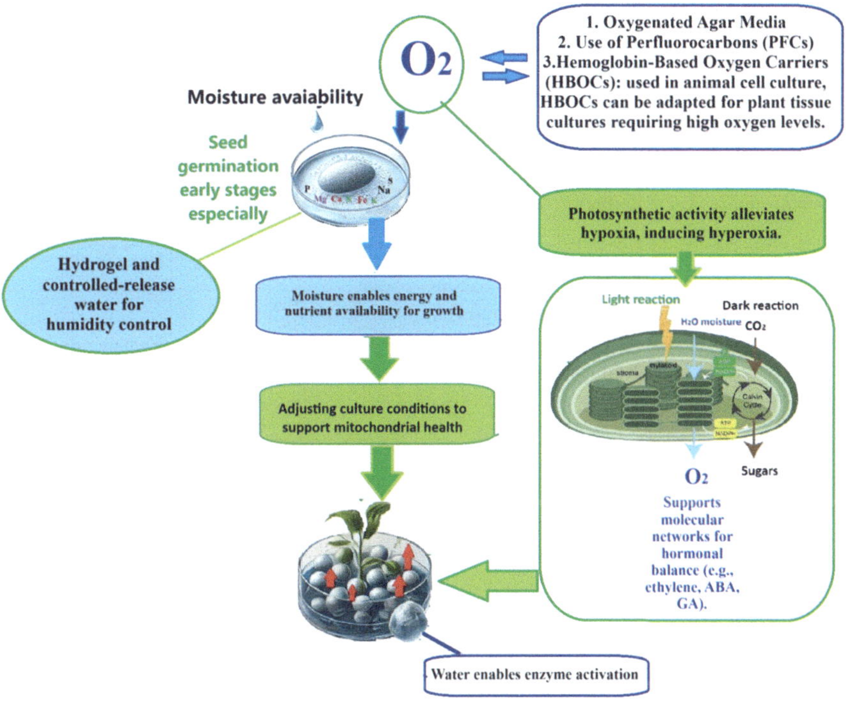

Fig. 1.3 Role of moisture and oxygen in plant in vitro culture development. (Source: Diagram prepared by authors)

Table 1.3 Adapted environmental factors to enhance seed germination

Plant object	Factor used to improve germination	Germination description	References
	Temperature		
Raspberry *Rubus idaeus* L.	Transversely cut seeds directly cultured at 24 °C	The germination rates ranged between 73% and 78%, with initial germination observed 4 days posttreatment Increasing of germination rate	Pergolotti et al. (2023)
Tree turmeric *Coscinium fenestratum (Gaertn.) Colebr.*	Complete darkness and an elevated temperature of 32 ± 2 °C	80% germination rate	Warakagoda and Subasinghe (2014)
Galax *Galax urceolata* (Poir.) Brummitt	Seeds subjected to cold stratification at 5 °C for a duration of 1 month	Highest germination percentage of 76%	Yang et al. (2013)
Dracula orchid *Dracula felix*	−10 °C in the dark	Seed viability was completely lost	Quijia-Lamiña et al. (2023)
Dracula orchid *Dracula felix*	Fluctuating light 22/17 °C, 12/12 h Light/Dark	Higher germination ratio	Quijia-Lamiña et al. (2023)
Dwarf spurge *Euphorbia nanus*	25 °C in the presence of light	66.66% germination rate	Catana et al. (2013)
Bamboo *Dendrocalamus membranaceus* Munro	Dark conditions, high temperatures around 30 °C	63.13% germination rate	Brar et al. (2013)
Hop *Humulus lupulus* L.	Stratification at 4 °C for 15 days	Increase of germination from 15% to 23%	Liberatore et al. (2018)
Alstroemeria *Alstroemeria ligtu* hybrid	Stratification in 5 ± 1 °C for 21 days plus soaking for 24 h in 100 mg/l GA_3	76.67% germination rate	Nasri et al. (2013)
Hairy St. John's Wort *Hypericum adenotrichum* Spach	18 °C conditions after stratification at 4 °C	Highest germination value 40%	Yamaner and Erdağ (2021)
Sea daffodil *Pancratium maritimum* L.	At 20–25 °C	Fast and high germination value over 70%	Alexopoulos et al. (2022)
Potato *Solanum tuberosum* L.	25 °C	Range of seeds germination for different cultivars - 70% to 86.66%	Dieme and Sy (2013)
Black plum *Vitex doniana*	25 °C–35 °C	High germination rate 80–87%	Haruna et al. (2024)

(continued)

Table 1.3 (continued)

Plant object	Factor used to improve germination	Germination description	References
Whorled sage *Salvia verticillata*	At −80 °C for 4 days, in dark at 20 ± 1 °C	57–67% germination rate	Tursun (2020), Yilmaz Gokdogan and Burun (2022)
Light			
Thale cress *Arabidopsis thaliana*	Monochromatic lights within the 300–780 nm wavelength range and extremely low fluence levels	Highest germination	Shinomura et al. (1996)
Galax *Galax urceolata* (Poir.) Brummitt	Light exposure at 97 μmol/m^2/s and minimal light at 17 μmol/m^2/s	Substantially enhanced seed germination rates, achieving 64% and 56%, respectively	Yang et al. (2013)
Greek mountain tea *Sideritis clandestina* subsp. *pelopponesiaca*	LED-BR-80 combined with 250 mg/L GA_3	Germination rate 40%, at day 55	Sarropoulou et al. (2024)
Genip *Genipa americana* L.	Stratification in the dark for 16 days	100% germination, with an average time of 30 days	de Souza et al. (2016)
Two-color sage *Salvia dichroantha* L.	In darkness, 20 °C	80% germination rate	Thanos and Doussi (1995), Yilmaz Gokdogan and Burun (2022)
Fruit-leaf sage *Salvia pomifera* ssp. *pomifera*	In darkness, 15 °C	70% germination rate	Thanos and Doussi (1995), Yilmaz Gokdogan and Burun (2022)
Algerian thyme *Thymus algeriensis* Boiss & Reut.	At ambient temperature (23 ± 2 °C) and continuous darkness	94.43% germination rate	Mehalaine et al. (2017)
White horehound *Marrubium vulgare* L.	At ambient temperature (23 ± 2 °C) and continuous darkness	57.76% germination rate	Mehalaine et al. (2017)
Saint Stephen's Peony *Paeonia corona-sancti-stefani.*	16/8 photoperiod 4 °C	High germination rate	Catana et al. (2013)
Greek mountain tea *Sideritis scardica*	5-year-old cold stratification 4 °C under LED-BGYOR-40 with 250 mg/L GA_3	Higher germination rates 80%	Sarropoulou et al. (2024)

(continued)

Table 1.3 (continued)

Plant object	Factor used to improve germination	Germination description	References
Sacred fig (*Ficus religiosa* L.)	24 h light	83.33% germination rate	Hesami et al. (2018)
Crispa orchid *Schomburgkia crispa* Lindl.	3000 K yellow LED	100% germination rate	Sorgato et al. (2021)
Pre-soaking with water and some additional components			
Genipap or jagua *Genipa americana* L.	Seeds were pre-soaked in distilled water for 48 h, followed by 16 days of stratification in the dark	High germination rate	de Souza et al. (2016)
Rosary pea *Abrus precatorius* Linn.	Seeds were immersed in water for durations of 2, 4, 6, 8, 10, and 12 h	Germination rates 60%, 60%, 90%, 50%, 50% and 30%, respectively	Qadir et al. (2012)
African camel's foot *Bauhinia thonningii* Schum.	Soaking in hot water for 10 min	40 and 53.3% germination for small and large seeds	Mwase and Mvula (2011)
Timroo *Zanthoxylum armatum* DC.	Pre-soaking seeds for 48 h gibberellic acid 200 ppm + kinetin 100 ppm	72.50% germination rate	Datt et al. (2017)
African locust bean tree *Parkia biglobosa*	Pre-soaking by (98%) H_2SO_4 for 5 min, then washed with sterile distilled water	87.76% germination rate	Abbas et al. (2018)
Kotschy's dragonhead *Dracocephalum kotschyi* Boiss	Soaking with hot water for 24 h	Germination rate more than 50%	Fattahi et al. (2011)
Snake bean tree *Bobgunnia madagascariensis*	Seeds were soaked in either cold or hot water for a duration of up to 24 h	Near 70% germination rate	Thokozani et al. (2011)
African baobab *Adansonia digitata* L.	75-day-old seed soaked in tap water for 1–10 days	No effect on the germination	Singh et al. (2010)

stratification, seeds exhibit increased functions of α-amylase and β-endo-mannanase, enzymes crucial for preparing substrates that facilitate rapid embryonic axis expansion upon light exposure (Goggin et al. 2011). For optimal in vitro germination of *Genipa americana* L. seeds, presaking in distilled water for 48 h followed by 16 days of dark stratification is necessary, with subsequent exposure to white fluorescent light (de Souza et al. 2016). Therefore, soaking in a 100 mg/L GA_3 solution and incubating at 20 °C under light conditions significantly improved germination rates (97.3%) (Payal et al. 2013). Additionally, halopriming with KNO_3 or applying gibberellic acid (GA_3) increases germination rates and reduces the mean germination time for crops such as *Cichorium intybus* (Tzortzakis 2009).

Determining the optimal concentration of growth regulators, such as cytokinins or gibberellins, is essential for promoting uniform and rapid seed germination across different species. For example, direct adventitious shoot regeneration from mature stem explants of *Withania somnifera* (a specific endemic plant) was effectively achieved via the combination of kinetin, 1-naphthalene acetic acid (NAA), and gibberellic acid (GA_3), with the best results (Sivanesan and Park 2015). The use of plant growth regulators, alongside treatments such as presoaking and temperature control, varies by plant species. The development of an effective in vitro seed germination protocol requires a deep understanding of the specific growth conditions of the chosen plant species, particularly for endemic or rare plants.

Experimental research examining the impact of temperature variations (25 °C, 22/11 °C, 27/15 °C, 29/19 °C, and 33/24 °C) and chilling at 10 °C on in vitro seed germination across diverse populations of *Calopogon tuberosus* var. *tuberosus* revealed that seeds from Michigan germinated more effectively under cooler conditions, whereas seeds from South Carolina and North Central Florida presented better germination rates under warmer temperatures. These findings highlight the importance of evaluating growth conditions and seed provenance to establish efficient protocols for in vitro seed germination (Kauth et al. 2011).

In *Arabidopsis*, seed germination is significantly influenced by phytochrome B (PhyB), the primary photoreceptor responsible for initiating germination in response to red light (Jiang et al. 2019). Current research is increasingly focused on the role of colored light in the biosynthesis of secondary metabolites under in vitro cultures (Gabr et al. 2022). Additionally, colored light may also affect the seed germination process (see Table 1.3). This emerging area of study could lead to further insights into optimizing light conditions for both seed germination and secondary metabolite production.

In vitro seed germination is a complex biological process influenced by both genetic (genotype) and physical factors (such as medium composition and environmental conditions). As a result, advanced mathematical techniques such as artificial neural networks are ideal for analyzing data and optimizing the conditions in this intricate system. A validation experiment on in vitro seed germination of industrial hemp (*Cannabis sativa* L.) revealed that the discrepancy between the experimental data and the optimized outcome was minimal (Hesami et al. 2021). Consequently, GRNN-NSGA-II demonstrated accurate predictions of seed germination and holds the potential for optimizing various factors involved in the in vitro culture of this versatile crop.

Another important factor that can regulate the in vitro germination process is moisture availability. Adequate moisture is essential for seed imbibition, as it initiates the physiological processes necessary for germination. The presence of water activates enzymes that breakdown stored nutrient compound reserves, providing the energy and nutrients required for seedling growth (Joshi 2018). In addition to moisture, oxygen is a critical factor in the seed germination process, as it facilitates the resumption of respiration and the subsequent reactivation of metabolism during seed imbibition (El-Maarouf-Bouteau et al. 2022). This leads to the generation of reducing power and ATP, which are essential for germination.

Hydrogel and controlled release water for humidity control are able to maintain optimal moisture availability during seed germination for in vitro culture (Gholizadeh et al. 2023; Garcia-Garcia et al. 2024). However, hydrogels offer certain advantages over agar in specific contexts: they have a superior water-holding capacity, providing a more consistent moisture environment, which is especially beneficial for seeds sensitive to drying (Oladosu et al. 2022; Sudheer et al. 2023). Additionally, some hydrogels release water gradually, helping to maintain optimal hydration levels for seeds over time. Hydrogels can also be applied directly as seed coatings, improving moisture availability at the seed level and increasing transplantation success (Kaur et al. 2023).

Oxygen plays a crucial role in regulating the molecular networks that control seed germination and dormancy through various mechanisms. It is essential for maintaining hormonal balance and particularly affects key hormones such as ethylene, abscisic acid (ABA), and gibberellins (GA). Oxygen levels significantly influence hormone signaling pathways, especially ABA sensitivity (González-Guzmán et al. 2021). The interaction between ABA and other phytohormones, including auxin, gibberellins (GA), cytokinin (CK), ethylene (ET), salicylic acid (SA), and jasmonic acid (JA), is critical for mitigating environmental stress (Rai et al. 2024). Therefore, the influence of environmental factors and their regulation must be carefully considered during the in vitro germination process.

Furthermore, mitochondria play crucial roles in reactive oxygen species (ROS) production under hypoxic conditions, impacting the germination process. The photosynthetic activity of seeds considerably alleviates hypoxic constraints and may even induce transient localized hyperoxia (Borisjuk and Rolletschek 2009). Abiotic conditions under in vitro culture must be optimized to prevent premature senescence, which is characterized by the degradation of chloroplasts and mesophyll tissues (Rodrigues et al. 2014). Additionally, the N-degron pathway is involved in the turnover of proteins that contribute to seed tolerance to hypoxia, highlighting the multifaceted role of oxygen in regulating in vitro germination processes (Corbineau 2022).

Oxygen regulation can be achieved through proper aeration techniques. This includes designing culture vessels that facilitate gas exchange and using oxygen-enriched media (Zobayed 2008) or bioreactors to maintain optimal oxygen levels, thereby supporting respiration and metabolic processes. Increasing mitochondrial function by adjusting culture conditions can also help manage reactive oxygen species (ROS) production, promoting healthier seedling development (Considine and Foyer 2021).

Moisture availability also plays a vital role in in vitro culture. Careful control of moisture levels in in vitro culture media is essential (Paul et al. 2012), and techniques such as the use of hydrogels or controlled-release water sources can help maintain optimal humidity for seed imbibition and germination. The hydrogen can function as a seed coating agent and (iv) improve the transplantation success rate (Kaur et al. 2023). Additionally, selecting appropriate substrates that retain moisture while allowing for proper drainage creates an ideal environment for seed germination. For example, embryos of *Orbignya phalerata* Mart. have demonstrated tolerance to desiccation, although a decrease in seed moisture content negatively impacts seedling length (da Silva et al. 2012). The use of humidity chambers or controlled environments to maintain high humidity levels can further support water absorption and promote the germination process.

Another crucial factor is the optimization of hormonal balance. The application of optimal concentrations of plant growth regulators (PGRs), including gibberellins, abscisic acid, and ethylene, plays a vital role in regulating seed germination and promoting seedling development (Patil et al. 2012; Ribalta et al. 2019; Pasternak and Steinmacher 2024). Ethylene is essential for overcoming dormancy in numerous species, with effective concentrations for stimulating the germination of dormant seeds typically ranging from 0.1 to 200 μL/L (Corbineau et al. 2014). For certain plant species, such as *Atropa acuminata* Royal ex Lindl., MS media without the addition of growth regulators does not induce any callus formation (Dar et al. 2021). Adjusting hormone ratios in the culture medium to align with the specific needs of the plant species can result in enhanced outcomes. Understanding and manipulating hormone signaling pathways through the addition of specific compounds can enhance abscisic acid sensitivity and other mechanisms essential for germination and growth.

Finally, monitoring and adjusting all the parameters are key to successful in vitro culture. It can optimize in vitro culture conditions, improve seed germination rates, promote healthy seedling growth, and ultimately increase plant production.

Consistently maintaining the right balance of moisture, oxygen, and suitable substrates prevents premature senescence, which is characterized by chloroplast and tissue degradation. The use of controlled humidity chambers, the selection of substrates that retain moisture without waterlogging, and the use of hydrogels or agar for seed coatings create an ideal environment for seedling growth and development.

Photosynthetic activity during germination can help alleviate hypoxia by creating transient hyperoxia conditions, reducing ROS accumulation (Borisjuk and Rolletschek 2009). Sheath oxygen (O_2) production helps reduce low-oxygen stress inside seeds, which can increase storage activity in the endosperm. This process creates better conditions for seeds to mature and germinate successfully (Brodersen and Kühl 2023). Adjusting techniques of in vitro culture cultivation, such as light and oxygen levels, helps manage ROS levels, thereby supporting healthy mitochondrial function (Rauf et al. 2024) and promoting seedling development.

Regular monitoring and fine-tuning of culture parameters ensure that in vitro *techniques* are optimized for the germination process and well-developed seedlings. Adjustments on the basis of observed growth and environmental response lead to improved seed germination rates, increased seedling development, and ultimately increased plant production.

1.6 Conclusion

Optimizing seed germination through in vitro culture techniques is a pivotal approach for promoting the cultivation of plants, conserving rare species, and addressing global agricultural challenges. By meticulously refining disinfection methods, nutrient media compositions, and environmental factors, researchers can significantly increase germination success and seedling vigor. These advancements are particularly valuable for rare and endangered plants, as illustrated by notable examples: Raspberry (*Rubus idaeus* L.), tree turmeric (*Coscinium fenestratum*), Dracula orchid (*Dracula felix*), Alstroemeria (*Alstroemeria ligtu hybrid*), and sacred fig (*Ficus religiosa*), with germination ratios ranging from 70 to 80%.

The data exemplify the impact of tailored environmental and hormonal interventions in achieving high germination rates for diverse plant species, including rare and medicinal ones. Moreover, the integration of modern biotechnological tools, such as molecular and genetic techniques, further expands the possibilities for fine-tuning germination protocols, thus increasing biodiversity conservation efforts.

As global challenges related to food security and environmental sustainability intensify, these methodologies bridge the gap between ecological preservation and agricultural innovation. By fostering the propagation of rare and endangered species, in vitro germination optimization contributes to both biodiversity and the sustainable utilization of plant resources, which are significantly impacted, ensuring long-term ecological and economic benefits.

Acknowledgments The authors express their gratitude to the BEWARE Fellowship Program for supporting the research presented in this study.

Competing Interests The authors have no relevant financial or non-financial interests to disclose.

References

Abbas MS, El-Shabrawi HM, Soliman AS, Selim MA (2018) Optimization of germination, callus induction, and cell suspension culture of African locust beans *Parkia biglobosa* (Jacq.). Benth J Genet Eng Biotechnol 16(1):191–201

Abubakar MS, Attanda ML (2022) Factors that cause seed dormancy. IntechOpen. https://doi.org/10.5772/intechopen.107004

Agogbua JU, Okoli BE (2022) Procedure for in vitro seed sterilization, germination and aseptic seedling establishment of *Zehneria capillacea* (Schumach) C. Jeffrey GSC Biol Pharm Sci 19(01):143–148

Ahammed GJ, Gantait S, Mitra M, Yang Y, Li X (2020) Role of ethylene crosstalk in seed germination and early seedling development: A review. Plant Physiol Biochem 151:124–131

Al-Ahmad H (2020) In vitro decoated seed germination and seedling development for propagation of wild mandrake (*Mandragora autumnalis* Bertol.). Plan Theory 9(10):1339

Alexopoulos AA, Mavrommati E, Kartsonas E, Petropoulos SA (2022) Effect of temperature and sucrose on in vitro seed germination and bulblet production of *Pancratium maritimum* L. Agronomy 12:2786

Ali F, Qanmber G, Li F, Wang Z (2021) Updated role of ABA in seed maturation, dormancy, and germination. J Adv Res 35:199–214
Arc E, Sechet J, Corbineau F, Rajjou L, Marion-Poll A (2013) ABA crosstalk with ethylene and nitric oxide in seed dormancy and germination. Front Plant Sci 4:63
Balde S, Ayessou NC, Sambou A, Niane K, Faye PG, Cisse M, Diop CM (2021) Effects of seed pretreatment on germination, growth and yield of *Momordica charantia* L. Agric Sci 12(1):39–57
Baskin CC, Baskin JM (1998). Seeds: ecology, biogeography, and evolution of dormancy and germination. Academic Press, San Diego. 666 pp.
Baskin CC, Baskin JM (2014) Seeds: ecology, biogeography, and evolution of dormancy and germination. Academic Press
Bernareggi G, Carbognani M, Mondoni A, Petraglia A (2016) Seed dormancy and germination changes of snowbed species under climate warming: the role of pre- and post-dispersal temperatures. Ann Bot 118(3):529–539
Bewley JD, Bradford KJ, Hilhorst HW, Nonogaki H (2013) Seed biology and technology. In: Seed science and technology. Springer, pp 1–400
Bhatt A, Caron MM, Chen X, Yu D, Niu Y (2022) Effect of temperature, light and storage on seed germination of *Salvia plebeia* R.Br., *Leonurus japonicus* Houtt., *Mosla scabra* (Thunb.) CYWu & HWLi and *Perilla frutescens* (L) Britton. J Appl Res Med Aromat Plant 29:100402
Borisjuk L, Rolletschek H (2009) The oxygen status of the developing seed. New Phytol 182:17–30
Bowen-O'Connor CA, Hubstenberger J, Killough C, VanLeeuwen DM, Hilaire RS (2007) *In vitro* propagation of *Acer grandidentatum* Nutt. In Vitro Cell Dev Biol-Plant 43(1):40–50
Brar J, Anand M, Sood A (2013) *In vitro* seed germination of economically important edible bamboo *Dendrocalamus membranaceus* Munro. Indian J Exp Biol 51:88–96
Breman E, Ballesteros D, Castillo-Lorenzo E, Cockel C, Dickie J, Faruk A, O'Donnell K, Offord CA, Pironon S, Sharrock S, Ulian T (2021) Plant diversity conservation challenges and prospects – the perspective of botanic gardens and the Millennium Seed Bank. Plants 10(11):2371
Brodersen KE, Kühl M (2023) Photosynthetic capacity in seagrass seeds and early-stage seedlings of *Zostera marina*. New Phytol 239(4):1300–1314
Bruegmann MM, Caraway V, Maunder M (2003) A safety net for Hawaii's rarest plants. Endangered Species Update, 20(1): S8+. Gale Academic OneFile. Accessed 21 Oct 2024
Carrera-Castaño G, Calleja-Cabrera J, Pernas M, Gómez L, Oñate-Sánchez L (2020) An updated overview on the regulation of seed germination. Plants 9(6):703
Catana R, Holobiuc I, Moldoveanu M (2013) *In vitro* seed germination in three rare taxa from the Romanian Carpathians flora. Muzeul Olteniei Craiova. Oltenia Studii şi comunicări Ştiinţele Nat 29(1):85–92
Chen D, Yuan Z, Luo XP, Yang L, Chen X, Zhang Z, Wang J, Hu X (2021) Cold stratification requirements for seed dormancy-break differ in soil moisture content but not duration for alpine and desert species. Plant Soil 471:393–407
Cheng Z, Wang Y, Zhang L (2018) Effect of mercuric chloride on seed germination and viability in tissue culture of plant species. Plant Growth Regul 86(2):295–303
Cheng J, Huang H, Liu W, Zhou Y, Han W, Wang X, Zhang Y (2022) Unraveling the effects of cold stratification and temperature on the seed germination of invasive *Spartina alterniflora* across latitude. Front Plant Sci 13:911804
Ciacka K, Krasuska U, Staszek P, Wal A, Zak J, Gniazdowska A (2020) Effect of nitrogen reactive compounds on aging in seed. Front Plant Sci 11:1011
Claeys H, De Bodt S, Inzé D (2014) Gibberellins and DELLAs: central nodes in growth regulatory networks. Trends in Plant Sci 19:231–239
Considine MJ, Foyer CH (2021) Oxygen and reactive oxygen species-dependent regulation of plant growth and development. Plant Physiol 186(1):79–92
Corbineau F (2022) Oxygen, a key signalling factor in the control of seed germination and dormancy. Seed Sci Res 32(3):126–136
Corbineau F (2024) Ethylene, a signaling compound involved in seed germination and dormancy. Plan Theory 13:2674

Corbineau F, Xia Q, Bailly C, El-Maarouf-Bouteau H (2014) Ethylene, a key factor in the regulation of seed dormancy. Front Plant Sci 5:539
Corlett RT (2016) Plant diversity in a changing world: status, trends, and conservation needs. Plant Divers 38(1):10–16
Cuba-Díaz M, Rivera-Mora C, Navarrete E et al (2020) Advances of native and non-native Antarctic species to *in vitro* conservation: improvement of disinfection protocols. Sci Rep 10:3845
Dar SA, Nawchoo IA, Tyub S, Kamili AN (2021) Effect of plant growth regulators on in vitro induction and maintenance of callus from leaf and root explants of *Atropa acuminata* royal ex Lindl. Biotechnol Rep 32:e00688
Datt G, Chauhan JS, Ballabha R (2017) Influence of pre-sowing treatments on seed germination of various accessions of Timroo (*Zanthoxylum armatum* DC.) in the Garhwal Himalaya. J Appl Res Med Aromat Plant 7:89–94
Debergh PC, Zimmerman RH (1991) Micropropagation: technology and application. Springer
Diantina S, Kartikaningrum S, McCormick AC et al (2020) Comparative *in vitro* seed germination and seedling development in tropical and temperate epiphytic and temperate terrestrial orchids. Plant Cell Tissue Organ Cult 143:619–633
Dieme A, Sy MO (2013) Environmental, morphological and physiological factors analyses for optimization of potato (*Solanum tuberosum* L.) microtuber *in vitro* germination. Adv Biosci Biotechnol 4(11):986–992
El-Maarouf-Bouteau H (2022) The seed and the metabolism regulation. Biology (Basel), 11(2), 168. https://doi.org/10.3390/biology11020168
El-Maarouf-Bouteau H, Bailly C (2008) Oxidative signaling in seed germination and dormancy. Plant Signal Behav 3(3):175–182
Ermis S, Özden E, Yıldırım E (2024) Seeds of resilience: physiology and mechanisms of Hardseededness. In: Seed biology – new advances. https://doi.org/10.5772/intechopen.1003847
Fadladeen LH, Toma RS, Mohammed MA, Ahmed A, Shaheen AA, Ahmed HB (2021) Response of red maple (*Acer rubrum* L.) micropropagation to different *in vitro* conditions. J Plant Prod 12(6):651–655
Fanzo J, Miachon L (2023) Harnessing the connectivity of climate change, food systems and diets: taking action to improve human and planetary health. Anthropocene 42:100381
Farooq MA, Ma W, Shen S, Gu A (2022) Underlying biochemical and molecular mechanisms for seed germination. Int J Mol Sci 23(15):8502
Fattahi M, Nazeri V, Sefidkon F et al (2011) The effect of pre-sowing treatments and light on seed germination of *Dracocephalum kotschyi* Boiss: an endangered medicinal plant in Iran. Hortic Environ Biotechnol 52:559–566
Finch-Savage WE, Leubner-Metzger G (2006) Seed dormancy and the control of germination. New Phytol 171(3):501–523, https://doi.org/10.1111/j.1469-8137.2006.01787.x
Fu Y, Ma L, Li J, Hou D, Zeng B, Zhang L, Liu C, Bi Q, Tan J, Yu X et al (2024) Factors influencing seed dormancy and germination and advances in seed priming technology. Plan Theory 13:1319
Funnekotter B, Mancera RL, Bunn E (2023) A simple but effective combination of pH indicators for plant tissue culture. Plan Theory 12:740
Gabr AMM, Fayek NM, Mahmoud HM, El-Bahr MK, Ebrahim HS, Sytar O, El-Halawany AM (2022) Effect of light quality and media components on shoot growth, rutin, and quercetin production from common buckwheat. ACS Omega 7(30):26566–26572
Gambor, OL, Miller RA, Ojima K (1968) Nutrient requirements of suspension cultures of soybean root cells. Experimental Cell Research, 50, 151–158. https://doi.org/10.1016/0014-4827(68)90403-5
Gamborg OL, Phillips GC (2013) Plant cell, tissue and organ culture: fundamental methods. Springer Science & Business Media
Gammoudi N, Nagaz K, Ferchichi A (2022) Establishment of optimized *in vitro* disinfection protocol of *Pistacia vera* L. explants mediated a computational approach: multilayer perceptron–multi–objective genetic algorithm. BMC Plant Biol 22:324

Garcia-Garcia A, Muñana-González S, Lanceros-Mendez S, Ruiz-Rubio L, Alvarez LP, Vilas-Vilela JL (2024) Biodegradable natural hydrogels for tissue engineering, controlled release, and soil remediation. Polymers 16(18):2599

Ge W, Bu H, Wang X, Xia Y, Martinez SA, Wang X, Qi W, Liu K, Du G (2020) Changes in endogenous hormone contents during seed germination of *Anemone rivularis* var. flore-minore. Glob Ecol Conserv 24:e01200

Ge N, Jia JS, Yang L, Huang RM, Wang QY, Chen C, Meng ZG, Li LG, Chen JW (2023) Exogenous gibberellic acid shortening after-ripening process and promoting seed germination in a medicinal plant *Panax notoginseng*. BMC Plant Biol 23(1):67

Geneve RL (1998) Seed dormancy in commercial vegetable and flower species. Seed Technol. Special Publication – A Symposium: Vegetable and Flower Seed Quality 20(2):236–250

George EF, Hall MA, De Klerk GJ (2008) Plant propagation by tissue culture, 3rd edn. Springer

Gholizadeh H, Landh E, Silva DM, Granata A, Traini D, Young PM, Fathi A, Maleknia SD, Abrams T, Dehghani F, Ong HX (2023) *In vitro* and *in vivo* applications of a universal and synthetic thermo-responsive drug delivery hydrogel platform. Int J Pharm 635(19):122777

Gilbert GS, Diaz A, Bregoff HA (2023) Seed disinfestation practices to control seed-borne fungi and bacteria in home production of sprouts. Food Secur 12(4):747

Goggin DE, Emery RJ, Powles SB, Steadman KJ (2010) Initial characterisation of low and high seed dormancy populations of *Lolium rigidum* produced by repeated selection. J Plant Physiol 167(15):1282–1288

Goggin DE, Powles SB, Toorop PE, Steadman KJ (2011) Dark-mediated dormancy release in stratified *Lolium rigidum* seeds is associated with higher activities of cell wall-modifying enzymes and an apparent increase in gibberellin sensitivity. J Plant Physiol 168(6):527–533

Gómez MV, Alcaraz CF, Cañizares AE (2020) Comparison of different sterilization methods for seed disinfection in in vitro cultures. Biotechnol Rep 28:e00573

González-Benito ME, González-Mas MC, Latorre J (2019) Hydrogen peroxide in plant tissue culture: a review. Plant Cell Tissue Organ Cult 139(3):441–453

González-Guzmán M, Gómez-Cadenas A, Arbona V (2021) Abscisic acid as an emerging modulator of the responses of plants to low oxygen conditions. Front Plant Sci 12:661789

Górnik K, Sas-Paszt L, Seliga Ł, Pluta S, Derkowska E, Głuszek S, Sumorok B, Mosa WFA (2023) The effect of different stratification and scarification treatments on breaking the dormancy of Saskatoon berry seeds. Agronomy 13(2):520

Guo W, Xing Y, Luo X, Li F, Ren M, Liang Y (2023) Reactive oxygen species: a crosslink between plant and human eukaryotic cell systems. Int J Mol Sci 24(17):13052

Gupta R, Chakrabarty SK (2013) Gibberellic acid in plants: still a mystery unresolved. Plant Signal Behav 8(9):e25504

Guzmán-Ortiz FA, Castro-Rosas J, Gómez-Aldapa CA, Mora-Escobedo R, Rojas-León A, Rodríguez-Marín ML, Falfán-Cortés RN, Román-Gutiérrez AD (2018) Enzyme activity during germination of different cereals: a review. Food Rev 35(3):177–200

Haruna M, Nakhooda M, Shaik S (2024) Development of seed germination and *in vitro* propagation protocols for *Vitex doniana*. S Afr J Bot 167:399–409

Hasnain A, Naqvi SAH, Ayesha SI, Khalid F, Ellahi M, Iqbal S, Hassan MZ, Abbas A, Adamski R, Markowska D, Baazeem A, Mustafa G, Moustafa M, Hasan ME, Abdelhamid MMA (2022) Plants *in vitro* propagation with its applications in food, pharmaceuticals and cosmetic industries; current scenario and future approaches. Front Plant Sci 13:1009395

Hesami M, Daneshvar MH, Yoosefzadeh-Najafabadi M (2018) Establishment of a protocol for *in vitro* seed germination and callus formation of *Ficus religiosa* L., an important medicinal plant. Jundishapur J Nat Pharm Prod 13(4):e62682

Hesami M, Pepe M, Monthony AS, Baiton A, Jones AM (2021) Modeling and optimizing *in vitro* seed germination of industrial hemp (*Cannabis sativa* L.). Ind Crop Prod 170:113753

Holdsworth M, Bentsink L, Soppe WJJ (2008) Molecular networks regulating *Arabidopsis* seed maturation, after-ripening, dormancy, and germination. New Phytol 179:33–54

Hong JK, Baek J, Park SR, Lee GS, Suh EJ (2023) A new protocol to mitigate damage to germination caused by black layers in maize (*Zea mays* L.) seeds. Agriculture 13:2147

Huan LVT, Tanaka M (2004) Callus induction from protocorm-like body segments and plant regeneration in *Cymbidium* (*Orchidaceae*). J Hortic Sci Biotechnol 79(3):406–410
Huang H, Ullah F, Zhou DX, Yi M, Zhao Y (2019) Mechanisms of ROS regulation of plant development and stress responses. Front Plant Sci 10:800
Hubert B, Leprince O, Buitink J (2024) Sleeping but not defenceless: seed dormancy and protection. J Exp Bot 75(19):6110–6124
Hussain MI, Iqbal MZ, Hussain A (2019) *In vitro* propagation of underutilized medicinal plants: a review. J Med Plants Res 13(7):124–136
Iliev I, Scaltsoyiannes A, Tsaktsira M, Gajdosova A (2010) Micropropagation of *Betula pendula* Roth cultivars by adventitious shoot induction from leaf callus. Acta Hortic 885:161–173
Isah T (2019) Stress and defense responses in plant secondary metabolites production. Biol Res 52:39
Jeevan Kumar SP, Chintagunta AD, Reddy YM, Rajjou L, Garlapati VK, Agarwal DK, Prasad SR, Simal-Gandara J (2021) Implications of reactive oxygen and nitrogen species in seed physiology for sustainable crop productivity under changing climate conditions. Curr Plant Biol 26:100197
Jia W, Wang Y, Mi Z, Wang Z, He S, Kong D (2022) Optimization of culture medium for in vitro germination and storage conditions of *Exochorda racemosa* pollen. Front Plant Sci 13:994214
Jiang Z, Xu G, Jing Y, Tang W, Lin R (2019) Phytochrome B and REVEILLE1/2-mediated signalling controls seed dormancy and germination in *Arabidopsis*. Nat Commun 7:12377
Joshi R (2018) Role of enzymes in seed germination. International Journal of Creative Research Thoughts (IJCRT), 6(2), 1481–1485
Kaur P, Agrawal RM, Pfeffer FM et al (2023) Hydrogels in agriculture: prospects and challenges. J Polym Environ 31:3701–3718
Kauth PJ, Kane ME, Vendrame WA (2011) Comparative *in vitro* germination ecology of *Calopogon tuberosus* var. tuberosus (*Orchidaceae*) across its geographic range. In Vitro Cell Dev Biol-Plant 47:148–156
Khan MIR, Fatma M, Per TS (2018) Hydrogen peroxide: a key player in plant abiotic stress responses. Plant Physiol Biochem 132:287–296
Khan MIR, Zain SM, Shafique MS (2019) Toxicity assessment of mercuric chloride on plant seeds and its alternatives in plant tissue culture. Environ Sci Pollut Res 26(30):31064–31075
Khan M, Ali S, Al Azzawi TNI, Saqib S, Ullah F, Ayaz A, Zaman W (2023) The key roles of ROS and RNS as signaling molecules in plant-microbe interactions. Antioxidants 12(2):268
Klupczyńska EA, Pawłowski TA (2021) Regulation of seed dormancy and germination mechanisms in a changing environment. Int J Mol Sci 22(3):1357
Koornneef M, Jorna ML, Der Swan DLCB-V, Karssen CM, Der Swan DLB-V (1982) The isolation of abscisic acid (ABA) deficient mutants by selection of induced revertants in non-germinating gibberellin sensitive lines of *Arabidopsis thaliana* (L.) Heynh. Theor Appl Genet 61:385–393
Koornneef M, Reuling G, Karssen CM (1984) The isolation and characterization of abscisic acid-insensitive mutants of *Arabidopsis thaliana*. Physiol Plant 61:377–383
Kulak V, Longboat S, Brunet ND, Shukla M, Saxena P (2022) *In vitro* technology in plant conservation: relevance to biocultural diversity. Plan Theory 11(4):503
Kumar A, Tripathi D, Singh V (2021) Use of hydrogen peroxide as a plant disinfectant: effectiveness and safety considerations. J Plant Pathol 103(1):23–35
Kunakhonnuruk B, Inthima P, Kongbangkerd A (2018) *In vitro* propagation of *Epipactis flava* Seidenf., an endangered rheophytic orchid: a first study on factors affecting asymbiotic seed germination, seedling development, and greenhouse acclimatization. Plant Cell Tissue Organ Cult 135:419–432
Lamine N'Doye A, Nahr SM, Ourèye SM (2012) Propagation of African Baobab (*Adansonia digitata* L., *Bombacoideae*, *Malvaceae*) germplasm through *in vitro* cloning. Adv Env Biol 6(10):2749–2757
Lamont BB, Pausas JG (2023) Seed dormancy revisited: dormancy-release pathways and environmental interactions. Funct Ecol 37:1106–1125

Lee SY, Park K, Jang BK, Ji B, Lee H, Baskin CC, Cho JS (2022) Exogenous gibberellin can effectively and rapidly break intermediate physiological dormancy of *Amsonia elliptica* seeds. Front Plant Sci 13:1043897

Li C, Na D, Yamei Z, Shasha W, Zhongjian L, Junwen Z (2021) A review for the breeding of orchids: current achievements and prospects. Hortic Plant J 7(5):2095–9885

Liberatore CM, Mattion G, Rodolfi M, Ganino T, Fabbri A, Chiancone B (2018) Chemical and physical pre-treatments to improve *in vitro* seed germination of *Humulus lupulus* L., cv. Columbus Sci Hortic 235:86–94

Lindsey BE, Rivero L, Calhoun CS, Grotewold E, Brkljacic J (2017a) Standardized method for high-throughput sterilization of *Arabidopsis* seeds. J Vis Exp 128:56587

Lindsey K, Pullen ML, Topping JF (2017b) Basic plant culture protocols. Humana Press

Luna B, Piñas-Bonilla P, Zavala G, Pérez B (2022) Effects of light and temperature on seed germination of eight *cistus* species. Seed Sci Res 32(2):86–93

Ma Z, Bykova NV, Igamberdiev AU (2017) Cell signaling mechanisms and metabolic regulation of germination and dormancy in barley seeds. Crop J 5:459–477

Maharjan S, Pradhan S, Thapa BB, Pant B (2019) In vitro propagation of endangered orchid, *Vanda pumila* Hook.f. through protocorms culture. Am J Plant Sci 10(7):1220–1232

Maiti SK (2013) Seeds: collection, storage, and treatment. In: Ecorestoration of the coalmine degraded lands. Springer, India

Malik R, Kaur J, Kumar D (2021) Challenges and advancements in the use of disinfectants for plant tissue culture: a review. J Plant Biochem Biotechnol 30(1):1–15

Mallon R, Onrubia M, Lota F, Corral P, Lorrain B, Michoux F (2014) *In vitro* propagation methods of the succulent plant *Euphorbia resinifera* and quantification of its major compounds. Planta Med 80:P2F2

Mangena P, Mokwala PW (2019) The influence of seed viability on the germination and *in vitro* multiple shoot regeneration of soybean (*Glycine max* L.). Agriculture 9(2):35

McCown BH, Lloyd G (1981) Woody Plant Medium (WPM)—A Mineral Nutrient Formulation for Microculture of Woody Plant Species. HortScience, 16, 453–453.

Meesakul P, Shea T, Fenstemacher R, Wong SX, Kuroki Y, Wada A, Cao S (2023) Phytochemistry and biological studies of endemic Hawaiian plants. Int J Mol Sci 24(22):16323

Mehalaine S, Menasria T, Bouguessa S, Yahia A (2017) *In vitro* seed germination of some Algerian medicinal plants and the effect of gibberellic acid (GA_3) on breaking dormancy. J Mater Environ Sci 8(6):2034–2039

Miguel S, Michel C, Biteau F, Hehn A, Bourgaud F (2020) In vitro plant regeneration and *agrobacterium*-mediated genetic transformation of a carnivorous plant, *Nepenthes mirabilis*. Sci Rep 10(1):17482

Milbau A, Graae BJ, Shevtsova A, Nijs I (2009) Effects of a warmer climate on seed germination in the subarctic. Ann Bot 104(2):287–296

Miransari M, Smith DL (2014) Plant hormones and seed germination. Environ Exp Bot 99:110–121

Mosoh DA, Khandel AK, Verma S et al (2023) Effects of sterilization methods and plant growth regulators on *in vitro* regeneration and tuberization in *Gloriosa superba* (L.). In Vitro Cell Dev Biol Plant 59:792–807

Murashige T, Skoog F (1962) A revised medium for rapid growth and bio assays with tobacco tissue cultures. Physiol Plant 15:473–497

Mwase WF, Mvula T (2011) Effect of seed size and pre-treatment methods of *Bauhinia thonningii* Schum. On germination and seedling growth. Afr J Biotechnol 10(26):5143–5148

Narsai R, Edwards JM, Roberts TH, Whelan J, Joss GH, Atwell BJ (2015) Mechanisms of growth and patterns of gene expression in oxygen-deprived rice coleoptiles. Plant J 82:25–40

Nasri F, Mortazavi SN, Ghaderi N, Javadi T (2013) Propagation in vitro of Alstroemeria ligtu hybrid through direct organogenesis from leaf base. J. Hortic. Res. 21(2), 71–76. https://doi.org/10.2478/johr-2013-0018

Nasril F, Ghaderi N, Mohammadi J, Mortazavi SN, Saba MK (2013) The effect of gibberellic acid and stratification on germination of *Alstroemeria* (*Alstroemeria ligtu* hybrid) seed under in vitro and in vivo conditions. J Ornamental Plants 3(4):221–228

Nawrot-Chorabik K, Sułkowska M, Osmenda M, Mohytych V, Surówka E, Latowski D (2022) The impact of biotic and abiotic stress factors on development of European ash tissue cultures. Forests 13:59

Nhut DT, Hien NTT, Don NT, Khiem DV (2007) *In vitro* shoot development of *Taxus wallichiana* Zucc., a valuable medicinal plant. In: Jain SM, Häggman H (eds) Protocols for micropropagation of Woody trees and fruits. Springer, Dordrecht

Notarnicola RF, Nicotra AB, Kruuk LEB, Arnold PA (2023) Effects of warming temperatures on germination responses and trade-offs between seed traits in an alpine plant. J Ecol 111:62–76

Nitsch C, Nitsch JP (1967) The induction of flowering in vitro in stem segments of Plumbago indica L.. *Planta 72*, 355–370. https://doi.org/10.1007/BF00390146

Okoroafor UE (2022) Microbial contamination in plant tissue culture and elimination strategies. Nigerian Agric J 53(2):348–355

Oladosu Y, Rafii MY, Arolu F, Chukwu SC, Salisu MA, Fagbohun IK, Muftaudeen TK, Swaray S, Haliru BS (2022) Superabsorbent polymer hydrogels for sustainable agriculture: a review. Horticulturae 8(7):605

Parasharami VA et al (2007) Neem (*Azadirachta indica* A. Juss.): a potential source for plant tissue culture and micropropagation. Plant Cell Biotechnol Mol Biol 8(1–2):95–100

Pasternak TP, Steinmacher D (2024) Plant growth regulation in cell and tissue culture *in vitro*. Plants 13(2):327

Patil JG, Ahire ML, Nikam TD (2012) Influence of plant growth regulators on *in vitro* seed germination and seedling development of *Digitalis purpurea* L. Asian Aus J Plant Sci Biotechnol 6(1):12–18

Paul S, Kumaria S, Tandon P (2012) An effective nutrient medium for asymbiotic seed germination and large-scale *in vitro* regeneration of *Dendrobium hookerianum*, a threatened orchid of Northeast India. AoB Plants 2012:plr032

Payal K, Maikhuri RK, Rao KS, Kandari LS (2013) Effect of gibberellic acid- and water-based pre-soaking treatments under different temperatures and photoperiods on the seed germination of *Allium stracheyi* Baker: an endangered alpine species of Central Himalaya. India Plant Biosyst 148(6):1075–1084

Pérez MT, Villalobos VM (2016) *In vitro* propagation of *Swietenia macrophylla* King (Meliaceae) using Gamborg's B5 medium. Tree Physiol 18(7):469–472

Pergolotti V, Marcellini M, Contreras E, Mezzetti B, Gambardella M, Capocasa F, Sabbadini S (2023) Standardization of an *in vitro* seed germination protocol compared to acid scarification and cold stratification methods for different raspberry genotypes. Horticulturae 9:153

Phuong Quynh T, Lan Xuan HT, Ai My T, Thanh Thao N, Phuong Thao N, Thien Quang N (2022) Effect of calcium hypochlorite on surface sterilization and seedling growth of Vietnamese coconut varieties. Vietnam J Biotechnol 20(4):663–673

Pierik RLM (1997) *In vitro* culture of higher plants. Springer Science & Business Media

Puglia GD (2024) Reactive oxygen and nitrogen species (RONS) signalling in seed dormancy release, perception of environmental cues, and heat stress response. Plant Growth Regul 103:9–32

Qaderi MM (2023) Environmental regulation of weed seed dormancy and germination. Seeds 2:259–277

Qadir M, Khan F, Khan SS (2012) Effect of pre-treatment on the germination of *Abrus precatorius* Linn. by soaking in water. Indian J Appl Pure Biol 27(1):105–108

Quijia-Lamiña PH, Baquero LE, Kane ME, Zettler LW (2023) *In vitro* seed germination and seedling development of *Dracula felix* (Luer) Luer—an orchid native to Ecuador. Diversity 15:732

Radomir AM, Stan R, Florea A, Ciobotea CM, Bănuță FM, Negru M, Neblea MA, Sumedrea DI (2023) Overview of the success of *in vitro* culture for ex situ conservation and sustainable utilization of endemic and subendemic native plants of Romania. Sustainability 15:2581

Raghavan V (2005) Developmental biology of seed plants. Springer

Rai MK et al (2009) *In vitro* propagation of *Shorea robusta* from cotyledonary nodes. Plant Cell Rep 28(2):237–244

Rai MK, Kalia RK, Singh R, Gangola MP, Dhawan AK (2011) Developing stress-tolerant plants through *in vitro* selection—an overview of the recent progress. Environ Exp Bot 71(1):89–98
Rai A, Kumar G, Mushtaq KD, Sadiya MC, Kumar P, Shanti K, Martínez-Andújar C, Martínez-Melgarejo AA (2024) Unlocking nature's stress buster: abscisic acid's crucial role in defending plants against abiotic stress. Plant Stress 11:100359
Rathore TS, Tandon P, Shekhawat NS (1991) *In vitro* regeneration of pitcher plant (*Nepenthes khasiana* Hook f.) – a rare insectivorous plant of India. J Plant Physiol 139:246–248
Rauf A, Khalil AA, Awadallah S, Khan SA, Abu-Izneid T, Kamran M, Hemeg HA, Mubarak MS, Khalid A, Wilairatana P (2024) Reactive oxygen species in biological systems: pathways, associated diseases, and potential inhibitors—a review. Food Sci Nutr 12:675–693
Ray S, Vijayan J, Sarkar RK (2016) Germination stage oxygen deficiency (GSOD): an emerging stress in the era of changing trends in climate and rice cultivation practice. Front Plant Sci 7:671
Ree JF, Powell C, Folgado R, Pence VC, Walters C, Maschinski J (2024) Development of a micropropagation protocol for the ex situ conservation of Nuttall's scrub oak (*Quercus dumosa*). Plants 13(8):1148
Reed RC, Bradford KJ, Khanday I (2022) Seed germination and vigor: ensuring crop sustainability in a changing climate. Heredity 128(6):450–459
Rezaei H, Mirzaie-Asl A, Abdollahi MR, Tohidfar M (2023) Comparative analysis of different artificial neural networks for predicting and optimizing *in vitro* seed germination and sterilization of petunia. PLoS One 18(5):e0285657
Ribalta FM, Pazos-Navarro M, Edwards K, Ross JJ, Croser JS, Ochatt SJ (2019) Expression patterns of key hormones related to *Pisum sativum* L. embryo physiological maturity shift in response to accelerated growth conditions. Front Plant Sci 10:1154
Rodrigues SP, Picoli EA, Oliveira DC, Carneiro RG, Isaias RM (2014) The effects of *in vitro* culture on the leaf anatomy of *Jatropha curcas* L. (Euphorbiaceae) Biosci J 30(6):1933–1941
Rodríguez-Gacio MC, Matilla-Vázquez MA, Matilla AJ (2009) Seed dormancy and ABA signaling: the breakthrough goes on. Plant Signal Behav 4(11):1035–1049
Roja G, Heble MR (1996) Isolation and tissue culture of *Pterocarpus santalinus*. J Plant Res 109(2):151–154
Rolli E, Brunoni F, Bruni R (2014) An optimized method for in vitro propagation of African baobab (*Adansonia digitata* L.) using two-node segments. Plant Biosyst 148(6):1075–1084
Rosbakh S, Hülsmann L, Weinberger I, Bleicher M, Poschlod P (2019) Bleaching and cold stratification can break dormancy and improve seed germination in *Cyperaceae*. Aquat Bot 158:103128
Salgotra RK, Chauhan BS (2023) Genetic diversity, conservation, and utilization of plant genetic resources. Genes 14(1):174
Sano N, Marion-Poll A (2021) ABA metabolism and homeostasis in seed dormancy and germination. Int J Mol Sci 22(10):5069
Sarropoulou V, Grigoriadou K, Maloupa E, Chatzopoulou P (2024) Enhancement of *in vitro* seed germination, growth, and root development in two *Sideritis* species through GA3 application and diverse LED light conditions. Seeds 3:411–435
Schreiber L, Riederer M, Schönherr J (2008) The plant cuticle: composition, properties, and uses. Springer
Sen MK, Henā Ā, Jamal M, Nasrin S (2013) Sterilization factors affect seed germination and proliferation of *Achyranthes aspera* cultured *in vitro*. Environ Exp Biol 11:119–123
Sharma P, Kaur S (2018) Effect of different growth media on seed germination of underutilized legumes. Int J Chem Stud 6(5):934–938
Shinomura T, Nagatani A, Hanzawa H, Kubota M, Watanabe M, Furuya M (1996) Action spectra for phytochrome A- and B-specific photoinduction of seed germination in *Arabidopsis thaliana*. Proc Natl Acad Sci 93(15):8129–8133
Si Y, Haxim Y, Wang L (2022) Optimum sterilization method for *in vitro* cultivation of dimorphic seeds of the succulent halophyte Suaeda aralocaspica. Horticulturae 8:289
Siddiqui MH, Al-Whaibi MH, Ghabrial SA (2020) Role of hydrogen peroxide in plant disease resistance. Physiol Mol Biol Plants 26(5):1085–1096

da Silva V, Vinícius M, de Fátima S, Guimarães Silva J, Rubio Neto F, Pereira P, Dionísio F (2012) The influence of moisture on the *in vitro* embryo germination and morphogenesis of babassu (*Orbignya phalerata* Mart.). Acta Sci Agron 34(4):453

Silva JT, Winarto B, Dobránszki J, Zeng S (2015) Disinfection procedures for *in vitro* propagation of anthurium. Folia Hortic 27(1):14–23

Silva JT, Zeng S, Wicaksono A, Kher MM, Kim HH, Hosokawa M, Dewir YH (2017) *In vitro* propagation of African violet: a review. S Afr J Bot 112:501–507

Singh S, Rai S, Khan S (2010) *In vitro* seed germination of *Adansonia digitata* L.: an endangered medicinal tree. Nanobiotech Univ 1(2):107–112

Sivanesan I, Park SW (2015) Optimizing factors affecting adventitious shoot regeneration, in vitro flowering, and fruiting of *Withania somnifera* (L.) Dunal. Ind Crop Prod 76:323–328

Smýkal P, Vernoud V, Blair MW, Soukup A, Thompson RD (2014) The role of the testa during development and in the establishment of dormancy of the legume seed. Front Plant Sci 5:351

Song Y, Zhu J (2019) The roles of metabolic pathways in maintaining primary dormancy of *Pinus koraiensis* seeds. BMC Plant Biol 19(1):550

Song Y, Gao X, Wu Y (2021) Key metabolite differences between Korean pine (*Pinus koraiensis*) seeds with primary physiological dormancy and no-dormancy. Front Plant Sci 12:767108

Sorgato JC, Mudolon ED, Guimarães F, Soares JS, Ribeiro LM (2021) Light sources on the germination and initial *in vitro* establishment of *Schomburgkia crispa* Lindl., a species of the Brazilian Cerrado. Ciência Rural 51(3):e20190022

de Souza RR, de Oliveira Paiva PD, da Silva RR, dos Reis MV, Nery FC, Paiva R (2016) Optimization of jenipapo *in vitro* seed germination process. Ciência E Agrotecnol 40(6):658–664

Stawska M, Oracz K (2019) PhyB and HY5 are involved in the blue light-mediated alleviation of dormancy of *Arabidopsis* seeds possibly via the modulation of expression of genes related to light, GA, and ABA. Int J Mol Sci 20(23):5882

Steber CM, Cooney SE, McCourt P (1998) Isolation of the GA-response mutant sly1 as a suppressor of ABI1-1 in *Arabidopsis thaliana*. Genetics 149:509–521

Sudheer WN, Praveen N, Al-Khayri JM, Jain SM (2022) Role of plant tissue culture medium components. In: Rai AC, Kumar A, Modi A, Singh M (eds) Advances in plant tissue culture. Academic Press, pp 51–83

Sudheer S, Bandyopadhyay S, Bhat R (2023) Sustainable polysaccharide and protein hydrogel-based packaging materials for food products: a review. Int J Biol Macromol 248:125845

Tarraf W, De Carlo A (2024) *In vitro* biotechnology for conservation and sustainable use of plant genetic resources. Plan Theory 13:1897

Thanos CA, Doussi MA (1995) Ecophysiology of seed germination in endemic labiates of Crete. Isr J Plant Sci 43:227–237

Thokozani B, Zulu D, Sileshi C, Teklehaimanot Z, Gondwe D, Sarasan V, Stevenson P (2011) Seed germination and *in vitro* regeneration of the African medicinal and pesticidal plant, *Bobgunnia madagascariensis*. Afr J Biotechnol 10(32):5959–5966

Thompson K, Bakker JP, Bekker RM (2020a) Ecological seed banks: a key to plant diversity and ecosystem restoration. Ecol Manag Restor 21(1):1–9

Thompson K et al (2020b) Seed germination ecology. J Ecol 108(3):850–863

Tikendra L et al (2022) *Dendrobium* sp.: *in vitro* propagation of genetically stable plants and ethnomedicinal uses. In: Mérillon JM, Kodja H (eds) Orchids phytochemistry, biology and horticulture, reference series in phytochemistry. Springer, Cham

Toorop PE, van der Hage ER, Lullien-Pellerin V (2008) The effect of light on seed germination: a quantitative analysis. Funct Plant Biol 35(5):395–403

Tu KL, Yin Y, Yang L, Wang J, Sun Q (2023) Discrimination of individual seed viability by using oxygen consumption technique and headspace gas chromatography-ion mobility spectrometry. J Integr Agric 22(3):727–737

Tursun AÖ (2020) The effects of low-temperature applications on dormancy of *Salvia verticillata* L. and *Rumex crispus* L. seeds. Pak J Bot 52(6):1885–1890

Tzortzakis, N.G. (2009). Effect of pre-sowing treatment on seed germination and seedling vigour in endive and chicory. Hort. Sci. (Prague), 36(3), 117–125. https://doi.org/10.17221/28/2008-HORTSCI

Utami ESW, Hariyanto S (2019) *In vitro* seed germination and seedling development of a rare Indonesian native orchid *Phalaenopsis amboinensis* J.J.Sm. Scientifica (Cairo) 3:8105138

Van de Poel B, Smet D, Van Der Straeten D (2015) Ethylene and hormonal cross-talk in vegetative growth and development. Plant Physiol 169(1):61–72

Vishal B, Kumar PP (2018) Regulation of seed germination and abiotic stresses by gibberellins and abscisic acid. Front Plant Sci 9:838

Volpi J, Lenzi A, Ruffoni B, Savona M (2024) Pre-germinative treatments and *in vitro* germination of *Dianthus caryophyllus* and *Alstroemeria* spp. seeds. Seeds 3:493–504

Wang Y, Zhang M, Dong S, Liu YL, Li ZH (2021) The opposite roles of white light in regulating germination of fresh and aged seed in tobacco. Plants 10(11):2457

Wang Y, Sun X, Peng J, Li F, Ali F, Wang Z (2024a) Regulation of seed germination: ROS, epigenetic, and hormonal aspects. J Adv Res (online). https://doi.org/10.1016/j.jare.2024.06.001

Wang JY, Bu ZJ, Poschlod P, Yusup S, Zhang JQ, Zhang ZX (2024b) Seed dormancy types and germination response of 15 plant species in temperate montane peatlands. Ecol Evol 14(7):e11671

Warakagoda PS, Subasinghe S (2014) *In vitro* seed germination of *Coscinium fenestratum* (Gaertn.) Colebr. Ann Res Rev Biol 4(23):3549–3565

Watt MP, Blakeway FC, Mokotedi MEO, Jain SM (2003) Micropropagation of *eucalyptus*. In: Jain SM, Ishii K (eds) Micropropagation of Woody trees and fruits, forestry sciences, vol 75. Springer, Dordrecht

Wrobleski A, Ernst S, Weber T, Delach A (2023) The impact of climate change on endangered plants and lichen. PLOS Clim 2(7):e0000225

Xu J, Du G (2023) Seed germination response to diurnally alternating temperatures: comparative studies on alpine and subalpine meadow populations. Global Ecol Conserv 44:e02503

Yamaner Ö, Erdağ B (2021) *In vitro* seed germination and seedling development of an endemic medicinal plant *Hypericum adenotrichum* Spach. Eur J Sci Technol 23:440–447

Yan D, Duermeyer L, Leoveanu C, Nambara E (2014) The functions of the endosperm during seed germination. Plant Cell Physiol 55(9):1521–1533

Yang J, Lovett-Doust J, Lovett-Doust L (1999) Seed germination patterns in green dragon (*Arisaema dracontium, Araceae*). Am J Bot 86(8):1160–1167

Yang G, Shen X, Jackson R, Lu Z (2013) Factors affecting *in vitro* seed germination and shoot proliferation of Galax [*Galax urceolata* (Poir.) Brummitt]. Aust J Crop Sci 7(11):1766–1771

Yang E, Fan L, Yan J, Jiang Y, Doucette C, Fillmore S, Walker B (2018) Influence of culture media, pH, and temperature on growth and bacteriocin production of bacteriocinogenic lactic acid bacteria. AMB Express 8(1):10

Yasin M, Andreasen C (2016) Effect of reduced oxygen concentration on the germination behavior of vegetable seeds. Hortic Environ Biotechnol 57:453–461

Yilmaz Gokdogan E, Burun B (2022) Studies on seed germination and in vitro cultures of *Salvia* L. species from Turkish flora. Nat Prod Biotechnol 2(1):60–73

Zobayed S (2008) Aeration in plant tissue culture. In: Gupta SD, Ibaraki Y (eds) Plant tissue culture engineering, focus on biotechnology, vol 6. Springer, Dordrecht

2 Slow-Growth Strategies for Prolonged Storage for Conservation of Rare and Endangered Germplasm

Engin Tilkat, Veysel Süzerer, and Yelda Özden Çiftçi

Abstract

This chapter examines the implementation of slow-growth storage strategies as an essential approach to safeguard and maintain the genetic resources of rare and threatened plant species. In light of escalating ecological and anthropogenic pressures, preserving genetic diversity has become pivotal for sustaining ecosystem resilience and maintaining the continuity of vital biological resources. Slow-growth techniques reduce cellular metabolic activity, thereby maintaining the functional integrity of plant tissues under controlled conditions and extending the duration of in vitro conservation. This chapter systematically addresses the core principles of slow-growth methods, emphasizing the fine-tuning of nutrient media, temperature, light, and environmental parameters. Furthermore, it details the application of these strategies across a diverse array of plant taxa, discussing specific adaptations, success metrics, and the inherent challenges encountered in the process. Through a synthesis of theoretical insights and practical implementations, this chapter aims to contribute to a robust framework for the medium-term conservation of plant genetic variation, underscoring its significance for future research and sustainable resource management.

E. Tilkat (✉)
Faculty of Science and Literature, Department of Biology, Batman University, Batman, Türkey
e-mail: engin.tilkat@batman.edu.tr

V. Süzerer
Vocational School of Health Services, Department of Pharmacy Services, Bingöl University, Bingöl, Türkey

Y. Özden Çiftçi
Faculty of Basic Sciences, Department of Molecular Biology and Genetics, Gebze Technical University, Gebze, Kocaeli, Türkey

S. Gantait, P. Chmielarz (eds.), *Conservation of Plant Genetic Resources*,
https://doi.org/10.1007/978-981-96-8331-4_2

Keywords

Slow-growth conservation · Germplasm storage · Rare and endangered species · In vitro preservation · Genetic diversity management

2.1 Introduction

Plant genetic resources (PGRs), also known as germplasms, contain the genomic information of plants, the biological mechanisms involved in translating this information into phenotypes, and the organizational structures that maintain this genetic variation (Bretting and Duvick 1997). According to the Food and Agriculture Organization, these genetic resources include currently used, newly developed, and obsolete cultivars and their relatives, as well as wild species and varieties, including elite and existing breeding lines (Phillips 2003; FAO 2024). Today, only 30 plant species, mainly maize (*Zea mays*), wheat (*Triticum aestivum*), and rice (*Oryza sativa*), feed the world (Haussmann et al. 2004). However, the impacts of climate change pose a significant risk to agricultural output and global food stability. In addition, the rising need for food from the growing global population has made it compulsory to increase output and enhance food quality. In this context, plant breeders have turned to genetic resource diversity to develop new cultivars with high yields that are tolerant to biotic and abiotic stresses (Engels and Ebert 2024). At this point, PGR conserves plant genetic diversity and ensures that this genetic diversity is utilized in modern breeding and agriculture (Salgotra and Chauhan 2023). This genetic diversity is not only a resource for obtaining new variants against changing climatic conditions but also increases the chance of adaptation of the founder species to changing environmental conditions and ensures the stability of the ecosystem (Hoban et al. 2018). However, genetic diversity sources are threatened and lost, which is also known as genetic erosion, due to factors such as modern agricultural practices, climate change, and habitat loss (Rao 2015).

Both in situ and ex situ conservation strategies have been developed to protect these critically important genetic materials. In the in situ approach, germplasm is maintained by preserving the plant in its habitat (Mattana et al. 2012). However, in this conservation method, the plant is exposed to biotic stress factors such as bacteria, viruses, and fungal diseases and abiotic stress factors such as changing environmental parameters and natural disasters (Chauhan et al. 2019). These stress factors are eliminated via the ex situ method. Ex situ conservation involves the removal of germplasm from its natural habitat and preservation under controlled conditions, such as in botanical gardens, seed banks, or gene banks (Murray 2017). Approximately 2500 botanical gardens with more than six million plants, 1750 gene banks with more than 7.4 million genetic resources, and seed banks with 15,698 records make important contributions to ex situ conservation (Chen and Sun 2018; Yang et al. 2021; Yadav et al. 2024).

PGR can also be conserved via biotechnological methods, including the application of plant tissue culture techniques for short- and medium-term conservation

(Muthoni et al. 2019). Plant tissue culture involves the in vitro production of new tissues and plants from plant explants under aseptic and defined conditions (Thorpe 2007). Among these methods, slow-growth storage techniques aim to prolong the subculture interval by reducing metabolic activity and thus the growth of germplasms through changes in both the storage medium and conditions. In this way, the subculture time is extended. Slow-growth storage is one of the most effective methods of plant conservation, particularly when seed production is limited or absent and seeds are characterized by recalcitrance and low germination rates (Chauhan et al. 2019; Benelli et al. 2022).

Along with slow-growth storage techniques and other conservation methods, the development of sequencing technology and bioinformatics has also paved the way to utilize genetic diversity more efficiently and to prove the use of PGRs for sustainable agriculture (Hanson et al. 2024).

2.1.1 The Role and Importance of the PGR

PGRs are integral to maintaining ecological balance and supporting human livelihoods. They contribute to essential ecosystem services such as soil formation, nutrient cycling, and water purification. Diverse plant species have unique root structures and nutrient requirements, which help maintain soil health and prevent erosion (Tilman et al. 2014).

In agriculture, the PGR provides the genetic diversity necessary for crop improvement. They offer traits that can be harnessed to increase yield, nutritional quality, and resistance to biotic and abiotic stresses. For example, wild relatives of crops have been utilized to introduce beneficial traits into cultivated varieties, thereby increasing their resilience and adaptability (Salgotra and Chauhan 2023). Moreover, PGRs are vital for food security and economic development. They enable the development of new crop varieties that can withstand changing climatic conditions and emerging pests and diseases. Therefore, the conservation and sustainable use of PGRs are critical to ensuring a stable and diverse food supply for the growing global population.

2.1.2 Rationale for Slow-Growth Storage Approaches

Despite their importance, PGRs are under threat from factors such as habitat loss, climate change, and genetic erosion—the loss of genetic diversity within a species. Genetic erosion can lead to decreased crop yields, increased vulnerability to pests and diseases, and reduced adaptability to environmental changes, ultimately threatening global food security (Ebert and Engels 2020).

To mitigate these risks, effective conservation strategies are essential. One such approach is slow-growth storage, an in vitro technique designed to preserve plant germplasm by controlling the growth and development of plantlets. This method economizes storage space and labor, reduces costs, and prolongs the timing between subcultures. Additionally, slow-growth storage lowers the risk of germplasm loss

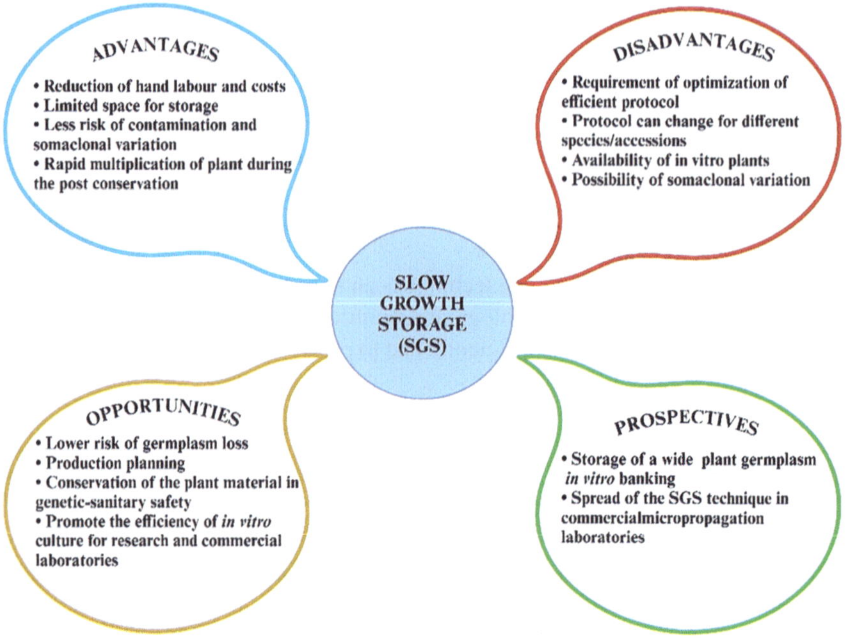

Fig. 2.1 Advantages, disadvantages, opportunities, and prospects of in vitro slow-growth storage (SGS) for plant germplasm conservation. (Adapted from Benelli et al. 2022 and created by authors)

through handling errors, such as contamination problems, and decreases the risk of genetic instability due to the reduction in the number of subcultures (Gopal and Chauhan 2010; Benelli et al. 2022). Implementing slow-growth storage approaches is crucial in mitigating the increased risk of genetic erosion. By maintaining plant viability and genetic stability over extended periods, these methods ensure the availability of diverse PGRs for future agricultural and ecological applications (Reed et al. 2011). Slow-growth storage techniques offer a strategic approach to the conservation of plant germplasms, providing a balance between prolonged viability and minimal maintenance. However, these methods also present challenges that require careful optimization, including protocol adjustments for different species and potential somaclonal variation risks. A summary of these advantages, limitations, opportunities, and future prospects is provided in Fig. 2.1 (Benelli et al. 2022).

2.2 Foundations of Slow-Growth Conservation Strategies

Slow-growth conservation strategies involve the deliberate reduction of metabolic activities in plant tissues to increase their viability during storage. This approach is particularly beneficial for the medium-term conservation of vegetatively propagated species and those with recalcitrant seeds that are not amenable to traditional seed banking methods

2.2.1 Defining Slow-Growth: Principles and Mechanisms

The concept of slow-growth storage refers to the deliberate reduction in plant tissue growth under controlled in vitro conditions. This approach is designed to lower metabolic activity while maintaining the viability and genetic stability of stored plant material. By carefully adjusting environmental and nutritional factors, researchers can extend storage periods and reduce the need for frequent subculturing.

Several key principles govern slow-growth strategies, each targeting specific physiological processes:

Temperature reduction Lowering the storage temperature decreases enzymatic activities and metabolic rates, effectively slowing plant growth.

Nutrient Limitation Modifying the culture medium by reducing nutrient availability helps restrict growth while preserving tissue viability.

Osmotic Regulation Adjusting the osmotic potential of the medium with compounds such as mannitol or sorbitol controls water uptake and cellular expansion, preventing excessive growth.

Growth Regulator Adjustment Altering the levels of plant growth regulators (e.g., cytokinins and auxins) modulates cell division and elongation, further contributing to slow-growth conditions.

The interaction of these factors forms the foundation of effective slow-growth storage strategies. Figure 2.2 provides a visual summary of these regulatory mechanisms, categorizing the four major approaches used to control in vitro plant storage. These strategies collectively contribute to the long-term preservation of plant genetic resources. Slow-growth techniques provide a flexible and efficient method for conserving plant germplasm, ensuring its long-term viability for breeding, research, and restoration programs. Researchers can develop customized protocols tailored to specific plant species and conservation goals by understanding and applying these principles.

2.3 Optimizing Conditions for Effective Conservation

Effective slow-growth storage relies on the precise optimization of nutrient media composition to regulate plant metabolism while maintaining genetic stability. By carefully adjusting the chemical and physical properties of the culture medium, plant tissues can be maintained for extended periods without compromising

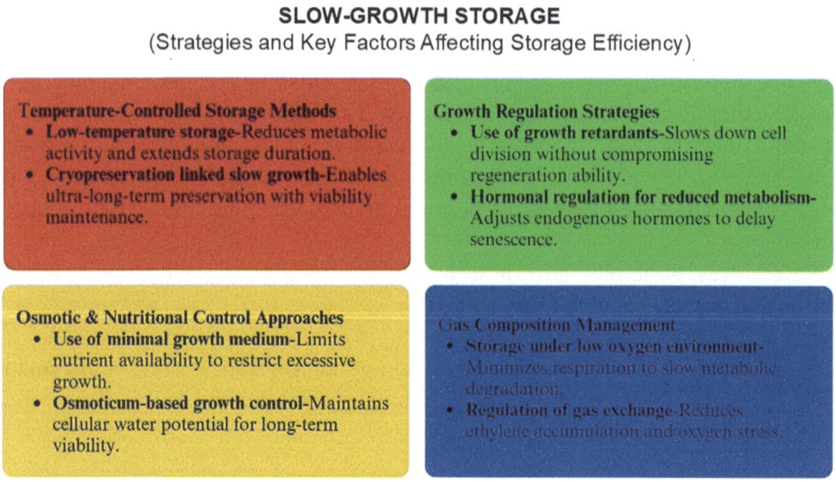

Fig. 2.2 Strategies and key factors influencing slow-growth storage (*This diagram categorizes four major approaches for regulating* in vitro *plant storage by modifying temperature, growth, osmotic conditions, and gaseous composition. These strategies collectively enhance the long-term preservation of plant genetic resources.*). (Source: Prepared by authors)

viability. The critical factors influencing the success of slow-growth conservation include carbon sources, mineral composition, plant growth regulators, and innovative additives. In addition to nutrient composition, environmental parameters such as osmotic balance, pH, and photoperiod play crucial roles in prolonging the storage duration of in vitro cultures. Recent advancements in tree biotechnology have highlighted the importance of species-specific adjustments to these variables, significantly improving conservation outcomes for recalcitrant species (Patel et al. 2025). By integrating these optimization strategies, researchers can increase the efficiency of in vitro conservation protocols, ensuring the long-term preservation of plant genetic resources.

2.3.1 Nutrient Media Formulation and Adjustments

The formulation and adjustment of nutrient media play pivotal roles in optimizing slow-growth conservation strategies for plant germplasms. This approach involves tailoring the chemical composition of the culture media to regulate metabolic activity, ensuring long-term viability while maintaining genetic stability. The key factors influencing the effectiveness of nutrient media include the selection of carbon sources, the use of osmotically active compounds, and the balance of plant growth regulators.

Carbon Sources and Osmotic Regulation In plant tissue culture, carbon sources are fundamental components of nutrient media and serve two primary functions:

providing energy for cellular metabolism and regulating osmotic potential. Sucrose is the most widely used carbon source because of its high efficiency in supporting in vitro plant growth and its availability (Pérez-Jiménez et al. 2021). However, alternative carbon sources, such as mannitol and sorbitol, are frequently employed in specialized protocols to induce osmotic stress, thereby modulating metabolic activity and influencing morphogenic responses in cultured tissues (Swedlund and Locy 1993; De Diego et al. 2013; Chen et al. 2016).

Mannitol is widely used in plant tissue culture to induce osmotic stress, effectively reducing water availability and slowing down cellular metabolism. This makes it particularly valuable for protocols involving long-term conservation and stress adaptation studies. However, plant tissues exhibit high sensitivity to mannitol, with growth rates declining significantly at low concentrations. Growth inhibition plateaus at relatively high concentrations, suggesting that mannitol induces a combination of osmotic and biotic stress responses, which may trigger secondary metabolite production and stress tolerance mechanisms (De Diego et al. 2013). These effects make mannitol essential for controlling metabolic activity in extended in vitro storage systems and for stress physiology research.

Sorbitol plays a dual role in plant tissue culture by functioning both as a carbon source and an osmotic regulator, making it a versatile alternative to sucrose. Unlike mannitol, which primarily acts as an osmotic stressor, sorbitol can also be metabolized by certain plant species and contribute to their growth and development (Swedlund and Locy 1993). Studies have demonstrated that sorbitol enhances the growth of embryogenic calli more effectively than sucrose does in some species, leading to increased regenerative potential. Additionally, research on *Prunus persica* indicates that the use of sorbitol as the primary carbon source significantly improves the uniformity of adventitious shoot regeneration (Chen et al. 2016). In some in vitro culture systems, sorbitol serves as an osmotic pressure regulator, preventing excessive water uptake by tissues while minimizing metabolic interference (Pérez-Jiménez et al. 2021). Its selective absorption and utilization make it a preferred choice for species that naturally accumulate polyols as part of their carbon metabolism.

In addition to regulating growth rates, osmotic stress induced by mannitol and sorbitol can mitigate oxidative stress, which is a major concern during prolonged storage. The accumulation of reactive oxygen species (ROS) in in vitro cultures can lead to oxidative damage and reduced viability. Mannitol and sorbitol have been shown to act as antioxidants, scavenging free radicals and thereby maintaining cell viability under stress conditions (Yaseen et al. 2013). This property makes them valuable for cryopreservation and in vitro conservation protocols, where oxidative stress is a limiting factor for tissue survival.

The selection of an appropriate carbon source and its concentration is crucial for optimizing in vitro culture protocols. While sucrose remains the standard choice owing to its efficiency and widespread applicability, mannitol and sorbitol offer unique advantages in specific applications. Mannitol is useful in inducing osmotic stress, reducing metabolic rates, and enhancing stress tolerance, whereas sorbitol supports both osmotic regulation and carbon metabolism, making it particularly beneficial for species that utilize polyols naturally. Given the species-specific responses to different carbon sources, empirical optimization is essential for designing successful tissue culture protocols (Kadota and Niimi 2004; Pérez-Jiménez et al. 2021). Understanding the interplay between carbon source availability, osmotic potential, and metabolic activity is key to improving the efficiency and success of in vitro plant propagation and conservation strategies.

Mineral Composition and Macro-/Micronutrient Balance The mineral composition of nutrient media is another critical determinant of tissue viability and growth under slow-growth conditions. Macronutrients, such as nitrogen, phosphorus, and potassium, must be optimized to support basal metabolic functions without promoting excessive growth. Micronutrients, including zinc, iron, and manganese, play essential roles in enzymatic activities and maintaining cellular homeostasis. Adjusting the concentrations of these nutrients can help extend the conservation period of germplasm by limiting excessive cell division while sustaining cellular functions (Rout et al. 2000; Yaseen et al. 2013).

Role of Plant Growth Regulators Plant growth regulators, especially cytokinins and auxins, are crucial in shaping the dynamics of in vitro growth. The balance between these two types of regulators significantly affects the growth behavior of in vitro cultures. Understanding this relationship is vital for optimizing growth conditions in laboratory settings. Cytokinins, such as 6-benzylaminopurine (BAP) or 6-benzyladenine (BA), promote shoot proliferation, whereas auxins, such as indole-3-butyric acid (IBA), stimulate root formation. In slow-growth conservation, adjusting the ratio of these growth regulators is crucial to achieve minimal growth rates, thereby extending the preservation period of plant cultures. Specifically, a relatively high concentration of auxins relative to that of cytokinins has been associated with reduced shoot elongation, contributing to prolonged conservation. For example, Lata et al. (2010) demonstrated that manipulating the concentrations of growth regulators effectively reduced growth rates in *Podophyllum peltatum*, thereby extending the conservation period. Additionally, recent advancements have introduced novel cytokinins, such as meta-topolin (mT), which have shown promise in increasing genetic stability and reducing somaclonal variations during in vitro propagation. Studies have demonstrated that mT can improve genetic fidelity in regenerated plants. For example, Ahmad et al. (2022) investigated the effects of mT on the micropropagation system of *Lagerstroemia speciosa*, assessing photosynthetic parameters and oxidative stress during acclimation, along with the genetic fidelity of regenerated plants. The use of mT has been associated with a decrease in somaclonal variations. Nowakowska and Pacholczak (2020) compared the effects of mT and BA on the micropropagation of *Daphne mezereum* L., focusing on shoot proliferation, biochemical changes, and genetic stability. These findings suggest that integrating meta-topolin into in vitro propagation protocols may increase genetic fidelity, offering a valuable tool for plant conservation and propagation efforts.

Innovative Additives for Enhanced Conservation Recent advancements in plant tissue culture have led to the integration of innovative additives, such as antioxidants and elicitors, to increase the efficacy of nutrient media under slow-growth conditions. Oxidative stress, which is caused primarily by the accumulation of reactive oxygen species (ROS), can significantly impact the viability of stored tissues. To counteract this, antioxidants such as ascorbic acid (vitamin C) and glutathione play crucial roles in mitigating oxidative damage, thereby increasing the long-term survival of plant tissues (Gallie 2013; Akram et al. 2017). Although these compounds have been widely studied for their role in abiotic stress responses, their ability to reduce ROS accumulation in plant cells suggests their potential effectiveness in slow-growth conservation systems as well.

Ascorbic acid functions as a major redox buffer, acting as a cofactor for enzymes involved in photosynthesis, hormone biosynthesis, and ROS detoxification (Gallie 2013; Akram et al. 2017). Additionally, glutathione plays a central role in redox homeostasis by interacting with the ascorbate-glutathione (AsA-GSH) cycle, a key defense mechanism against oxidative stress (Foyer and Noctor 2011; Hasanuzzaman et al. 2019). This pathway has been shown to be crucial for maintaining cellular stability in plants exposed to environmental stress, and its role in reducing oxidative stress suggests its potential for improving the viability of plant germplasms under slow-growth storage.

In addition to antioxidants, the integration of elicitors, such as salicylic acid, has shown promise in enhancing the resilience of plant tissues against oxidative stress during conservation. These bioactive compounds function by triggering defense responses, leading to the upregulation of antioxidant enzymes and the activation of stress adaptation pathways. While elicitor-induced stress tolerance has primarily been demonstrated in abiotic stress studies, the underlying mechanisms may also be beneficial for improving the stability of in vitro-stored tissues (Rout et al. 2000; Benelli et al. 2022). By modulating oxidative stress responses, elicitors provide an additional layer of protection, complementing the effects of antioxidants in maintaining cellular integrity during prolonged conservation.

In conclusion, optimizing nutrient media composition with the strategic incorporation of antioxidants and elicitors is essential for enhancing the effectiveness of slow-growth storage techniques. The combined use of ascorbic acid, glutathione, and defense-stimulating compounds such as salicylic acid may help increase the viability of in vitro-stored germplasm, reduce oxidative stress, and potentially improve overall conservation efficiency. Further experimental validation is needed to confirm the direct benefits of these compounds specifically under slow-growth storage conditions, ensuring the long-term preservation of valuable plant genetic resources for agricultural and ecological applications.

2.3.2 Temperature, Light, and Humidity Optimization

Optimizing environmental conditions—specifically temperature, light, and humidity—is crucial for the effective in vitro conservation of plant germplasms. These factors significantly influence the metabolic activities, growth rates, and overall viability of plant tissues maintained under laboratory conditions.

Temperature Regulation Temperature profoundly affects the metabolic rate of plant tissues. Lowering the storage temperature can decelerate metabolic processes, thereby increasing the viability of conserved samples. For example, in the in vitro conservation of *Carica pubescens*, a storage temperature of 8 °C effectively suppressed growth over a 6-month period while maintaining plantlet viability upon regeneration (Rahayu and Habibah 2016). Similarly, for *Glycyrrhiza glabra*, cultures incubated at 10 °C under low light intensity presented optimal survival rates (Srivastava et al. 2013). These findings underscore the importance of tailoring temperature settings to the specific thermal preferences of each species to minimize stress and prolong conservation periods. Figure 2.3 illustrates the impact of slow-growth storage on *Fraser photinia* microshoots after 3 months at 15 °C. The preservation of shoot viability under these conditions underscores the importance of temperature modulation in medium-term conservation.

Light Intensity and Quality Light is a critical factor influencing photosynthetic activity and morphogenesis in plant tissues. Adjusting light intensity and quality can effectively control growth rates during in vitro conservation. Research indicates that

Fig. 2.3 View of 3-month-old *Fraser photinia* microshoots stored in vitro at 15 °C under slow-growth storage conditions (Bar: 0.95 cm). (Source: Authors)

low light intensities, such as 2.5 μmol/m^2/s, are beneficial for maintaining slow growth in certain species (Srivastava et al. 2013). Additionally, the spectral quality of light, including the use of specific wavelengths, can impact morphogenesis and should be optimized on the basis of the species' requirements (Gautam et al. 2024). The implementation of light-emitting diode (LED) technology allows precise control over light spectra, offering a means to fine-tune light conditions to meet the needs of the conserved plant material (Gago-Calderón et al. 2023).

Humidity Control Maintaining appropriate humidity levels is essential to prevent desiccation and ensure the integrity of plant tissues during in vitro conservation. Relative humidity (RH) levels between 30% and 50% are generally recommended to balance moisture needs and reduce the risk of microbial contamination (Berger et al. 2024). Moreover, controlling RH fluctuations is vital, as rapid changes can lead to physiological stress and compromise tissue viability. The use of culture vessels with appropriate sealing and gas exchange properties can help maintain stable humidity levels, thereby supporting the longevity and health of conserved germplasms (Gautam et al. 2024).

2.3.3 Additional Environmental Parameters and Their Impact

In addition to temperature, light, and humidity, several other environmental factors significantly influence the success of in vitro plant conservation. Optimizing these parameters is essential for maintaining plant viability, genetic stability, and overall health during conservation efforts.

Carbon Dioxide Concentration Carbon dioxide (CO_2) is a critical component of photosynthesis, and its concentration within the culture environment can markedly affect plant growth and development. Elevated CO_2 levels have been shown to increase photosynthetic rates, leading to improved biomass accumulation and vigor in cultured plantlets. Thus, implementing controlled CO_2 enrichment in tissue culture systems can be beneficial for the in vitro conservation of various plant species (Solárová and Pospíšilová 1997).

Ventilation and Gas Exchange Proper ventilation and gas exchange are essential for in vitro plant tissue culture, particularly for slow-growth conservation. Maintaining adequate gas exchange prevents the accumulation of ethylene and other potentially harmful gases that can lead to physiological disorders such as leaf epinasty, chlorosis, and reduced growth rates (Jackson et al. 1991). The incorporation of gas-permeable membranes or ventilation systems in culture vessels facilitates efficient gas exchange, ensuring a stable microenvironment conducive to long-term plant preservation (Kozai et al. 1997). In slow-growth conservation, regulating metabolic activity is crucial for extending subculture intervals and minimizing genetic instability. While factors such as temperature, light intensity, and nutrient composition are commonly adjusted to achieve this, gas exchange plays an equally critical role. Proper ventilation maintains optimal oxygen and carbon dioxide levels, which are essential for sustaining photosynthesis and respiration. Although improved ventilation has been shown to increase in vitro plant growth by reducing ethylene accumulation and promoting gas exchange, its direct impact on extending storage duration and preserving genetic stability in slow-growth conservation remains underexplored. Further research is necessary to optimize ventilation strategies tailored specifically for long-term in vitro germplasm preservation.

Osmotic Pressure and Water Availability The osmotic potential of the culture medium significantly influences water uptake and cellular turgor, directly affecting plantlet growth and survival. The concentration of osmotic agents, such as mannitol or sorbitol, can be adjusted to modulate osmotic pressure, thereby controlling growth rates and enhancing stress tolerance during conservation. For example, mannitol and sorbitol are commonly used as osmotic regulators during in vitro plant cultures to lower the water potential of the medium and restrict plant development, mimicking water deficit conditions (Wijerathna-Yapa and Hiti-Bandaralage 2023). Moreover, the use of osmotic agents such as mannitol and sorbitol plays a crucial role in controlling water availability and reducing metabolic activity during slow-growth storage. Recent studies on tree conservation strategies emphasize the need for species-specific osmotic regulation to maintain tissue viability over extended storage periods (Patel et al. 2025). Careful calibration of osmotic conditions is particularly important for species sensitive to water stress. Increased osmotic pressure directly interferes with plant metabolism, leading to reduced sprout and shoot growth development, as observed in certain species (Chauhan et al. 2019).

pH of the Culture Medium The pH of the culture medium affects nutrient availability, enzyme activity, and overall cellular metabolism. Most plant species thrive in media with pH values ranging from 5.5 to 6.0. Deviations from this optimal range can impair nutrient uptake and physiological functions, leading to suboptimal growth or even tissue necrosis. Regular monitoring and adjustment of the pH of the medium are therefore imperative for successful in vitro conservation (Chauhan et al. 2019). Studies have shown that medium pH is essential for many aspects of explant development and growth, with sensitivity or tolerance to pH variations differing among species (Chen et al. 2014). The pH of the culture medium plays a fundamental role in maintaining enzymatic activity and nutrient solubility, which are essential for plant tissue survival. Studies on tree biotechnology have demonstrated that fine-tuning pH levels enhances the stability of in vitro-stored germplasm, preventing tissue necrosis and oxidative stress (Patel et al. 2025).

Photoperiod Regimes The duration of light exposure, or photoperiod, plays a critical role in regulating developmental processes such as flowering, dormancy, and organogenesis. Tailoring photoperiod regimes to meet the specific requirements of conserved species can promote desired developmental pathways and improve conservation outcomes (Arcidiacono et al. 2021). For example, certain long-day plants may require extended light periods to maintain vegetative growth and prevent premature flowering. Research has demonstrated that a 16-h light and 8-h dark photoperiod is optimal for shoot regeneration in some species, whereas continuous darkness can adversely affect growth (Yaacob et al. 2014). Additionally, studies on *Cannabis sativa* have shown that specific photoperiods can significantly influence in vitro growth and development, underscoring the importance of photoperiod optimization in tissue culture (Moher et al. 2021). In addition, photoperiod regulation is key in controlling in vitro developmental processes such as dormancy, organogenesis, and vegetative growth. Recent advances suggest that adjusting light exposure on the basis of species-specific requirements enhances conservation efficiency, particularly for tree species with long generation cycles (Patel et al. 2025).

2.4 Applications of Slow-Growth Storage Across Plant Species

Slow-growth storage is a pivotal in vitro conservation strategy employed across various plant taxa, including orchids, forest trees, and tropical plants. This method involves modulating environmental and nutritional conditions to decelerate plant growth, thereby extending the intervals between subcultures and reducing labor and resource inputs. The following sections are dedicated to the specific applications, challenges, and successful protocols associated with slow-growth storage in these plant categories. Various plant species have been successfully conserved via slow-growth storage, with protocols tailored to their specific physiological and genetic characteristics. Table 2.1 provides an overview of in vitro slow-growth conservation protocols for selected plant species, detailing key parameters such as medium composition, temperature, light conditions, and storage duration.

2.4.1 Storage Protocols for Selected Plant Species

Ornamental Plants and Orchids The ornamental plant industry relies on species diversity and the consistent availability of high-quality propagation materials. In vitro plant tissue culture techniques play crucial roles in the propagation and conservation of many plant species, including ornamental plants. Moreover, during periods when there is little or no demand for such plants, the slow-growth storage technique allows their preservation without affecting their viability or regrowth potential (Da Silva et al. 2018). Shoots of *Dianthus ingoldbyi* were successfully preserved in a dark and cold environment for up to 6 months without losing their regeneration ability (Arda et al. 2016). While slow-growth storage is widely utilized for ornamental plants, it is particularly valuable for species with specialized propagation requirements and long reproductive cycles, such as orchids. Orchids, renowned for their ecological and ornamental value, often face threats due to habitat loss and overcollection. In vitro conservation through slow-growth storage offers a viable means to preserve orchid germplasm. For example, a study on *Epipactis flava* demonstrated that supplementing Murashige and Skoog (MS) media with 2% man-

Table 2.1 In vitro slow-growth conservation protocols for selected and endangered plant species

Plant species	Medium composition	Storage condition (temperature, light, duration)	Key findings	References
Fraser photinia	QL medium with 15 g/L sucrose, 15 g/L mannitol, and 1 mg/L BAP	4 °C, fully dark, 15 months	The incorporation of sucrose and mannitol into the medium significantly enhanced shoot proliferation during 15 months of storage under dark and cold conditions	Akdemir et al. (2010)
Solanum tuberosum	MS medium with 20 g/L sucrose and addition of 40 g/L sorbitol	8 °C, 16 h/8 h, 18 months	Low temperature, reduction of sucrose, and addition of sorbitol effectively slowed growth, facilitating medium-term conservation	Gopal and Chauhan (2010)
Apricot	MS medium with 58.4 mM (both sucrose and mannitol)	5 °C, 16 h/8 h, 7 months	The addition of mannitol and sucrose provided a 100% survival rate for 7 months storage	Marino et al. (2010)
Thymus vulgaris	MS medium with addition 1 mg/L KIN 0.3 mg/L GA_3	4 °C, fully dark, 12 months	Storage for up to 12 months was achieved under cold and dark conditions	Ozudogru et al. (2011)
Thymus moroderi	MS medium with addition 1.5% mannitol and 1.5% sucrose	4 °C, fully dark, 7 months	The combined effect of osmotically active compounds present in the culture medium, along with dark and cold conditions, enabled storage	Marco-Medina and Casas (2012)
Castanea sativa	WPM medium with 0.44 μM BAP	8 °C, 16 h/8 h, 48 months	A low temperature during slow-growth storage positively affected shoot survival rates over extended preservation periods	Capuana and Di Lonardo (2013)
Musa spp.	MS medium with addition 4% mannitol	23 °C, 12 h/12 h, 16 months	The addition of mannitol extended the storage duration up to 16 months	Irish et al. (2013)

(continued)

Table 2.1 (continued)

Plant species	Medium composition	Storage condition (temperature, light, duration)	Key findings	References
Epidendrum chlorocorymbos	MS medium with addition 1% mannitol and 1% sorbitol	25 °C, 16 h/8 h, 12 months	The addition of mannitol and sorbitol reduced growth rates, allowing for extended storage without subculturing. High recovery rates were observed post storage	Lopez-Puc (2013)
Pistacia lentiscus L.	MS Medium with 1 mg/L BAP	4 °C, Fully dark, 12 months	The storage process was observed for up to 12 months under low temperature and dark conditions	Koç et al. (2014)
Allium sativum	¼ MS medium with addition 45 g/L sucrose	25 °C, 16 h/8 h, 7 months	Quarter-strength MS medium combined with a high concentration of sucrose enabled storage for up to 7 months	Pardo et al. (2014)
Prunus mahaleb L.	½ MS medium without sucrose	25 °C, 16 h/8 h, 5 months	Reducing MS salt concentration and eliminating sucrose from the nutrient medium proved effective for mid-term storage periods of up to 5 months	Sota and Kongjika (2014a)
Ziziphus jujuba	MS medium with addition 1 mg/L BAP and 0,05 mg/L IBA	4 °C, Fully dark, 10 months	The combination of low temperature (4 °C) and a reduced light regime is the most effective method for medium-term preservation	Sota and Kongjika (2014b)
Genipa americana	½ MS medium	25 °C, 12 h/12 h, 5 months	Reducing MS medium to half strength had a positive effect on plant preservation	Lédo et al. (2015)

(continued)

Table 2.1 (continued)

Plant species	Medium composition	Storage condition (temperature, light, duration)	Key findings	References
Cynara cardunculus var. *scolymus*	MS medium supplemented with 1.5% sorbitol and reduced cytokinin levels	4 °C, 12 h/12 h, 12 months	The protocol was effective for conserving both wild and elite genotypes, with high recovery rates post-storage	Tavazza et al. (2015)
Asparagus racemosus	MS medium supplemented with 2% mannitol and 1.5% sucrose	25 °C, 16 h/8 h, 6 months	Room temperature and reduced sucrose levels enabled storage for up to 6 months	Thakur et al. (2015)
Senecio macrophyllus	½ MS medium with addition 1,5% mannitol and 1.5% sorbitol	10 °C, Fully dark, 9 months	Storage for up to 9 months in a dark, cold environment with a nutrient medium supplemented with mannitol and sorbitol	Trejgell et al. (2015)
Dianthus ingoldbyi	MS medium with addition 0,5 mg/L NAA	4 °C, fully dark, 6 months	The cold and dark condition storage technique used as an effective method for in vitro conservation	Arda et al. (2016)
Simmondsia chinensis	MS medium with addition 1 mg/L BAP	5 °C, fully dark, 9 months	Low temperature decreases both the number and height of shootlets while significantly promoting new branch regeneration. No necrosis was observed at low temperatures for up to nine months of storage. Therefore, we conclude that a temperature of 5 °C is optimal for the healthy storage of in vitro-grown jojoba shootlets	Bekheet et al. (2016)
Ananas comosus	1/2 MS medium with 30 g/L sucrose	21 °C, 12 h/12 h, 24 months	The shoot cultures have been renewed every 24 months over a span of 10 years	Da Silva et al. (2016)

(continued)

Table 2.1 (continued)

Plant species	Medium composition	Storage condition (temperature, light, duration)	Key findings	References
Manihot esculenta	MS medium with addition 20 g/L mannitol	16 °C, fully dark, 16 months	The osmotic preservation of secondary somatic embryos of cassava on 2% mannitol at 16 °C is the most effective slow-growth method	Opabode et al. (2016)
Carica pubescens	1/2 MS medium with addition 2 ppm BAP	8 °C, 16 h/8 h, 6 months	Shoots exhibited growth during six months of in vitro storage and maintained their viability in the regeneration medium	Rahayu and Habibah (2016)
Pyrus communis	MS medium with reduced nitrogen and sucrose	4 °C, fully dark, 24 months	Low temperature and darkness conditions effectively slowed growth, facilitating long-term conservation. High recovery rates were observed post-storage	Arbeloa et al. (2017)
Phoenix dactylifera	MS medium with addition 10 mg/L 2,4-D, 3 mg/L 2iP, 6 mg/L ABA, 102 g/L sucrose, 3 g/L AC	15 °C, fully dark, 12 months	Short-term in vitro preservation of date palm shoot-tip and callus cultures under slow-growth storage conditions at 15 °C for 12 months, utilizing sucrose as an osmotic agent and abscisic acid (ABA) as a growth inhibitor	El-Dawayati (2017)
Prunus avium Gisela®5	MS medium with increased sucrose (45/60 g/L)	4 °C, fully dark, 16 months	Increased sucrose concentration, along with dark and cold conditions, enabled storage for up to 16 months	Ozudogru et al. (2017)
Ullucus tuberosus	1/2 MS medium with 30 g/L mannitol	17 °C, 16 h/8 h, 18 months	The addition of mannitol was found to be effective in the storage of microshoots for 18 months	Hammond Hammond et al. (2019)

(continued)

Table 2.1 (continued)

Plant species	Medium composition	Storage condition (temperature, light, duration)	Key findings	References
Ribes nigrum	MS medium with decreased sucrose and addition %2 mannitol) and PGR	4 °C, 10 h/14 h, 18 months	The reduction of sucrose, along with the addition of PGR and mannitol, was effective in storage for 18 months	Turdiyev et al. (2020)
Vanilla planifolia	MS medium with addition 3 mg/L ABA	22 °C, 16 h/8 h, 4 months	The most effective treatment for highest survival in in vitro conservation	Bautista-Aguilar et al. (2021)
Fragaria ananassa	1/4 MS medium with addition PGR	25 °C, 16 h/8 h, 9 months	Cultures can be successfully preserved in vitro for nine months using a 1/2 strength MS medium	Kadhim and Abdulhussein (2021)
Vitis vinifera	3/4 MS, 5.5% sorbitol	5 °C, fully dark, 12 months	Maintained its green color and health after 12 months of storage	Al Mousa and Hassan (2022)
Berula erecta	MS medium with addition 40 g/L sucrose and 20 g/L sorbitol	4 °C, fully dark, 6 months	A high concentration of carbon source, sorbitol, and cold conditions preserved the regeneration ability of the shoots	Ambrožič-Dolinšek et al. (2022)
Tetraclinis articulata (Vahl) Masters	SH medium with addition PGR	4 °C, fully dark, 28 weeks	Explant survival and growth remained above 90% after 28 weeks of cold storage at a low temperature	Juan-Vicedo et al. (2022)
Hancornia speciosa	MS medium with decreased sucrose and addition 5 g/L sorbitol	25 °C, 16 h/8 h, 4 months	The reduced sucrose amount and added sorbitol enabled the storage of nodal shoots for 120 days	Oliveira and Aloufa (2022)
Saccharum officinarum	MS medium with addition 4 mg/L Paclobutrazol	28 weeks	The addition of paclobutrazol ensured the survival of shoots for 60 weeks while maintaining their multiplication ability	Azizi et al. (2023)

(continued)

Table 2.1 (continued)

Plant species	Medium composition	Storage condition (temperature, light, duration)	Key findings	References
Rubus fruticosus L. cvs Chester and Thornfree	DKW medium with addition PGR	4 °C, fully dark, 5 months	Liquid culture in TIS using the ElecTIS bioreactor consistently enhances the recovery of shoot cultures from long-term conservation in SGS, significantly accelerating their return to an optimal level of shoot proliferation	Elazab et al. (2023)
Solanum habrochaites	MS medium supplemented with 60 g/L sorbitol	22–25 °C, 16 h/8 h, 9 months	The nutrient medium enriched with sorbitol proved to be the most effective formulation for slow-growth storage	Miroshnichenko et al. (2023)
Epipactis flava	MS medium supplemented with 2% mannitol	25 °C, 16 h/8 h, 24 weeks	Mannitol effectively reduced growth rates, allowing for extended storage without subculturing. Post-storage recovery was successful	Linjikao et al. (2024)
Malus domestica ('Yanfu-6' and 'Qingzhen-1')	MS medium with 4.5% sucrose or 0.5% mannitol; temperature reduced to 12 °C	12 °C, 14 h/10 h, 9–12 months	Lowering the temperature to 12 °C extended the conservation period to 9 months for 'Yanfu-6' and over 12 months for 'Qingzhen-1'. High recovery rates were observed	Lu et al. (2024)
Ribes janczewskii	WPM medium with addition 20 g/L mannitol	4 months	Shoots remained viable for 4 months with the addition of mannitol for in vitro slow-growth conservation	Nurtaza et al. (2024)

DKW Driver and Kuniyuki Walnut medium, *QL* Quoirin and Lepoivre medium, *MS* Murashige and Skoog medium, *SH* Schenk and Hildebrandt medium, *WPM* woody plant medium, *PGR* plant growth regulator, *2IP* 2-isopentenyladenine, *2,4-D* 2,4-dichlorophenoxyacetic, *ABA* abscisic acid, *BAP* 6 benzylaminopurine, *IBA* indole-3-butyric acid, *AC* activated charcoal nitol effectively reduced growth rates, allowing for 24 weeks of storage without subculturing. Poststorage recovery was successful, indicating the efficacy of the protocol for medium-term conservation (Linjikao et al. 2024).

Forest Trees Conserving forest tree germplasms presents significant challenges because of their long generation times, slow growth, and, in some cases, recalcitrant seed behavior (Gantait et al. 2016). Traditional ex situ conservation methods, such as seed banking, are often ineffective for these species. Consequently, in vitro slow-growth storage techniques have been explored as promising alternatives for long-term germplasm preservation. These techniques involve modifying culture conditions, such as adjusting temperature, osmotic stress, and growth regulators, to extend subculture intervals while maintaining genetic stability (Guan et al. 2016; Reeves et al. 2023; FAO 1994). The application of in vitro slow-growth storage in forest species varies depending on the biological characteristics of the species and their responsiveness to tissue culture conditions. While these methods are not yet widely implemented, several tree species have shown promising results under slow-growth storage conditions:

Pinus radiata: Studies have demonstrated that cold storage can improve the initiation and viability of somatic embryos in *Pinus radiata*, highlighting the potential for in vitro conservation of this species. Temperature manipulation was found to increase SE efficiency, suggesting its role in the extended preservation of viable germplasm (Guan et al. 2016; Reeves et al. 2023).

Alnus glutinosa: Protocols have been established for the in vitro conservation of *Alnus glutinosa* via slow-growth storage techniques. These approaches have enabled the long-term maintenance of genetic resources with minimal subculturing, making them a viable strategy for preserving this species (Lidder and Sonnino 2011).

Eucalyptus species: Various *Eucalyptus* species have been successfully maintained under in vitro slow-growth storage conditions. These techniques provide an effective means for preserving genetic diversity, particularly in breeding programs where maintaining clonal fidelity is essential (Padayachee et al. 2009).

Populus species: Several studies have demonstrated the efficacy of in vitro slow-growth storage for *Populus* species. These techniques allow for the conservation of clonal material while reducing the frequency of subculturing, thereby lowering the risk of somaclonal variation. Cryopreservation and slow-growth storage methods have also been explored for maintaining genetic stability over extended periods (Son et al. 1991; Jokipii et al. 2004; Chauhan et al. 2019).

These examples highlight the applicability of in vitro slow-growth storage for forest species with challenging seed storage requirements or those requiring vegetative propagation. While further optimization is necessary to standardize protocols across a broader range of forest tree species, current research suggests that these methods provide an effective means for ex situ conservation, ensuring the long-term preservation of valuable genetic resources.

Tropical Plants Tropical plants, including numerous fruit species, often exhibit sensitivity to conventional conservation methods. Slow-growth storage can be categorized into four major approaches used to regulate in vitro plant storage by modifying temperature, growth, osmotic conditions, and gaseous composition. These strategies significantly increase the long-term preservation of plant genetic resources. A comprehensive review of in vitro conservation techniques over the past decade has emphasized the importance of tailoring storage conditions to the specific physiological needs of each species. Successful protocols have been established for various tropical fruit trees, facilitating the preservation of their genetic diversity (Benelli et al. 2022).

Medicinal Plants Medicinal plants hold significant therapeutic value and are often subject to overharvesting, leading to a decline in their natural populations. In vitro slow-growth storage offers a viable method for conserving these valuable species. For example, protocols have been developed for the conservation of *Rauvolfia serpentina*, a medicinal plant known for its antihypertensive properties. By adjusting the culture medium and environmental conditions, researchers have successfully maintained viable cultures over extended periods, ensuring the preservation of this species for future medicinal use (Rohani 2020).

Endangered Species The conservation of endangered plant species is crucial for maintaining biodiversity. In vitro slow-growth storage has been effectively utilized to preserve these species, providing a valuable tool for their conservation. For example, a validated in vitro slow-growth conservation protocol has been established for globe artichoke (*Cynara cardunculus* var. scolymus), an endangered species. This protocol serves as a cost-effective approach to preserving both wild and elite genotypes, contributing to the conservation of genetic diversity within the species (Tavazza et al. 2015). Additionally, several studies have reported successful slow-growth storage techniques for critically endangered species. For example, *Dianthus ingoldbyi* Turrill, a critically endangered plant, was successfully preserved under in vitro slow-growth conditions via modified nutrient media and growth retardants (Arda et al. 2016). Similarly, cryopreservation techniques have been effectively applied to endangered species such as *Tuberaria major*, *Thymus lotocephalus*, and *Plantago algarbiensis* (Benelli et al. 2022). These approaches significantly extend conservation periods while maintaining genetic stability. Recent advancements emphasize the integration of cryopreservation techniques, such as encapsulation–dehydration and droplet vitrification, alongside slow-growth storage strategies to optimize the long-term conservation of endangered plant germplasms (Coelho et al. 2020). The application of these complementary techniques provides a more robust strategy for preserving species that are vulnerable to habitat destruction and climate change.

Fruit Trees Fruit trees are vital for global agriculture and food security, contributing significantly to human nutrition and economic sustainability. However, many fruit tree species face challenges in conservation because of their recalcitrant seeds and long generation times, which limit their ability to be stored via conventional seed banking methods. Given these limitations, in vitro slow-growth storage techniques have emerged as effective alternatives for preserving fruit tree germplasm. In vitro slow-growth storage techniques have been applied to various fruit tree species to address these conservation challenges. These approaches enable the long-term maintenance of valuable genetic material while minimizing the frequency of subculturing, thereby reducing the risk of somaclonal variation. For example, research has focused on developing routine protocols to maintain cultures for various fruit tree genotypes, ensuring their preservation and availability for future breeding programs (Arbeloa et al. 2017). By optimizing culture conditions such as temperature, osmotic regulation, and growth retardants, these techniques allow for the successful conservation of both wild and elite fruit tree cultivars. The application of in vitro slow-growth storage is species specific and adaptable to various environmental and genetic factors, making it a versatile tool in plant biotechnology. These examples illustrate the broad applicability and effectiveness of in vitro slow-growth storage techniques across a range of plant species, underscoring their critical role in global conservation efforts (Benelli et al. 2022). As climate change and habitat loss continue to threaten plant biodiversity, refining and expanding the use of in vitro conservation strategies for fruit trees will ensure their long-term survival and sustainable utilization.

2.5 Challenges and Technological Innovations

Slow-growth storage is a vital in vitro conservation strategy for preserving plant genetic resources. However, its implementation presents several environmental and biological challenges. Recent technological innovations offer potential solutions to these issues.

2.5.1 Challenges in Slow-Growth Storage

Contamination Risks Maintaining aseptic conditions over extended storage periods is challenging. Contaminants can outcompete plant tissues, leading to culture loss. Regular monitoring and stringent sterilization protocols are essential to mitigate this risk (Reed and DeNoma 2016).

Genetic Stability Prolonged in vitro culture can induce somaclonal variations, resulting in genetic and phenotypic changes. With the application of slow-growth storage techniques, this risk is reduced; however, genetic fidelity, which requires optimizing culture conditions and periodic assessment of genetic stability, should also be ensured in germplasm (Cruz-Cruz et al. 2013).

Physiological Deterioration Extended storage may lead to physiological disorders, such as reduced vigor or viability. Adjusting environmental factors such as light, temperature, and nutrient composition is crucial for maintaining tissue health (Benelli et al. 2022).

Species-Specific Responses Different species exhibit varying responses to slow-growth conditions, necessitating tailored protocols. The development of species-specific guidelines can be resource intensive and time-consuming (Chauhan et al. 2019).

2.5.2 Technological Innovations

Synthetic Seed Technology Encapsulation of somatic embryos or shoot tips in alginate beads allows easier handling and storage. This method facilitates short- to medium-term conservation and simplifies germplasm exchange (Yücesan 2019).

Cryopreservation Advances While traditionally used for long-term storage, recent improvements in cryopreservation techniques, such as vitrification, offer complementary options for species that do not respond well to slow-growth methods (Popova et al. 2015).

Molecular Markers for Genetic Fidelity Utilizing molecular markers to assess genetic stability ensures that the preserved germplasm remains true-to-type, aiding in the early detection of somaclonal variations.

Addressing the challenges in slow-growth storage through these technological innovations enhances the effectiveness of in vitro conservation efforts, ensuring the preservation of plant biodiversity for future generations.

2.5.3 Technical Advances in Medium-Term In Vitro Storage

Recent technological advancements have addressed some of the challenges associated with medium-term in vitro storage:

Optimized Culture Media Adjusting the composition of culture media, including the use of growth retardants and osmotic agents, has been effective in reducing growth rates, thereby extending subculture intervals. For example, minimal media and growth retardants have been used to achieve slow-growth storage (Chauhan et al. 2019).

Temperature Modulation Storing cultures at relatively low temperatures (e.g., 15–18 °C) has been successful in slowing metabolic processes, thus prolonging the viability of in vitro cultures. This approach has been particularly effective for subtropical and tropical plants (Reed and Denoma 2016).

Synthetic Seed Technology Encapsulation of somatic embryos or shoot tips in alginate beads allows easier handling and storage. This method facilitates medium-term conservation and simplifies germplasm exchange (Tarraf and De Carlo 2024).

Molecular Tools for Genetic Assessment The use of molecular markers enables the assessment of genetic stability in stored cultures, ensuring the integrity of germplasm over extended storage periods (Cruz-Cruz et al. 2013).

By addressing environmental and biological constraints through these technical advancements, medium-term in vitro storage has become a more reliable and efficient method for conserving plant genetic resources.

2.6 Conclusions and Future Research Directions

The advancement of slow-growth conservation methodologies has markedly enhanced the preservation of plant genetic resources, offering a viable solution for safeguarding biodiversity. By meticulously adjusting environmental and nutritional parameters, these in vitro techniques have successfully prolonged subculture intervals, thereby diminishing the frequency of maintenance and conserving both labor and resources. Despite these achievements, fully harnessing the potential of slow-growth conservation necessitates ongoing research and its seamless integration into comprehensive global conservation frameworks.

Progress in Conservation Methodologies In recent years, substantial progress has been made in refining in vitro conservation protocols. Modifications in culture media composition, precise regulation of temperature and light intensity, and the strategic application of growth retardants have collectively contributed to decelerating plant tissue growth, facilitating the development of medium-term storage solutions. These refined protocols have been effectively implemented across a diverse array of species, encompassing those with recalcitrant seeds and challenges in vegetative propagation. For example, the use of minimal growth maintenance media has enabled the successful conservation of various plant species under controlled laboratory conditions (Chauhan et al. 2019). Furthermore, the incorporation of molecular tools has significantly bolstered the capacity to monitor genetic stability throughout the storage period, thereby ensuring the fidelity and integrity of conserved germplasms. Techniques such as DNA fingerprinting and marker-assisted selection have proven instrumental in identifying and mitigating somaclonal variations, which is crucial for maintaining the genetic authenticity of plant genetic resources (Cruz-Cruz et al. 2013).

Potential for Broader Applications and Cross-Sector Implications Expanding the application of slow-growth conservation methods holds considerable promise for enhancing global biodiversity preservation initiatives. By developing and implementing species-specific protocols, it becomes feasible to encompass a broader spectrum of plant taxa, particularly those that are endangered or possess significant ecological and economic value. Moreover, integrating these in vitro methods with other conservation strategies, such as cryopreservation and seed banking, can establish a more robust and comprehensive approach to germplasm preservation (Cruz-Cruz et al. 2013). Cross-sector collaboration among research institutions, botanical gardens, and agricultural organizations is pivotal for the successful implementation and scaling of these conservation strategies. Such partnerships facilitate the exchange of knowledge, the sharing of resources, and the development of standardized protocols, thereby increasing the effectiveness and reach of conservation efforts. Additionally, public awareness campaigns and educational initiatives play a vital role in promoting the importance of plant conservation, fostering community involvement, and garnering support for biodiversity preservation endeavors (Tarraf and De Carlo 2024).

Competing Interests The authors have no relevant financial or non-financial interests to disclose.

References

Ahmad N, Fatima N, Faisal M, Alatar AA, Pathirana R (2022) Photosynthetic parameters and oxidative stress during acclimation of crepe-myrtle (*Lagerstroemia speciosa* (L.) Pers.) in a meta-topolin-based micropropagation system and genetic fidelity of regenerated plants. Plan Theory 11(9):1163

Akdemir H, Kaya E, Ozden Y (2010) In vitro proliferation and minimum graft storage of Fraser photinia: influences of different medium, sugar combinations, and culture vessels. Sci Hortic 126:268–275

Akram NA, Shafiq F, Ashraf M (2017) Ascorbic acid—a potential oxidant scavenger and its role in plant development and abiotic stress tolerance. Front Plant Sci 8:613

Al Mousa R, Hassan N (2022) Effect of osmotic regulators on in vitro conservation of grape. J Genet Environ Resour Conserv 10:60–66

Ambrožič-Dolinšek J, Oblonšek S, Ciringer T (2022) Slow growth storage of *Berula erecta* in vitro – Effect of sucrose, sorbitol, and temperature. Agric Sci 19:33–41

Arbeloa A, Marín JA, Andreu P, García E, Lorente P (2017) In vitro conservation of fruit trees by slow growth storage. Acta Hortic 1155:101–106

Arcidiacono M, Catalano C, Motisi A, Sajeva M, Carimi F, Carra A (2021) Influence of culture conditions on in vitro asymbiotic germination of *Anacamptis longicornu* and *Ophrys panormitana* (Orchidaceae). Plants 10(11):2543

Arda H, Dayan S, Kartal Ç, Güler N (2016) In vitro conservation of critically endangered *Dianthus ingoldbyi* Turrill under slow growth conditions. Trakya Univ J Nat Sci 17:47–54

Azizi AAA, Andarini YN, Supriati Y, Roostika I, Sabda M, Mastur RH (2023) Sugarcane germplasm conservation using in vitro minimal growth technique and the regeneration capacity to provide sustainable genetic material. AIP Conf Proc 2972:070007

Bautista-Aguilar JR, Iglesias-Andreu LG, Martínez-Castillo J, Ramírez-Mosqueda MA, Ortiz-García MM (2021) In vitro conservation and genetic stability in *Vanilla planifolia* Jacks. HortScience 56:1494–1498

Bekheet SA, Matter MAHS, Taha EAAA (2016) In vitro conservation of jojoba (*Simmondsia chinensis*) shootlet cultures using osmotic stress and low temperature. Middle East J 5:396–402

Benelli C, Tarraf W, Izgu T, De Carlo A (2022) In vitro conservation through slow-growth storage technique of fruit species: An overview of the last 10 years. Plan Theory 11:3188

Berger J, Essah E, Blanusa T (2024) The impact of plants on the humidity of naturally-ventilated Office indoor environments. J Build Eng 86:108814

Bretting PK, Duvick DN (1997) Dynamic conservation of plant genetic resources. Adv Agron 61:1–51

Capuana M, Di Lonardo S (2013) In vitro conservation of chestnut (*Castanea sativa*) by slow growth. In Vitro Cell Dev Biol Plant 49:605–610

Chauhan R, Singh V, Quraishi A (2019) In vitro conservation through slow-growth storage. In: Faisal M, Alatar A (eds) Synthetic seeds. Springer, Cham, pp 457–478

Chen G, Sun W (2018) The role of botanical gardens in scientific research, conservation, and citizen science. Plant Divers 40:181–188

Chen CC, Bates R, Carlson J (2014) Effect of environmental and cultural conditions on medium pH and explant growth performance of Douglas-fir (*Pseudotsuga menziesii*) shoot cultures. F1000 Research 3:298

Chen X, Li D, Yang Y, Wu H (2016) Mannitol and sorbitol improve uniformity of adventitious shoot regeneration in *Echinacea purpurea* L. Moench J Biomed Sci Eng 9:58–64

Coelho N, Gonçalves S, Romano A (2020) Endemic plant species conservation: Biotechnological approaches. Plan Theory 9(3):345

Cruz-Cruz CA, González-Arnao MT, Engelmann F (2013) Biotechnology and conservation of plant biodiversity. Resources 2:73–95

Da Silva RL, Ferreira CF, da Silva Ledo CA, de Souza EH, da Silva PH, de Carvalho Costa MAP, Souza FVD (2016) Viability and genetic stability of pineapple germplasm after 10 years of in vitro conservation. Plant Cell Tissue Organ Cult 127:123–133

Da Silva DPC, Ozudogru EA, Dos Reis MV, Lambardi M (2018) In vitro conservation of ornamental plants. Ornam Hortic 24(1):28–23

De Diego N, Sampedro MC, Barrio RJ, Saiz-Fernández I, Moncaleán P, Lacuesta M (2013) What is stress? Dose-response effects in commonly used in vitro stress assays. Physiol Plant 147:341–352

Ebert AW, Engels JMM (2020) Plant biodiversity and genetic resources matter! Plan Theory 9:1706

Elazab D, Capuana M, Ozudogru EA, Anichini M, Lambardi M (2023) Use of liquid culture with the ElecTIS bioreactor for faster recovery of blackberry (*Rubus fruticosus* L.) shoots from conservation at 4°C. Horticulturae 9:680

El-Dawayati MM (2017) In vitro conservation of date palm shoot-tip explants and callus cultures under minimal growth conditions. In: Methods in molecular biology, vol 1638. Springer, Berlin/Heidelberg, pp 49–59

Engels JM, Ebert AW (2024) How can we strengthen the global genetic resources' conservation and use system? Plan Theory 13:702

FAO (1994) Chapter three: cryopreservation and in vitro storage. In: Forest genetic resources conservation in Romania. Retrieved from https://www.fao.org/4/t2114e/t2114e04.htm

FAO (2024) Seeds and plant genetic resources: a basis for life. Food and Agriculture Organization of the United Nations. Retrieved from https://www.fao.org/agriculture/crops/core-themes/theme/seeds-and-plant-genetic-resources/en/

Foyer CH, Noctor G (2011) Ascorbate and glutathione: the heart of the redox hub. Plant Physiol 155:2–18

Gago-Calderón A, Redrado-Salvatierra GM, Andres-Diaz JR, Barcelo-Muñoz M, Barcelo-Munoz A (2023) Effect of the adaptation of LED lighting in in vitro chambers on the environmental conditions of temperature and humidity of the plant cultures. IEEE Sustainable Smart Lighting World Conference & Expo (LS18), Mumbai, India, pp 1–6

Gallie DR (2013) L-Ascorbic acid: a multifunctional molecule supporting plant growth and development. Scientifica 2013:795964

Gantait S, Kundu S, Wani SH, Das PK (2016) Cryopreservation of forest tree seeds: A mini-review. J For Environ Sci 32:311–322

Gautam N, Faroda P, Agarwal T, Harish, Gupta AK (2024) Effect of microclimatic physical factors on in vitro morphogenesis of plants: a systematic review. Discov Agric 2:13

Gopal J, Chauhan N (2010) Slow growth in vitro conservation of potato germplasm at low temperature. Potato Res 53:141–149

Guan Y, Li SG, Fan XF, Su ZH (2016) Application of somatic embryogenesis in woody plants. Front Plant Sci 7:938

Hammond SD, Viehmannova I, Zamecnik J, Panis B, Cepkova PH (2019) Efficient slow-growth conservation and assessment of clonal fidelity of *Ullucus tuberosus* Caldas microshoots. Plant Cell Tissue Organ Cult 138:559–570

Hanson J, Lusty C, Furman B, Ellis D, Payne T, Halewood M (2024) Opportunities for strategic decision making in managing ex situ germplasm collections. Plant Genet Resour 22:195–200

Hasanuzzaman M, Bhuyan MHMB, Anee TI, Parvin K, Nahar K, Mahmud JA, Fujita M (2019) Regulation of ascorbate-glutathione pathway in mitigating oxidative damage in plants under abiotic stress. Antioxidants 8:384

Haussmann BIG, Parzies HK, Presterl T, Sušić Z, Miedaner T (2004) Plant genetic resources in crop improvement. Plant Genet Resour 2:3–21

Hoban S, Kallow S, Trivedi C (2018) Implementing a new approach to effective conservation of genetic diversity, with ash (*Fraxinus excelsior*) in the UK as a case study. Biol Conserv 225:10–21

Irish BM, Goenaga R, Reed BM (2013) Amending storage vessel and media increase transfer interval of *Musa* spp. tissue culture plantlets. J Agric Univ Puerto Rico 97:1–14

Jackson MB, Abbott AJ, Belcher AR, Hall KC, Butler R, Cameron J (1991) Ventilation in tissue cultures and effects of poor aeration on ethylene and carbon dioxide accumulation, oxygen depletion and explant development. Ann Bot 67:229–237

Jokipii S, Ryynänen L, Kallio PT, Aronen T, Häggman H (2004) A cryopreservation method maintaining the genetic fidelity of a model forest tree, *Populus tremula* L. × *Populus tremuloides* Michx. Plant Sci 166:799–806

Juan-Vicedo J, Serrano-Martínez F, Cano-Castillo M, Casas JL (2022) In vitro propagation, genetic assessment, and medium-term conservation of the coastal endangered species *Tetraclinis articulata* (Vahl) Masters (Cupressaceae) from adult trees. Plants 11:187

Kadhim ZK, Abdulhussein MAA (2021) Minimal media strength for in vitro conservation of strawberry (*Fragaria ananassa*) cultures. Basrah J Agric Sci 34:1–9

Kadota M, Niimi Y (2004) Influences of carbon sources and their concentrations on shoot proliferation and rooting of 'Hosui' Japanese pear. HortScience 39:1681–1683

Koç İ, Akdemir H, Onay A et al (2014) Cold-induced genetic instability in micropropagated *Pistacia lentiscus* L. plantlets. Acta Physiol Plant 36:2373–2384

Kozai T, Kubota C, Jeong BR (1997) Environmental control for large-scale production of plants through in vitro techniques. Plant Cell Tissue Organ Culture 51(1):49–56

Lata H, Moraes RM, Bertoni B, Pereira AMS (2010) In vitro germplasm conservation of *Podophyllum peltatum* L. under slow growth conditions. In Vitro Cell Dev Biol Plant 46:22–27

Lédo ADS, Almeida CS, Silva AVC, Francisco J, Junior DS, Araujo AGD (2015) Medium-term conservation and regrowth of the tropical multipurpose species Genipa americana. Afr J Agric Res 10:3885–3892

Lidder P, Sonnino A (2011) Biotechnologies for the management of genetic resources for food and agriculture. Commission on Genetic Resources for Food and Agriculture, Background Study Paper No. 52. Food and Agriculture Organization (FAO). Retrieved from http://www.fao.org/4/mb387e/mb387e.pdf

Linjikao J, Inthima P, Limmongkon A, Kongbangkerd A (2024) Mannitol and sorbitol concentration optimization for effective *Epipactis flava* Seidenf. in vitro slow growth storage. In Vitro Cell Dev Biol Plant 60:496–507

Lopez-Puc G (2013) An effective in vitro slow growth protocol for conservation of the orchid *Epidendrum chlorocorymbos* Schltr. Trop Subtrop Agroecosyst 16:61–68
Lu X, Sun P, Liu R et al (2024) In vitro slow-growth conservation, acclimatization, and genetic stability of virus-free apple plants. Hortic Adv 2:30
Marco-Medina A, Casas JL (2012) Polyamine content during minimal growth storage of Thymus moroderi explants. Biol Plant 56:590–594
Marino G, Negri P, Cellini A, Masia A (2010) Effect of carbohydrates on in vitro low-temperature storage of shoot cultures of apricot. Sci Hortic 126:434–440
Mattana E, Fenu G, Bacchetta G (2012) Regional responsibility for plant conservation: The 2010 GSPC Target 8 in Sardinia. Plant Biosyst 146:649–653
Miroshnichenko TM, Ivchenko TV, Bashtan NO, Mozgovska HV (2023) Medium composition for in vitro mid-term storage of *Solanum habrochaites* test-tube plants. Veg Melon Growing 74:6–18
Moher M, Jones M, Zheng Y (2021) Photoperiodic Response of In Vitro *Cannabis sativa* Plants. HortScience 56(1):108–113
Murray BG (2017) Plant diversity, conservation and use. In: Encyclopedia of applied plant sciences, 2nd edn. Elsevier, pp 289–308
Muthoni J, Shimelis H, Melis R (2019) Long-term conservation of potato genetic resources: methods and status of conservation. Aust J Crop Sci 13:717–725
Nowakowska K, Pacholczak A (2020) Comparison of the effect of meta-topolin and benzyladenine during *Daphne mezereum* L. Micropropagation. Agronomy 10(12):1994
Nurtaza A, Dyussembekova D, Islamova S et al (2024) In vitro conservation and genetic diversity analysis of rare species *Ribes janczewskii*. Sci Rep 14:31117
Oliveira KS, Aloufa MAI (2022) Slow growth in vitro culture for conservation of *Hancornia speciosa* Gomes. Floresta 52:7–16
Opabode JT, Ajibola OV, Oyelakin OO, Akinyemiju OA (2016) Somatic embryogenesis and genetic uniformity of cassava plants regenerated from secondary somatic cotyledons preserved in osmotic agents. Plant Tissue Cult Biotechnol 26:47–54
Ozudogru EA, Kaya E, Kirdok E (2011) Development of protocols for short-, medium-, and long-term conservation of thyme. Acta Hortic 918:43–50
Ozudogru EA, Benelli C, Dradi G et al (2017) Effect of culture container and carbohydrate content on in vitro slow growth storage of the cherry rootstock 'Gisela®5'. Acta Physiol Plant 39:94
Padayachee K, Watt M, Edwards N, Mycock D (2009) Cryopreservation as a tool for the conservation of *Eucalyptus* genetic variability: concepts and challenges. South For J For Sci 71:165–170
Pardo A, Rivero S, Alvarado G (2014) In vitro conservation of garlic microbulbs (Allium sativum L.). Bioagro 26:115–122
Patel AK, Sharma I, Choudhary M, Shekhawat NS, Rai MK (2025) Exploring plant tissue culture for in vitro conservation of trees: recent progress. In: Kalia RK, Pathak R (eds) Tree biology and biotechnology. Springer, Singapore, pp 1–15
Pérez-Jiménez M, Guevara-Gázquez A, Carrillo-Navarro A, Cos-Terrer J (2021) How carbon source and seedcoat influence the in vitro culture of peach (*Prunus persica* L. Batsch) immature seeds. HortScience 56:136–137
Phillips PWB (2003) Ethics and biosafety; ownership of plant genetic resources. In: Encyclopedia of applied plant sciences. Elsevier, pp 289–295
Popova E, Shukla M, Kim HH, Saxena PK (2015) Plant cryopreservation for biotechnology and breeding. In: Al-Khayri J, Jain S, Johnson D (eds) Advances in plant breeding strategies: breeding, biotechnology and molecular tools. Springer, Cham
Rahayu E, Habibah N (2016) Optimization of in vitro conservation protocol of *Carica pubescens* Lenne & K. Koch through medium concentration, temperature, and irradiation duration decrease. Biosaintifika J Biol Educ 8:85–93
Rao BR (2015) Genetic diversity, genetic erosion, conservation of genetic resources, and cultivation of medicinal plants. In: Genetic diversity and erosion in plants: case histories. Springer, p 357
Reed BM, DeNoma J (2016) Medium-term in vitro storage of pear as a complementary germplasm preservation technique. Acta Hortic 1113:251–256

Reed BM, Sarasan V, Kane M, Bunn E, Pence VC (2011) Biodiversity conservation and conservation biotechnology tools. In Vitro Cell Dev Biol Plant 47:1–4

Reeves C, Tikkinen M, Aronen T, Krajnakova J (2023) Application of cold storage and short in vitro germination for somatic embryos of *Pinus radiata* and *P. sylvestris*. Plants 12:2095

Rohani RA (2020) It's all about technique: in vitro methods for storing medicinal plant germplasm. United Plant Savers, Medicinal Plant Conservation

Rout GR, Samantaray S, Das P (2000) In vitro manipulation and propagation of medicinal plants. Biotechnol Adv 18:91–120

Salgotra RK, Chauhan BS (2023) Genetic diversity, conservation, and utilization of plant genetic resources. Genes 14:174

Solárová J, Pospíšilová J (1997) Effect of carbon dioxide enrichment during in vitro cultivation and acclimation to ex vitro conditions. Biol Plant 39:23–30

Son SH, Chun YW, Hall RB (1991) Cold storage of in vitro cultures of hybrid poplar shoots (*Populus alba* L. × *P. grandidentata* Michx.). Plant Cell Tissue Organ Cult 27:161–168

Sota L, Kongjika E (2014a) The effect of nutrient media in micropropagation and in vitro conservation of wild population of mahaleb cherry (*Prunus mahaleb* L.). J Microbiol Biotech Food Sci 3:453–456

Sota L, Kongjika E (2014b) Slow growth in vitro conservation of *Ziziphus jujuba* Mill. Agric For 60:27–37

Srivastava MRI, Purshottam DK, Srivastava AK, Misra P (2013) In vitro conservation of *Glycyrrhiza glabra* by slow growth culture. Int J Biol Res 3(1):49–58

Swedlund B, Locy RD (1993) Sorbitol as the primary carbon source for the growth of embryogenic callus of maize. Plant Physiol 103:1339–1346

Tarraf W, De Carlo A (2024) In vitro biotechnology for conservation and sustainable use of plant genetic resources. Plants 13:1897

Tavazza R, Rey NA, Papacchioli V, Pagnotta MA (2015) A validated slow-growth in vitro conservation protocol for globe artichoke germplasm: a cost-effective tool to preserve from wild to elite genotypes. Sci Hortic 197:135–143

Thakur S, Tiwari KL, Jadhav SK (2015) In vitro approaches for conservation of *Asparagus racemosus* willd. In Vitro Cell Dev Biol-Plant 51:619–625

Thorpe TA (2007) History of plant tissue culture. Methods Mol Biol 6:9–28

Tilman D, Isbell F, Cowles JM (2014) Biodiversity and ecosystem functioning. Annu Rev Ecol Evol Syst 45:471–493

Trejgell A, Kamińska M, Tretyn A (2015) In vitro slow growth storage of *Senecio macrophyllus* shoots. Acta Physiol Plant 37:234

Turdiyev TT, Kovalchuk IY, Kabylbekova BZ, Chukanova NI, Frolov SN (2020) In vitro germplasm cold storage of fruit and berry plants of Kazakhstan. Eurasian J Biosci 1219:1213–1219

Wijerathna-Yapa A, Hiti-Bandaralage J (2023) Tissue culture-a sustainable approach to explore plant stresses. Life 13(3):780

Yaacob JS, Mahmad N, Mat Taha R, Mohamed N, Mad Yussof AI, Saleh A (2014) Optimization of culture conditions (sucrose, pH, and photoperiod) for in vitro regeneration and early detection of somaclonal variation in ginger lime (*Citrus assamensis*). Scientific World J 2014:262710

Yadav B, Pandit DL, Banjade D, Mehata DK, Bhattarai S, Bhandari S, Ghimire P, Yadav P, Paudel P (2024) Insights into the germplasm conservation and utilization: implications for sustainable agriculture and future crop improvement. Arch Agric Environ Sci 9:180–193

Yang X, Baskin CC, Baskin JM, Pakeman RJ, Huang Z, Gao R, Cornelissen JH (2021) Global patterns of potential future plant diversity hidden in soil seed banks. Nat Commun 12:7023

Yaseen M, Ahmad T, Sablok G, Standardi A, Hafiz IA (2013) Role of carbon sources for in vitro plant growth and development. Mol Biol Rep 40:2837–2849

Yücesan B (2019) Synseed: a new trend in seed technology. In: Faisal M, Alatar A (eds) Synthetic Seeds. Springer, Cham

Cryopreservation of Cell Suspension Cultures and Protoplasts

3

Jana Krajňáková and Lloyd Donaldson

Abstract

Cryopreservation is a vital technique for the long-term conservation of plant cell suspension cultures and protoplasts, ensuring the stability of valuable genetic resources. This chapter examines recent advances in cryopreservation methods, focusing on their applications in woody species and biotechnological processes. While suspension cultures serve as a sustainable source of secondary metabolites, their cryopreservation is limited by cellular heterogeneity, oxidative stress, and post-thaw viability issues. Protoplasts, although promising for genetic transformation, face additional hurdles owing to their sensitivity to cryogenic stress and ice formation. Standardized protocols, including slow cooling, vitrification, and encapsulation–dehydration, have been developed, yet species-specific optimization remains necessary. Advances in omics technologies, particularly transcriptomics, and metabolomics, provide deeper insights into cryotolerance mechanisms, whereas microscopy techniques play crucial roles in evaluating membrane integrity, intracellular ice formation, and post-thaw recovery. High-resolution imaging methods, such as cryo-scanning electron microscopy and fluorescence microscopy, offer valuable tools for assessing structural damage and optimizing freezing protocols. Despite progress, improving post-thaw regeneration and metabolic stability remains a challenge. Integrating AI-driven monitoring and nanotechnology-based warming approaches could increase cryopreservation efficiency. Future research should refine protocols, improve genetic stability, and expand cryobank accessibility to support biodiversity conservation and commercial applications.

J. Krajňáková (✉) · L. Donaldson
Forest Genetics and Biotechnology, Scion, Rotorua, New Zealand
e-mail: jana.krajnakova@scionresearch.com

S. Gantait, P. Chmielarz (eds.), *Conservation of Plant Genetic Resources*,
https://doi.org/10.1007/978-981-96-8331-4_3

Keywords

Cryopreservation · Microscopic techniques · Post-thaw · Secondary metabolites · Woody species

3.1 Introduction

The cultivation of isolated plant cells in a chemically defined culture medium has a long history, dating back to 1902 with the work of Gottlieb Haberlandt, who made the first attempts to culture single isolated plant cells (Hartel 2003). Currently, plant cells and tissues can be cultivated in vitro for various applications, including plant breeding, propagation, germplasm preservation, and molecular biology (Lelu-Walter et al. 2013; Rosvall 2019; Rodríguez et al. 2022).

Approximately 45% of the estimated 374,000 plant species worldwide are woody, including trees, shrubs, and lianas. This suggests that there are approximately 168,000 woody plant species globally, with approximately 64,100 known tree species (Fazan et al. 2020; Cazzolla Gatti et al. 2022). Woody plant species are distributed across various global ecosystems, including tropical forests, temperate zones, boreal forests, and arid and semiarid regions. These species belong to two major clades, gymnosperms and angiosperms, which diverged approximately 300–350 million years ago (van der Kooi and Ollerton 2020). Molecular phylogenetic studies indicate that this split occurred as early seed plants evolved distinct reproductive strategies. Gymnosperms, including conifers, cycads, and ginkgos, retained exposed seeds, whereas angiosperms, which later became dominant, developed flowers and enclosed seeds within fruits.

Owing to the economic importance of woody plant species and the threats posed by natural and anthropogenic factors, various conservation approaches, including in situ and ex situ strategies, have been developed to preserve biodiversity. Cryopreservation techniques fall under ex situ conservation strategies within in vitro methods designed to ensure the long-term storage of diverse biological materials (Gonzalez-Arnao et al. 2014; Potter et al. 2017). Approaches for the in vitro conservation of woody plant germplasms have been reviewed by González-Arnao et al. (2017), Corredoira et al. (2017), Li et al. (2018), Normah et al. (2019), and Ballesteros et al. (2024).

Owing to their difficulty in regenerating whole plants under in vitro conditions, woody species are often classified as recalcitrant. This challenge persists despite extensive manipulation of culture conditions, growth regulators, and explant sources. Recalcitrance is particularly problematic in adult trees and genetically superior clones, where somatic tissues frequently lose their ability to dedifferentiate and form embryos (Bonga et al. 2010; Wendling et al. 2014a, b; Bonga 2017).

This chapter examines different types of cryopreservation protocols for suspension cultures of woody plant species. It covers various plant cell types, including calli, embryogenic tissue, and nucellar cells of *Citrus* species. These suspension cultures serve as essential starting materials for cryopreservation protocols aimed at producing bioactive compounds or facilitating plant regeneration. Protoplast cultures have been established in many woody plant species, particularly with advancements in regeneration technologies such as somatic embryogenesis (Reed and Bargmann 2021). However, practical applications remain limited due to inconsistent regeneration success. *Citrus* sp. stands out as an exception, with well-documented protoplast regeneration and somatic hybridization applications (Omar et al. 2016; Liu et al. 2024). Despite advancements in cryopreservation techniques for herbaceous protoplasts, no studies have focused on the cryopreservation of protoplasts from woody species, highlighting a significant research gap and the need for further investigation.

Callus and embryogenic tissue are two distinct types of plant tissue commonly used in tissue culture and regeneration studies. While both originate from dedifferentiated plant cells, they differ in structure, function, and regenerative potential. A callus is an unorganized mass of dedifferentiated cells that form in response to wounding or hormonal induction. It consists of heterogeneous cell types, including parenchyma-like cells, vacuolated cells, and nondividing cells, and appears as an amorphous, friable, or compact mass with irregular cellular organization. The callus lacks polarity and often includes both embryogenic and non-embryogenic cells. It has limited regenerative capacity unless it undergoes further differentiation into organogenic or embryogenic pathways. More commonly, calluses rather than direct plant regeneration are used for studies of somaclonal variation. It is induced and maintained in media with high auxin-to-cytokinin ratios (e.g., 2,4-D, NAA), requiring continuous subculturing for sustained growth. Owing to their undifferentiated nature, calluses are widely employed in genetic transformation, secondary metabolite production, and mutagenesis studies (Ikeuchi et al. 2013, 2016; Feher 2019).

In contrast, embryogenic tissue consists of highly organized and actively dividing cells capable of forming somatic embryos. These cells exhibit a dense cytoplasm, prominent nuclei, and minimal vacuolation, resembling early zygotic embryo cells. In gymnosperms, embryogenic tissue forms proembryogenic masses (PEMs) that exhibit distinct polarity and developmental organization, progressing through different embryogenic stages (Filonova et al. 2000; von Arnold et al. 2019). Unlike calli, embryogenic tissue has high regenerative potential, forming somatic embryos that develop into whole plants without further differentiation. It requires relatively low auxin concentrations for induction and often benefits from abscisic acid (ABA) for somatic embryo maturation (von Arnold 2008). Owing to its ability to generate uniform plants, embryogenic tissue is primarily used in large-scale clonal propagation and forestry breeding programs (Park et al. 2016). A key distinguishing feature of gymnosperms is their well-defined suspensor structure during SE, a trait not commonly observed in angiosperms (Smertenko and Bozhkov 2014).

3.2 General Aspects of Cryopreservation

Cryopreservation enables the long-term storage of plant cells, tissues, and organs at ultralow temperatures, typically in liquid nitrogen (−196 °C) or vapor-phase nitrogen (−150 °C). It halts metabolic activity and cellular degradation, preserving plant material indefinitely without causing genetic or physiological changes (Benson 2008; Reed 2008).

At temperatures below −130 °C, molecular movement slows or stops, preventing enzymatic activity and aging (Nagel et al. 2024). However, maintaining viable cells, tissues, or organs requires careful water management. Water is essential for membrane and protein function, but freezing between 0 °C and − 40 °C results in the formation of ice crystals that expand, rupture membranes, and cause cell death. Preventing ice formation is achieved by increasing the solute concentration, which disrupts water molecule bonding or, by ultrafast cooling, which halts molecular motion and induces vitrification. In this glass-like state, ice crystallization is avoided, preserving cell viability (Wang et al. 2021b; Nagel et al. 2024).

A plant cell can be considered a watery solution enclosed by a membrane within a rigid cell wall. To remove water without damaging cellular structures, two primary mechanisms are used: osmosis and cryoprotection. During natural cold acclimation, plants accumulate cryoprotectants such as sugars, proteins, and amino acids. As temperatures drop below zero, ice forms in intercellular spaces, creating an osmotic gradient that draws water out of cells. This dehydration concentrates the cytoplasm, increasing resistance to freezing. Many temperate plants can supercool to −15 °C without freezing, whereas some woody species can reach −40 °C before spontaneous ice nucleation occurs. This temperature threshold determines the tree line in mountainous regions. However, in some species, even cold temperatures without ice or excessive dehydration can be lethal. Cryopreservation techniques counteract these damaging effects (Reed 2011; Reed et al. 2016; Normah et al. 2019).

The success of cryopreservation depends on increasing the solute concentration, optimizing cooling rates, and enhancing plant tolerance to dehydration and freezing (Sakai and Engelmann 2007). Nagel et al. (2024) identified five fundamental methods to increase the cellular solute concentration: air-drying, metabolic adjustment, osmotic dehydration, and the use of penetrating molecules. These methods form the foundation of all cryopreservation techniques that are described briefly below.

Slow cooling, controlled-rate cooling or two-step cooling protects plant cells by regulating osmotic balance and inducing freeze dehydration (Reed and Uchendu 2008). The cells were pretreated with cryoprotectants and gradually cooled to intermediate temperatures (−35 °C to −40 °C) before immersion in liquid nitrogen (LN). Ice nucleation occurs at approximately −9 °C, promoting extracellular ice formation while removing intracellular water to prevent lethal ice crystal formation. The process consists of five key stages:

(i) Preconditioning and cold acclimation—enhances survival by adjusting the osmotic balance, applying cold acclimation, or modifying in vitro culture conditions.

(ii) Pretreatment with cryoprotectants—reduces cellular damage by introducing protective compounds.
(iii) Slow cooling—gradual temperature reduction at 0.1 °C to 5 °C per min. Samples are held at intermediate temperatures (−35 °C to −40 °C) before immersion in LN to ensure optimal dehydration.
(iv) Rapid warming—quick rewarming (40–45 °C for 1–2 min) prevents ice recrystallization, which can damage cells. Slow warming can cause structural stress due to ice expansion and contraction.
(v) Viability and recovery assessment—evaluated through direct regrowth and viability stains such as fluorescein diacetate and Evans blue.

This method is best suited for unorganized cell clumps, where preserving cell integrity is more important than maintaining tissue structure.

Vitrification prevents ice formation by transitioning water into an amorphous glassy state via highly concentrated cryoprotectants (CPAs). Unlike ice crystallization, vitrification avoids cellular damage by increasing the solution viscosity and enabling supercooling below −70 °C. This technique stabilizes tissues, prevents solute imbalance, and halts molecular activity, preserving viability over time (Sakai and Engelmann 2007; Sakai et al. 2008; Engelmann et al. 2008). Since most plants cannot naturally vitrify due to their high water content, effective cryoprotectants must have strong glass-forming ability, induce dehydration, and minimize toxicity. PVS2 (30% glycerol, 15% ethylene glycol, 15% DMSO, 0.4 M sucrose) is widely used for plant shoot tips, whereas PVS3 (50% glycerol, 50% sucrose) serves as a DMSO-free alternative, requiring longer exposure times (Sakai and Engelmann 2007). Vitrification techniques are widely applicable because of their simplicity, efficiency, and reproducibility. They are successfully used for cryopreserving cell suspensions, embryogenic tissues, shoot tips, and seeds of both woody and herbaceous species from temperate and tropical regions (Sakai and Engelmann 2007; Gonzalez-Arnao et al. 2014, 2017; Normah et al. 2019).

Encapsulation-dehydration increases the solute concentration in plant tissues through osmotic dehydration and protection. Explants are encapsulated in calcium alginate beads, forming artificial seeds (Palanyandy et al. 2020). These beads are exposed to high sucrose concentrations and dried under sterile airflow or with silica gel to reduce the water content to 20–30%. Once dehydrated, they are immersed in liquid nitrogen (LN) for long-term storage (Benelli et al. 2013; González-Arnao et al. 2017).

Table 3.1 compares essential aspects to consider when one of the three cryopreservation methods is selected: slow cooling, vitrification, and encapsulation-dehydration or any modifications of these methods.

3.3 Cryopreservation of Suspension Cultures

The cryopreservation of suspensions or dedifferentiated cell cultures was recently reviewed by Nausch and Buyel (2021) and earlier by Schumacher et al. (2021), Heine-Dobbernack et al. (2008), Grout (2007), Reinhoud et al. (2000), Schrijnemakers and Van Iren (1995), and (Grout 1995b). The application of

Table 3.1 Comparison of key aspects in choosing a cryopreservation method

Aspect	Slow cooling	Vitrification	Encapsulation-dehydration
Cooling rate	0.3 °C to 1 °C per min (gradual)	Rapid (~10,000 °C per min)	Moderate (~1 °C per min, with hydrogel or polymer coating)
Ice crystal formation	Possible (if cooling not controlled properly)	None (transitions to glass-like state)	Minimized due to controlled diffusion of CPA
Cryoprotectant (CPA) concentration	Low (~5–10%)	High (~15–50%)	Medium (~10–20%)
Toxicity risk	Lower toxicity risk	Higher toxicity risk (due to high CPA concentration)	Lower toxicity due to controlled CPA exposure
Thawing process	Slow, controlled thawing	Rapid warming to prevent devitrification	Controlled thawing, often with stepwise CPA removal
Survival rate	Moderate (lower for delicate cells)	Higher (better for delicate cells)	High (better structural integrity for larger samples)
Equipment required	Requires programmable freezer, could be replaced with Mr. Frosty™	Programmable freezer is not needed	Requires biocompatible encapsulation materials
Best for	Embryos, embryogenic tissue, cells, shoot tips	Embryos, small tissues, shoot tips, meristems	Shoot tips, somatic embryos
Challenges	Risk of intracellular ice, lower survival rates	Cryoprotectant toxicity, risk of thermal cracking	Complex preparation, potential CPA diffusion issues

vitrification methods to plant cell suspension cryopreservation was discussed by Reinhoud et al. (1995), while alginate encapsulation was covered by Schumacher et al. (2021). The combination of encapsulation and vitrification was explored by Wang and Perl (2006) and Yin et al. (2012). Nausch and Buyel (2021) systematically analyzed 50 research articles, providing in-depth analysis of slow cooling, vitrification, and encapsulation dehydration techniques across various species, including model plants such as *Arabidopsis*, tobacco, and rice, which are widely used for secondary metabolite production (Eibl et al. 2018; Gubser et al. 2021). Their review identified key parameters that maintain viability and promote fast, vigorous regrowth. In contrast, reviews such as Schumacher et al. (2021), Wang and Perl (2006), Yin et al. (2012), and Schrijnemakers and Van Iren (1995) focus on step-by-step guidance and practical recommendations for executing specific cryopreservation methods.

Table 3.2 summarizes woody gymnosperm and angiosperm species where suspension cultures have been cryopreserved or stored long-term, including studies where gymnosperm embryogenic tissue was maintained in suspension long-term.

Table 3.2 List of selected woody species from the gymnosperm and angiosperm clades in which suspension cultures of different origins were cryopreserved or long-term stored

Plant species	Clade	Purpose	Cultivation	Culture type	Method	No cultivars/ cell lines	Main aims/ achievements	References
Acer pseudoplatanus L.	ANG	NA	SC	Callus	TSC	1	Use of CPA, essential to use the cells in lag phase	Sugawara and Sakai (1974)
	ANG	NA	SC	Callus	TSC	1	Use of CPA, DMSO, and glycerol, small meristematic cells survived better	Nag and Street (1975a)
	ANG	NA	SC	Callus	TSC	1	Importance of CPA	Nag and Street (1975b)
	ANG	NA	SC	Callus	TSC	1	CPA—mannitol, cells taken at lag and early exponential phase, effect of inoculation density	Withers and Street (1977)
	ANG	NA	SC	Callus	TSC	1	Ultrastructural modifications in thawed cells	Withers and Davey (1978)
	ANG	NA	SC	Callus	TSC	1	Importance of the cell cycle stage	Withers (1978)
	ANG	NA	SC	Callus	TSC	1	Description of freezing unit and routine cryo-preservation method	Withers and King (1980)

(continued)

Table 3.2 (continued)

Plant species	Clade	Purpose	Cultivation	Culture type	Method	No cultivars/ cell lines	Main aims/ achievements	References
	ANG	NA	SC	Callus	TSC	1	Mannitol and sorbitol used as CPAs, ultrastructural studies of cells	Pritchard et al. (1986c)
	ANG	NA	SC	Callus	TSC	1	Water potential, osmotic adjustment, metabolic changes	Pritchard et al. (1986a)
	ANG	NA	SC	Callus	TSC	1	Osmotic stress significantly improved cryo-preservation outcomes	Pritchard et al. (1986b)
	ANG	NA	SC	Callus	Cold storage	1	Cold storage in a phosphate-free medium at 5 °C for 6 months	Boisson et al. (2012)
Corylus avellana L.	ANG	SM	SEM	Callus	DES	1	First cryo-preservation of *C. avellana* callus, use of ABA	Mojarrad Nanas et al. (2023)
Citrus deliciosa Tan.	ANG	REG	SC	Nucellar cells, callus	TSC	1	−0.5 °C/min to −40 °C, pretreatment with 0.15 M sucrose and 5% DMSO 50%–60% survival rates	Aguilar et al. (1993)

Citrus deliciosa Tan *Citrus reshni* Hort. Ex Ten. *Citrus sinensis* Osb. *C. sinensis* (L.) Osb. var. Hamlin *Citrus aurentifolia* Swing.	ANG	REG	SC	Nucellar cells, callus	TSC	6	Programmable and non-programmable freezing method CPAs: 0.15 M sucrose and 5%–15% DMSO 5 out 6 tested citrus varieties survived cryopreservation	Engelmann et al. (1994)
Citrus sinensis Osb.	ANG	REG	SC	Nucellar cells, callus	TSC	1	Slow cooling to −40 °C at 0.5 °C/min CPAs: 5% DMSO and 1.2 M sucrose 70% post-thaw viability	Kobayashi et al. (1990)
	ANG	REG	SC	Nucellar cells, callus	TSC	1	CPAs: 2 or 3 M glycerol and 0.4 M sucrose	Sakai et al. (1991a)
Citrus sinensis Osb. var. *brasiliensis* Tanaka	ANG	REG	SC	Nucellar cells, callus	VITRIF	1	PVS2 applied in 2 steps an average survival rate of 80%. The vitrification procedure required only 10 min.	Sakai et al. (1990)

(continued)

Table 3.2 (continued)

Plant species	Clade	Purpose	Cultivation	Culture type	Method	No cultivars/ cell lines	Main aims/ achievements	References
	ANG	REG	SC	Nuclear cells	VITRIF	1	PVS2, at 25 °C for 3 min achieved high survival rates (up to 90%) Method was applied to other citrus cultivars, demonstrating broad applicability	Sakai et al. (1991b)

C. sinensis cv Valencia cv. Anliucheng cv. Newhall *C. reticulata* cv. Ponkan *C. aurantium* cv. spp. (sour orange) *C. paradisi* cv. Red Marsh. (capable of SE) *C. unshiu* cv. Guoqing No. 1, *C. reticulata* × *C. sinensis* cv. Murcott, *Microcitrus papuana*, *C. reticulata* × *C. paradisi* cv. Minneola, *Fortunella hindsii* cv. Suichuan, Meiwa. (recalcitrant to SE)	ANG	REG	SC	callus	VITRIF	12	Over 90% viability across all genotypes; cryopreserved cells maintained regenerative ability. No ploidy changes or sequence mutations were detected post-cryopreservation. DNA methylation patterns were altered, potentially influencing SE	Hao et al. (2002)

(continued)

Table 3.2 (continued)

Plant species	Clade	Purpose	Cultivation	Culture type	Method	No cultivars/ cell lines	Main aims/ achievements	References
Elaeis guineensis	ANG	REG	SC	ET	TSC	4	The best survival rates (38–81%) were obtained using CPAs 10% DMSO and 1 M glucose Somatic embryos were successfully regenerated from all tested cell lines	Chabrillange et al. (2000)
Picea abies (L.) Karst.	GYM	REG	SC	ET	TSC	6	Effect of culture period and cell density on the regrowth The best regrowth was obtained when sedimented cell volume (SCV) was adjusted to 50% (v/v) before sorbitol addition	Find et al. (1998)

Picea glauca (Moench) Voss	GYM	REG	SC		TSC	1	CPA: 0.4 M sorbitol +5% DMSO pretreatment, only meristematic and some upper suspensor cells survived. Following freezing (−35 °C, LN storage for 1 h) and thawing, only embryogenic region cells remained viable, leading to re-growth and new somatic embryos	Kartha et al. (1988)
Picea sitchensis (Bong.) Carr.	GYM	REG	SC	ET	TSC	NA	Pre-culture with CPA 0.4 M sorbitol, slow cooling at 0.5 °C/min to −40 °C. The cryopreserved cells exhibited regrowth rates similar to non-frozen cultures	Find et al. (1993)

(continued)

Table 3.2 (continued)

Plant species	Clade	Purpose	Cultivation	Culture type	Method	No cultivars/ cell lines	Main aims/ achievements	References
	GYM	REG	SC	ET	TSC	4	Effect of culture period and cell density on the regrowth	Find et al. (1998)
Pinus caribaea Mor. var. *hondurensis*	GYM	REG	SC	ET	TSC	NA	CPA: sucrose and (DMSO) The cryopreserved embryogenic cells maintained their ability to regenerate into plants	Laine et al. (1992)
Populus hybrid	ANG	NA	SC	callus	TSC	NA	Max. survival (40%) achieved with 5.0% DMSO and 5.0% glycerol. Pre-growth with 1.0% proline was more effective	Binder and Zaerr (1980)
Taxus chinensis (Pilg.) Rehder	GYM	SM	SC	Callus	VITRIF	NA	Pretreatment with 3% mannitol/3 days the cells are incubated at 4 °C/3 h transfer into a vitrification solution containing 40% (w/v) ethylene glycol and 30% (w/v)	Kadkade et al. (1999)

	GYM	SM	SC	Callus	TSC	NA	No pre-culture step, after cryo, cultures maintained their paclitaxel production capability	Kim et al. (2001)
Taxus floridana Nutt.	GYM	SM	SC	Callus	TSC	1	Use of Mr. Frosty, DNA content monitored by flow cytometry	Skrlep et al. (2008)
	GYM	SM	SC	Callus	TSC	1	7 days old cells pre-cultured either on 0.5 or 0.33 M mannitol, pre-culturing 4 or 7 days, regrowth after 10–14 days	Mustafa et al. (2011)
Taxus x *media*	GYM	SM	SC	Callus	TSC	5	Use of Mr. Frosty, DNA content monitored by flow cytometry	Skrlep et al. (2008)
	GYM	SM	SC	Callus	TSC	5	7 days old cells pre-cultured on 0.5 M mannitol, pre-culturing 2 days, regrowth after 10–14 days	Mustafa et al. (2011)

(continued)

Table 3.2 (continued)

Plant species	Clade	Purpose	Cultivation	Culture type	Method	No cultivars/ cell lines	Main aims/ achievements	References
	GYM	SM	SC	Callus	Cold storage		Storage of SC for 180 days at 5 ± 1 °C, genetic and epigenetic variation measured	Li et al. (2013)
V. vinifera x *V. berlandieri*	ANG	REG	SC	ET	TSC	1	60% recovery obtained when using 0.25 M maltose and 5% DMSO as CPAs	Dussert et al. (1991)
V. vinifera x *V. berlandieri*						1	58% recovery	Dussert et al. (1992)
V. vinifera, rootstock 41B	ANG	REG	SC	ET	TSC	1	50% recovery 2 mg/l NAA improved recovery	
V. vinifera L.	ANG	REG	SC	ET	ENCAP +DEHYD	1	Stepwise sucrose preculture, encapsulation, air-drying to 20.6% water content, 78% recovery	Wang et al. (2002)

V. berlandieri x *V. rupestris*, rootstock 110 Richer rootstock 41B *V. vinifera*, cv. Sugarone cv. Red globe cv. Gammy cv. Chardonay	ANG	REG	SC	ET	ENCAP +VITRIF	6	First successful cryo-procedure using encapsulation/ vitrification procedure with PVS2	Wang et al. (2004)
V. vinifera, cv. Albariño cv. Tempranillo	ANG	REG	SC	ET	ENCAP +DEHYD	2	'Tempranillo' cells exhibited higher post-thaw regrowth than 'Albariño'	González-Benito et al. (2009)
V. berlandieri x *V. rupestris*, rootstock 110 Richer *V. vinifera*, cv. Riesling *V. vinifera*, cv. Tempranillo	ANG	REG	SC	ET	DEHYD +ENCAP	3	Stepwise sucrose, sucrose/glycerol solution, encapsulation, 43–78% recovery	Ben-Amar et al. (2013)
V. vinifera, cv. Sultana, cv. Red globe, cv. Merlot cv. Roobernet cv. Dauphine cv. Chardonay	ANG	REG	SC	ET	ENCAP +DEHYD	6	Post-thaw survival rates varied by cultivar, ranging from 0% to 40%	Vasanth and Vivier (2011)

(continued)

Table 3.2 (continued)

Plant species	Clade	Purpose	Cultivation	Culture type	Method	No cultivars/ cell lines	Main aims/ achievements	References
Vitis vinifera, cv. Tempranillo cv. Riesling *V. berlandieri* x *V. rupestris*, rootstock 110 Richer	ANG	REG	SC	ET	ENCAP + DEHYD ENCAP + VITRIF	3	Higher survival (81%, 62%, and 48% for different varieties) and regeneration rates (72%, 60%, and 54%) than ENCAP/ VITRIF. Cryogenic stress, particularly osmotic priming with CPAs, was crucial in enhancing survival and mitigating oxidative damage	Ben-Amar et al. (2024)

Abbreviations: *ANG* angiosperms, *GYM* gymnosperms, *REG* regeneration, *SM* secondary metabolite, *SC* suspension culture, *SEM* semi-solid medium, *ET* embryogenic tissue, *TSC* two steps slow cooling, *VIT* vitrification, *ENCAP* encapsulation, *DEHYD* dehydration, *DES* desiccation, *CPA* cryoprotective agent

Gymnosperm embryogenic tissue in liquid culture achieves higher multiplication rates than semisolid media do (Gupta and Timmis 2005; Denchev and Grossnickle 2019). However, variations among lines affect homogeneity, and excessive agitation can impair embryo development (Mamun et al. 2018; Sun et al. 2010). Ensuring protocol suitability for a wide range of lines is crucial for maintaining genetic diversity and adaptability in clonal forest regeneration, particularly for long-lived gymnosperms (Park et al. 2016; Park 2014). These differences explain the disparity in cryopreservation protocols for suspension cultures of angiosperms and gymnosperms. While angiosperms have been cryopreserved via various methods, only slow-cooling protocols have been applied to gymnosperms. Slow-cooling protocols accounted for 65% of the reviewed articles, whereas vitrification and encapsulation-dehydration each represented 14%. The remaining studies focused on cold storage.

Although woody plant species are generally recalcitrant to in vitro techniques and cryopreservation, pioneering studies on *Acer pseudoplatanus* suspension cultures in the 1970s and 1980s played a crucial role in the development of cryopreservation techniques. These studies, particularly those involving slow cooling, provided fundamental insights into freezing tolerance (Withers 1985; Steponkus et al. 1995).

A major breakthrough in cryopreservation was the refinement of freezing and thawing protocols. Early research revealed that an optimal freezing rate of 1–2 °C per min before plunging into liquid nitrogen prevented intracellular ice formation, whereas thawing at 40 °C significantly improved survival rates (Sugawara and Sakai 1974; Nag and Street 1975b; Withers and Street 1977).

Freeze tolerance in *Acer pseudoplatanus* cells was found to depend on the cell cycle stage, with early G1-phase cells showing greater survival than G2- or stationary-phase cells (Withers 1978). A 7-day subculturing regime helped remove highly vacuolated, freeze-sensitive cells, improving viability. Synchronization studies involving nitrate starvation and low-density inoculation confirmed that G1-phase cells presented the highest tolerance, although the mechanism involved remains unclear. However, synchronization is not practical for routine cryopreservation due to technical limitations.

Cryoprotectants are critical for preventing ice formation and cell damage. The combination of 5% DMSO and 10% glucose was highly effective across multiple species, significantly improving survival and post-thaw regrowth (Nag and Street 1975a, b). Studies of frozen-thawed *A. pseudoplatanus* cells revealed mitochondrial and plastid dilation, indicating structural stress, but controlled freezing with cryoprotectants minimized this damage, enabling metabolic recovery (Withers and Davey 1978).

To further increase survival, osmotic preconditioning with mannitol has been widely adopted. Supplementing the culture medium with 3.3–6% (w/v) mannitol was found to increase freeze tolerance by reducing the vacuolar volume and promoting the formation of multiple small vacuoles (Pritchard et al. 1986b). A mixture of 5% DMSO and 10% glycerol was identified as an optimal cryoprotectant formulation (Withers 1985).

Pritchard and coworkers (1986a, b, c) demonstrated that osmotic stress improves cryopreservation success, particularly in sycamore cells. Preconditioning reduced the cell size and vacuolation, increasing cryotolerance. While sycamore cells adjust metabolically through increased protein levels and respiration, soybean cells primarily accumulate proline, which does not necessarily improve post-thaw survival. These findings suggest that osmotic stress treatments should be tailored to individual species for maximum effectiveness.

The cryopreservation of *Vitis* embryogenic suspension cultures has been extensively studied to ensure long-term storage and viability while maintaining genetic stability and embryogenic potential. Various techniques, including two-step freezing, encapsulation-dehydration, and encapsulation-vitrification, have been developed to optimize survival and post-thaw regrowth (Bi et al. 2017; Carra et al. 2019, 2024).

Pioneering studies by Dussert et al. (1991, 1992) demonstrated that two-step freezing is superior to rapid freezing for *Vitis* embryogenic suspensions. The method involved slow cooling at −0.5 °C per min to −40 °C, followed by immersion in liquid nitrogen (LN). When 5% DMSO and 0.25 M maltose were used as cryoprotectants, the survival rates exceeded 60%. Subsequent studies optimized post-thaw culture conditions, emphasizing the importance of semisolid media with naphthoxyacetic acid to support recovery.

Wang et al. (2002) developed an encapsulation-dehydration method for *Vitis vinifera* suspension cultures, which achieved 78% survival and a sevenfold increase in embryogenesis. The cryopreserved cells regenerated normally, maintaining genetic stability. Wang et al. (2004) introduced encapsulation-vitrification for suspension cultures of multiple *Vitis* cultivars. The cells were precultured in sucrose (up to 0.75 M), encapsulated, and dehydrated with PVS2. Viability ranged from 42% to 82%, with *the variety Red Globe* showing the highest survival. Post-thaw, the cells resumed embryogenesis, with increased soluble sugars and proteins improving freeze tolerance. Wang et al. (2005) optimized *Agrobacterium*-mediated transformation in *Vitis vinifera* by refining selection conditions and utilizing cryopreserved embryogenic cell suspensions.

González-Benito et al. (2009) optimized encapsulation-dehydration for suspension cultures of Albariño and Tempranillo, achieving 50% post-thaw viability and 100% bead regrowth. Tempranillo showed greater regrowth, confirming the reliability of the method for long-term conservation. Vasanth and Vivier (2011) refined encapsulation-vitrification using sucrose-treated embryogenic tissue from anthers. Survival rates vary (0–40%), with secondary embryogenesis occurring faster than primary embryogenesis. Cultures transferred to NN medium supplemented with glutamine and activated charcoal yielded the greatest number of embryos (31–53 per bead).

Ben-Amar et al. (2024) compared encapsulation-dehydration and encapsulation-vitrification, finding higher survival and regeneration in encapsulation-dehydration (81% vs. 62% survival). Cryopreservation rejuvenated senescent cells, increasing embryogenesis through increased SERK1, BBM, and WOX gene expression. Osmotic priming improved survival and reduced oxidative stress, reinforcing cryopreservation as both a conservation tool and a method to increase cell viability.

Owing to their economic importance and adaptability to in vitro and biotechnological applications (Germanà et al. 2020), *Citrus* cryopreservation technologies have been extensively reviewed by De Carlo et al. (2011), providing step-by-step protocols. Slow-freezing methods were first applied for the cryopreservation of *Citrus sinensis* suspension cultures (Kobayashi et al. 1990; Sakai et al. 1991a) and *Citrus deliciosa* (Aguilar et al. 1993; Engelmann et al. 1994). Histological analysis confirmed greater survival in meristematic cells, whereas cryopreserved cultures presented enhanced somatic embryogenesis (Aguilar et al. 1993). Engelmann et al. (1994) developed a cost-effective cryopreservation method using a nonprogrammable freezing approach (Nalgene Mr. Frosty™), 0.15 M sucrose, and 5–15% DMSO. Five out of the six *Citrus* varieties survived, with 10–15% DMSO yielding the best results, indicating that this approach is effective for long-term germplasm conservation.

Sakai et al. (1990) successfully cryopreserved *Citrus sinensis* nucellar cells via vitrification with PVS2 (30% glycerol, 15% ethylene glycol, 15% DMSO). The cells were incubated in 60% PVS2 at 25 °C for 5 min and then chilled at 0 °C for 3 min before immersion in liquid nitrogen (−196 °C). The method achieved 80% survival, with regrowth starting in 2 days and full plant regeneration. The 10-min process eliminated the need for slow freezing, making vitrification a fast and effective conservation technique. The method was further optimized using PVS2 with 0.4 M sucrose, achieving up to 90% survival with a 3-min PVS2 treatment at 25 °C. Post-thaw growth resumed within 3 days, and the plants regenerated within 3 months. This method has proven highly efficient and broadly applicable to other citrus cultivars (Sakai et al. 1991b).

Cell growth in suspension culture can be monitored via the total cell number (TCN), packed cell volume (PCV), fresh weight (FW), dry weight (DW), cell viability, and medium conductivity (Mohamad Puad and Abdullah 2018). At least two methods are recommended for tracking growth until the stationary phase (Mustafa et al. 2011), with additional indicators, including carbon source consumption, nutrient depletion, oxygen uptake, and carbon dioxide production. PCV, the ratio of cryopreservation medium volume to cell volume, determines the minimum cell density threshold for post-cryopreservation regrowth (Mustafa et al. 2011). PCVs of 20–30% are optimal for slow freezing and encapsulation-dehydration, achieving ~80% regrowth, whereas 40–50% PCVs are preferred for vitrification (Nausch and Buyel 2021). In *Picea abies*, increasing the PCV from 20% to 50% tripled regrowth rates (Find et al. 1998), whereas in *Vitis vinifera*, increasing the PCV from 10% to 70% improved regrowth from 50% to 75%, although excessively high PCVs reduced viability (Dussert et al. 1991).

In angiosperm cultures, higher cell densities generally improve survival post-cryopreservation (Bajaj 1976; Nag and Street 1975a). However, in Find's study (Find et al. 1998), cultures adjusted to high densities exhibited low regrowth. While low regrowth at suboptimal densities may result from fewer cells being plated, poor recovery at supra-optimal densities suggests that other factors, such as injurious compounds released by damaged cells, impair regrowth. High PCVs may also limit cryoprotectant efficiency, reducing survival.

Cell morphology plays a crucial role in cryotolerance. In both angiosperms and gymnosperms, small, dense cells exhibit higher survival rates, whereas large vacuolated cells are more susceptible to cryoinjury (Aguilar et al. 1993; Kristensen et al. 1994). In angiosperms, surviving cells are typically found at the periphery of small aggregates, with cultures showing enhanced SE after cryopreservation (Aguilar et al. 1993; Wang et al. 2002; Ben-Amar et al. 2024). In gymnosperms, viable cells capable of regrowth are located within the embryogenic region of proembryogenic masses (PEMs) (Kartha et al. 1988; Kristensen et al. 1994). Cryotolerance in gymnosperm cultures appears to be closely linked to changes in the cellular composition during cultivation, emphasizing the importance of selecting the optimal growth stage for cryopreservation.

Ben-Amar et al. (2024) studied embryogenic suspension cultures of *Vitis* and reported that cryopreservation promoted somatic cell rejuvenation, leading to increased embryogenic competence. Histological and microscopic analyses revealed significant morphological changes, with cryopreserved cells forming well-organized structures and displaying enhanced embryo development. Gene expression analysis revealed increased transcript levels of the SERK1, BBM, and WOX genes, which are associated with embryogenic competence, along with stress response genes such as ChitIV and LEA.

Early studies on *Picea glauca* cryopreservation revealed differences in survival across cell types (Kartha et al. 1988). FDA staining revealed that after 0.4 M sorbitol +5% DMSO pretreatment, only meristematic and upper suspensor cells survived. Following freezing (−35 °C, LN storage for 1 h) and thawing, only embryogenic region cells remained viable, leading to regrowth and new somatic embryos. The loss of most suspensor cells before freezing suggests that these cells are not directly involved in somatic embryo formation. Kristensen et al. (1994) provided the first detailed description of the survival, morphology, and development of embryogenic and suspensor cells following cryopreservation of an embryogenic suspension culture of *Picea sitchensis*.

The cultivation of plant cells in batches follows a typical sequence of growth phases (Mohamad Puad and Abdullah 2018): (1) lag phase, where cells adapt to the medium with limited proliferation; (2) acceleration phase, where proliferation begins but has not yet reached its peak; (3) exponential phase, where cells proliferate at their maximum rate; (4) deceleration phase, where division slows owing to high density or nutrient depletion; and (5) stationary phase, where proliferation stops as nutrients are depleted or maximum density is reached.

Cells in the exponential phase contain smaller vacuoles, resulting in a higher cytosol-to-vacuole ratio and increased solute concentration, which may act as natural cryoprotectants, increasing the resistance of these cells to freezing stress (Sugawara and Sakai 1974; Withers 1978; Withers and Davey 1978; Reinhoud et al. 2000). For example, *Acer pseudoplatanus* cryopreserved during the lag or stationary phase showed <5% regrowth, whereas >50% regrowth was observed when the cells were cryopreserved in the exponential phase (Withers 1978).

For *Picea abies* and *Picea sitchensis*, regrowth after cryopreservation is strongly influenced by the culture period prior to freezing (Find et al. 1998). Unlike angiosperm cultures, which survive best when collected in the late lag or exponential phase, conifer cultures exhibit maximum regrowth when collected during the stationary phase. This difference reflects distinct proliferation patterns: angiosperm cultures mainly consist of single cells and small aggregates (Withers and Street 1977), whereas conifer cultures contain structured proembryogenic masses (PEMs) at various developmental stages (Filonova et al. 2000; von Arnold 2008; von Arnold et al. 2020).

The development of new cryopreservation protocols requires aligning plant characteristics with the appropriate method. Controlled-rate freezing is commonly used for temperate species, whereas vitrification and encapsulation-dehydration are better suited for tropical and desiccation-sensitive plants. Assessing laboratory resources and optimizing key factors, such as cryoprotectant exposure, cooling rates, and post-thaw recovery, are essential for achieving high survival rates and long-term storage success (Reed 2011).

Kim and Popova (2023) suggested classifying plant materials by size, structure, and response to cryoprotectants to refine protocol development. Their approach reduces trial-and-error by testing present protocols and selecting the most effective. Kim and Popova (2023) emphasized the importance of optimizing regrowth conditions, including osmotic balance, light exposure, and growth regulators, to improve post-cryopreservation recovery and expand applications to cryopreservation-sensitive species.

3.4 Cryopreservation of Protoplasts

Protoplasts have long been valuable tools in plant science, facilitating genotypic diversity through hybridization, cell biology research, transcriptional assays, and transient transformation. The ability of these strains to bypass long life cycles and overcome the challenges of transforming recalcitrant genotypes makes them particularly useful for gene editing and applied research (Jiang et al. 2013; Pasternak et al. 2020; Reed and Bargmann 2021; Yue et al. 2021; Reyna-Llorens et al. 2023; Mukundan et al. 2025). However, further advancements are needed to fully understand and optimize protoplast regeneration.

To analyze research trends in protoplast regeneration, Yue et al. (2021) examined 779 related articles and reported that protoplasts are most commonly derived from Solanaceae, Poaceae, and Brassicaceae due to their relatively high regeneration efficiency and economic importance. The optimization of the plant source, culture incubation, protoplast density, basal medium, growth regulators, and supplements is species and sometimes variety specific.

In gymnosperms, somatic embryo regeneration from protoplasts isolated from embryogenic cells of *Pinus taeda* and *Picea glauca* has been achieved. Among angiosperms, protoplast regeneration has been reported in at least 12 woody species across six families (Reed and Bargmann 2021), including *Ulmus americana* (American elm), *Ulmus edulis* (passion fruit), *Malus pumila* (apple), *Morus indica* (mulberry), and *Phellodendron amurense* (Amur cork tree). Despite these successes, woody species remain challenging for protoplast regeneration, and further research is needed to refine culture conditions and improve efficiency.

Citrus is a leading model system for protoplast regeneration, primarily because of its strong ability for somatic embryogenesis (Omar et al. 2016). Well-established methods exist for protoplast isolation from embryogenic calli and suspension cultures, followed by plant regeneration (Tusa et al. 1990; Dutt et al. 2010; Grosser and Gmitter 2011). Protoplast fusion techniques have been used extensively in *Citrus*, generating somatic hybrids from over 500 parental combinations, including more than 300 hybrids (Grosser et al. 1992; Tusa et al. 1990; Grosser and Gmitter 2011). Additionally, hundreds of diploid Citrus cybrid plants have been regenerated via SE as a by-product of protoplast fusion (Grosser et al. 1996). Protoplasts have also been instrumental in genetic transformation, particularly in *Citrus*, which has led to protoplast-mediated transformation among woody fruit trees. The robust SE of citrus has enabled successful transformation and plant recovery (Omar et al. 2008; Germanà et al. 2020), further confirming its role as a model system.

The cryopreservation of protoplasts gained popularity during the initial wave of research on plant regeneration from protoplasts (Steponkus 1985; Withers 1985; Bajaj 1989; Grout 1995b). The unique properties of protoplasts, particularly their lack of intercellular connections, make them ideal for studying cryopreservation at the cellular level. Cryo-microscopy studies have provided precise observations of ice formation and nucleation events (Steponkus et al. 1983; Steponkus et al. 1984).

Owing to their high susceptibility to intracellular ice formation, osmotic stress, and structural damage, protoplasts require specialized freezing protocols. The use of cryoprotectants such as DMSO, mannitol, and sucrose improves survival rates, although species-specific variability remains a limitation. Slow-cooling techniques combined with rapid thawing minimize cellular damage while preserving viability and regenerative potential. Studies have confirmed that cryopreserved protoplasts can retain their morphogenetic ability, making them valuable for both fundamental research and plant breeding (Bajaj 1989; Grout 1995a).

Although cryopreservation of plant protoplasts is feasible, success varies significantly across species and has been largely limited to a few model species. To date, no studies have reported the cryopreservation of protoplasts isolated from woody species. The key limiting factors include cryoinjury mechanisms, particularly intracellular ice formation, and variability in survival rates, especially in hybrid protoplasts. The optimization of cryoprotectants and freezing-thawing protocols is essential for improving outcomes (Bajaj 1989; Grout 1995a).

3.5 Suspension Cultures and Protoplasts for the Production of Secondary Metabolites

Despite extensive research on secondary metabolites, current estimates suggest that only approximately 6% of higher plants (from an estimated 300,000 to 500,000 species) have been systematically studied for their pharmacological potential, whereas only 15% have been evaluated for phytochemical composition (Espinosa-Leal et al.

2018). This indicates significant opportunities for further exploration in this field, especially with woody species (Arora et al. 2022).

The cultivation of plant suspension cultures in controlled environments, along with advancements in transformation techniques, has highlighted the importance of optimizing biosynthetic pathways for high-value secondary metabolites (Eibl et al. 2021). This approach has proven more efficient and attractive than relying on intact plants for metabolite production (Murthy et al. 2014; Santos et al. 2016; Donini and Marusic 2019; Gubser et al. 2021; Popova et al. 2021; Xu et al. 2025).

Currently, more than 50 bioactive small molecules are produced from plant cell suspension cultures for applications in cosmetics, food, and feed products and as active pharmaceutical ingredients (Wilson and Roberts 2012; Eibl et al. 2018).

A notable example is paclitaxel, a widely used breast cancer drug originally isolated from the bark of the endangered Pacific yew tree (*Taxus brevifolia*). To address sustainability concerns, paclitaxel is now produced in cell suspension cultures, particularly *T. chinensis*, which has become the leading industrial method for paclitaxel production, offering a sustainable alternative to bark extraction (Yin et al. 2025). Companies such as Phyton Biotech and Samyang Biopharm have commercialized the process, with yields increasing from 1–3 mg/L to 565 mg/L through genetic manipulation, optimized culture conditions, and metabolic inducers (Yin et al. 2025). Recent studies suggest that *Corylus avellana* (hazelnut) cell cultures present a promising alternative for paclitaxel production, addressing challenges in *Taxus* cultivation (Gallego et al. 2017; Ferrucci et al. 2023).

Given the high value of *Taxus* cell cultures, cryopreservation techniques have been developed for long-term storage, although reports on their application remain limited. Only two studies have documented cryopreservation of *T. chinensis*: one using two-step freezing with a programmable freezer (Kim et al. 2001) and another employing vitrification, patented by Phyton Biotech (Kadkade et al. 1999). The most valuable cell lines are likely preserved via proprietary methods not publicly available in scientific databases.

Kim et al. (2001) successfully cryopreserved *T. chinensis* cell cultures using 0.5 M DMSO and 0.5 M glycerol in a controlled freezing protocol, achieving 40% post-thaw viability. While biomass accumulation was lower in cryopreserved cultures than in control cultures, paclitaxel production remained unchanged. Skrlep et al. (2008) developed a simplified, cost-effective two-step freezing method for *T.* × *media* and *T. floridana* suspension cultures. In this method, plant cell suspensions are cryopreserved in a freezing container instead of a programmable freezer. The cells were preconditioned in mannitol-supplemented medium (*T.* × *media*: 0.5 M; *T. floridana*: 0.33–0.5 M), followed by exposure to a cryoprotectant mixture containing 2 M sucrose, 1 M glycerol, and 1 M DMSO. Slow cooling at −1 °C/min to −40 °C, followed by rapid immersion in liquid nitrogen, resulted in successful regrowth in three of the four *T.* × *media* lines and in *T. floridana*. Notably, the nuclear DNA content remained stable, reducing concerns about genetic variability in cryopreserved *taxus* cell lines. The success of this method across two *Taxus* species suggests its potential applicability to other species. Sahari Moghaddam et al. (2024) expanded *Taxus* conservation techniques by evaluating cold storage and cryopreservation of apical buds via encapsulation-dehydration in *T. baccata*. The results of these studies are not included in Table 3.2.

Cryobanking is essential for maintaining stable, high-quality plant cell lines used in the commercial production of bioactive molecules, ensuring consistent yields and product integrity. This method enables the long-term preservation of elite cell cultures, minimizing the need for continuous subculturing, which can cause genetic drift and variability in metabolite production (Yuorieva et al. 2023). To maintain biosafety and consistency, a structured master cell banking system should be established. The widely accepted three-tier system includes a research cell bank, a master cell bank (MCB), and a working cell bank (WCB), a standard approach for both plant and mammalian cell lines (Schmale et al. 2006).

Despite their high biotechnological potential, documented collections of plant cell strains are limited, with major repositories in the USA, Finland, Japan, Germany, Hungary, the Czech Republic, and Russia (Yuorieva et al. 2023). The largest and most accessible collections include the Plant Cell Culture Library at the University of Massachusetts Amherst (>1000 plant species, https://www.umass.edu/ials/pccl-database, accessed on 30 March 2025), the Leibniz Institute DSMZ-German Collection (>80 plant families) (Kreis 2019), the VTT Culture Collection (Finland, 23 species, http://culturecollection.vtt.fi, accessed on 30 March 2025), RIKEN BRC Plant Cultured Cell Resources (Japan, 32 species, https://epd.brc.riken.jp/en/pcellc, accessed on 30 March 2025), the University of Debrecen collection (Hungary) (Máthé et al. 2012) and the Institute of Experimental Botany collection (Czech Republic, >20 species, http://lpb.ueb.cas.cz/About.html, accessed on 30 May 2025), and the All-Russian Collection of Plant Cell Cultures at IPPRAS (Russia, storing over 1000 in vitro cultures and seeds, covering 457 species from 74 families, all preserved in liquid nitrogen for long-term conservation). These collections include cell strains with anticancer, antimicrobial, antioxidant, and insecticidal properties, as well as model species such as *Nicotiana tabacum* and *Arabidopsis thaliana* (Yuorieva et al. 2023).

3.6 Microscopic Studies

Microscopic evaluation of cryopreserved plant cell suspensions allows visualization of damage to cell membranes and organelles, thus informing optimization of freezing protocols for particular cell types, especially when combined with the recovery of live cell cultures post storage. There are considerable parallels between cryopreservation and cryo-fixation for microscopy, both aiming to preserve cell viability.

3.6.1 Light and Fluorescence Microscopy

Fluorescein diacetate (FDA) and propidium iodide (PI) staining have both been used to perform live/dead cell analysis on post-cryo cell cultures (Whaley et al. 2021). FDA stains live cells with intact membranes, whereas PI stains nonviable cells. This method is widely used for assessing post-thaw viability, but more modern live/dead cell assays based on membrane integrity or metabolic activity, including DAPI (with freeze permeabilization), ethidium bromide, Evans blue, SYTOX dyes, calcein AM, and autofluorescence, could be used (Baker and Mock 1994; Truernit and Haseloff 2008; Ciniglia et al. 2010; Rajeckaite et al. 2018; Donaldson 2020; Whaley et al. 2021) (Fig. 3.1). In addition to microscopy, this approach is

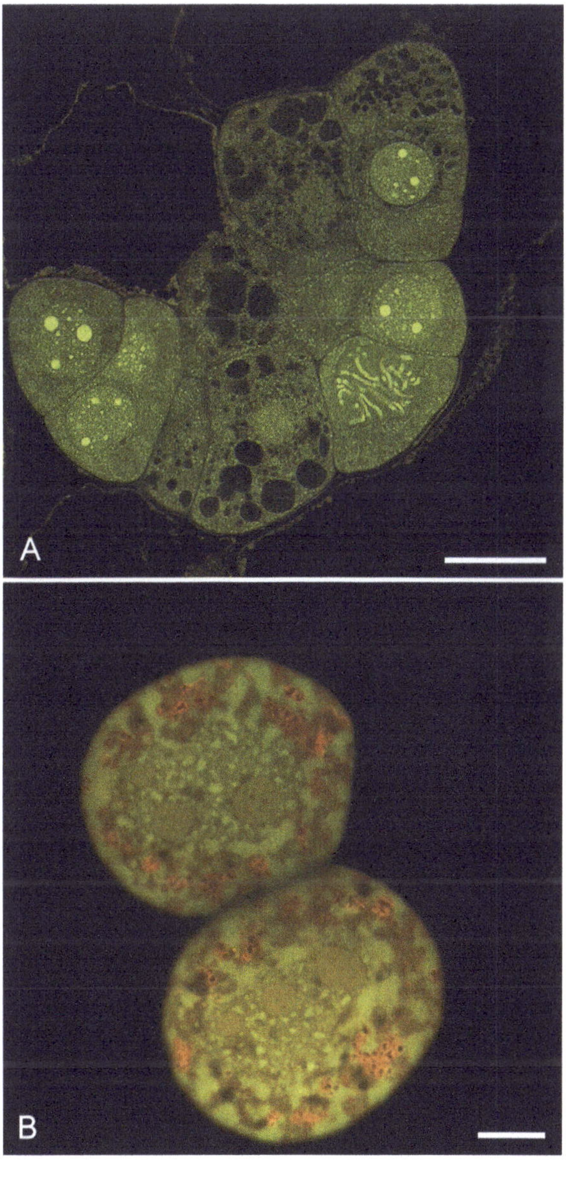

Fig. 3.1 (**a**) Embryogenic tissue of *Pinus radiata* showing dense cytoplasmic contents and active cell division. This image is a single confocal optical slice showing autofluorescence. Scale bar = 30 μm. (**b**) Pine protoplast. This image is a single confocal optical slice showing autofluorescence. (Image provided by Dr Sanjeeve Raikar, Scion. Scale bar = 10 μm)

amenable to flow cytometry for the detection/measurement of viable cells after cryopreservation (San Jose et al. 2015; Sliwinska et al. 2022).

In combination with differential interference contrast microscopy, Calcofluor white staining is used to detect cell wall regeneration in plant protoplasts (Neubauer et al. 2022). The latter can also be used to evaluate multinucleate cell formation in protoplast cultures (Fig. 3.2). Immunolocalization of cell wall components [arabinogalactan proteins (JIM13, JIM16), extensin (JIM20), xyloglucan (LM25), and pectins (LM20, LM5, LM6)] has also been used in relation to cell wall regeneration in protoplast cultures (Sala-Cholewa et al. 2024). There is scope to apply this technique to cultures from woody plants, especially in relation to amenability to the transformation of different varieties/clones.

Surprisingly, confocal fluorescence microscopy has not been widely used for the evaluation of cell damage related to cryopreservation, although it has been applied to natural freezing damage and the assessment of cryoprotectants (Kobayashi et al. 2014; Vidot et al. 2018). Confocal microscopy has been used to localize plant cells via CLEM (correlative light and electron microscopy) via the use of endogenous expression tags such as GFP (Sanchez Carrillo et al. 2023). Likewise, the application of super-resolution techniques to evaluate the cryopreservation of plant cells seems not to have been explored other than the application of cryofixation (Vidot et al. 2018). Both staining/expression tag and autofluorescence protocols need to be developed to investigate the effects of cryopreservation in live-cell or fixed-cell applications (Knapp et al. 2012; Vidot et al. 2018; Donaldson 2020).

Raman microscopy techniques have enabled studies of cryoprotectant distribution in cells prepared for cryopreservation, offering valuable insight into how these chemicals protect cells from freezing damage. Using cultured rice cells, Samuels et al. (2021) and Samuels et al. (2023) used Raman microscopy to study the distribution of DMSO labeled by deuteration. This work revealed that DMSO was unevenly distributed in cell compartments, accumulated in amyloplasts, and likely led to uneven cryoprotection. In cell types that did not contain amyloplasts, DMSO was distributed more evenly. The plasmolysis response of plant cells was substantially slower than the process of cryoprotectant permeation. Mixtures of cryoprotectants facilitate the permeation of individual components into living cells (Pearce et al. 2024).

3.6.2 Electron Microscopy (SEM and TEM)

Cryopreservation and cryo-electron microscopy share several common applications, and cryomicroscopy can be used both to determine the physical effects of freezing on cells/tissues and to provide insight into the freezing methodology. Compared with conventional chemical fixation, which often induces artifacts, cryomicroscopy aims to preserve cells in their natural state via rapid (high-pressure) freezing and vitrification (McDonald and Auer 2006). Chemical fixation of plant tissues via glutaraldehyde and osmium can often be limited by slow diffusion into tissues such as buds (Endoh et al. 2009) or cambium (Rensing and Samuels 2004).

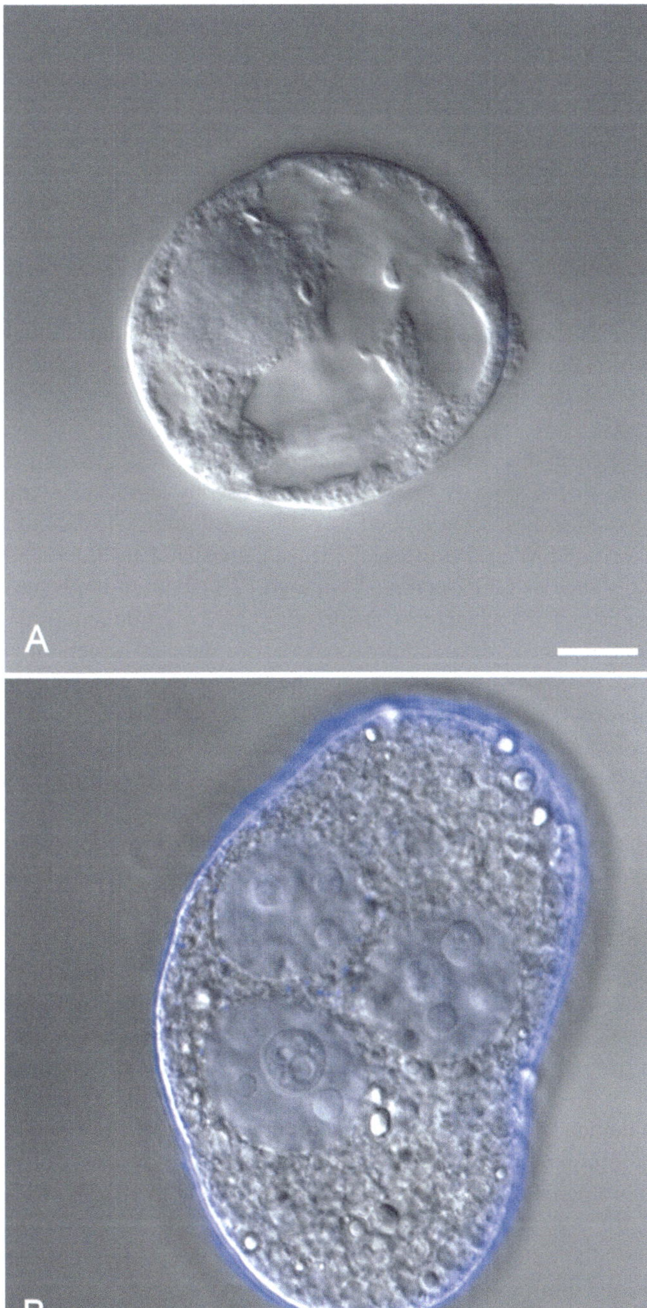

Fig. 3.2 Protoplasts of *Pinus radiata* stained with calcofluor white to show primary cell walls combined with differential interference contrast (DIC). (**a**) Pine protoplast at time 0 h showing absence of a primary cell wall. (**b**) Pine protoplast at time 18 h showing the presence of a developing cell wall. This cell is multinucleate. (Images provided by Dr Sanjeeve Raikar, Scion. Scale bars = 10 μm)

Cryo-fixation can also be limited by slow freezing rates associated with ice formation in tissues but is generally much more successful with cell suspensions, including protoplasts or cultured tissues (McDonald and Auer 2006; Sanchez Carrillo et al. 2023).

High-pressure freezing, a technique used to preserve cells for cryomicroscopy that parallels cryopreservation, is a state-of-the-art method for the cryofixation of cells. This technique does not work as well with organized tissues because of the slow rate of freezing but is suitable for cell suspensions and protoplasts (Sakai and Engelmann 2007; Sanchez Carrillo et al. 2023). In this technique, single cells or thin sections of organized tissue are frozen in liquid nitrogen at high pressure, which results in vitreous freezing, an approach that does not seem to have been used for cryopreservation.

The development of correlative cryomicroscopy, whereby frozen cells can be observed in the frozen state by fluorescence microscopy and then by transmission electron microscopy, has been applied to plant cell cultures. In this method, cell suspensions on a TEM grid are rapidly frozen, localized by fluorescence using expression tags such as GFP, sectioned via cryo-FIB (focused ion beam) and then examined via cryo-TEM (Sanchez Carrillo et al. 2023). While requiring advanced equipment and expertise, this technique is ideal for detailed structural analysis of plant cell suspensions and protoplasts, potentially enabling evaluation of freezing behavior to inform cryopreservation studies.

Cryomicroscopy has also been successfully applied to studies of frozen primary cell walls via the use of cryotomography to determine the three-dimensional architecture of cellulose (McDonald and Auer 2006; Nicolas et al. 2022). Secondary cell walls from the xylem have also been studied via methods involving cryosectioning and electron tomography (Reza et al. 2013; Sarkar et al. 2014; Fernando et al. 2023). Cell walls are generally amenable to freezing, and damage to the cytoplasm, especially the plasmalemma, limits cryopreservation (Yamada et al. 2002; Wen et al. 2010).

Serial block face microscopy allows 3D reconstruction of plant cells at high resolution via serial sectioning, block face imaging, and high-resolution backscattered electron microscopy. Although a relatively new technique as applied to plants (Mursalimov et al. 2021), this approach is well suited for the assessment of cryopreservation damage to cell and protoplast cultures. Cell cultures can potentially be prepared via high-pressure freezing and freeze substitution before and after cryopreservation protocols. These preparation methods are known to minimize fixation artifacts.

Microscopy plays a crucial role in evaluating the effects of cryopreservation on both suspension cultures and protoplasts. It helps in assessing membrane integrity and permeability, intracellular structures (vacuolization and cytoplasmic changes), ice crystal formation and intracellular damage, cell division, and post-thaw recovery.

3.7 Challenges and Future Trends

For cryopreservation of suspension cultures aimed at secondary metabolite production, maintaining a stable metabolic pathway is essential. However, callus-derived suspension cultures often exhibit heterogeneity in metabolite content and cell color (Heine-Dobbernack et al. 2008). This variability arises because traditional suspension cultures originate from dedifferentiated calli, which retain a mix of specialized cell types from the parent tissue and remain heterogeneous over time (Gubser et al. 2021).

To circumvent dedifferentiation, cambial meristematic cells (CMCs) have been isolated from *Taxus cuspidata*, *Panax ginseng*, *Ginkgo biloba*, and *Solanum lycopersicum* (Lee et al. 2010; Ochoa-Villarreal et al. 2015). CMC cultures from *T. cuspidata* exhibit lower growth variability, smaller aggregates, and higher paclitaxel yields than traditional suspension cultures do, while they are also more resistant to shear stress in bioreactors. Further optimization of CMC cultures across various species could increase the scalability and commercial viability of plant cell culture systems.

Advancing cryopreservation requires a deeper understanding of the molecular mechanisms underlying plant freezing tolerance, including cold acclimatization; the activation of cold-responsive genes; and the protective roles of transcription factors, noncoding RNAs, and epigenetic modifications. Specialized proteins such as antifreeze proteins (AFPs) and late embryogenesis abundant (LEA) proteins protect plant cells during freezing and thawing (Jahed et al. 2023; Bialoskorska et al. 2024). While temperate species naturally acclimate to cold through physiological and biochemical adaptations, tropical and subtropical species lack these mechanisms, making their cryopreservation more difficult (Volk 2010; Benson et al. 2013).

Cryopreservation exposes plant explants to multiple stressors, including extreme temperatures, dehydration, oxidative stress, osmotic imbalances, and chemical toxicity, which can lead to cellular damage, such as vacuolization, membrane rupture, autophagy, and lysis (Wesley-Smith et al. 2015; Ren et al. 2021). Freezing tolerance is closely linked to the ability to withstand these stresses. Reactive oxygen species (ROS)-induced oxidative stress damages plant cells, reducing their ability to survive cryopreservation and often leading to low recovery rates or complete failure. To mitigate its effects and improve recovery, future research should focus on understanding how plant explants respond to ROS-induced stress, elucidating the mechanisms behind oxidative stress resistance and tolerance, and applying antioxidants to enhance the cryopreservation of recalcitrant species, particularly endangered, endemic, and tropical plants. Advanced analytical techniques should be integrated into cryopreservation studies. Fluorescence probes offer high-resolution monitoring of ROS dynamics (Abo and Weerapana 2019; Grasso et al. 2023), whereas high-throughput omics technologies can identify key genes and proteins involved in cellular protection against ROS stress (Suzuki and Mittler 2006).

During cryopreservation, methylation-sensitive markers indicate that methylation patterns may shift, depending on the species, cryopreservation method, and tissue culture conditions (Malhotra et al. 2024). Despite these potential changes, comparisons of morphology, growth, and reproductive capacity between cryopreserved and non-cryopreserved plants confirm the maintenance of genetic integrity (Wang et al. 2021a). Further research using genome-wide profiling techniques with higher marker density is needed to better understand epigenetic stability during cryopreservation.

To ensure genetic integrity and long-term viability, strict hygiene standards must be maintained to prevent contamination in cryotanks (Bajerski et al. 2021). Cryopreservation should prioritize genetically valuable, high-quality material, preserving it under optimal conditions for maximum longevity while ensuring true-to-type regrowth. Standard operating procedures must document all cryopreservation and rewarming protocols while preventing mislabeling, such as tracking accessions with barcodes. Additionally, storage data should be securely managed in digital systems (Nagel et al. 2024).

Emerging cryopreservation technologies such as laser warming (Narida et al. 2023), nanowarming with magnetic iron oxide nanoparticles (Chen et al. 2023), vacuum infiltration vitrification (Nadarajan and Pritchard 2014), ultrasound-assisted freezing, and magnetic field applications have shown promise in improving warming rates, preventing recrystallization, and enhancing devitrification control (Chang and Zhao 2021). Additionally, ice-inhibiting molecules such as IBPs, synthetic polymers, nanomaterials, and hydrogels (Ma et al. 2021; Chopra et al. 2016) present exciting possibilities. However, rigorous risk assessment is needed to ensure genetic stability and the long-term storability of preserved accessions.

3.8 Conclusions

The cryopreservation of suspension cultures from woody species presents unique challenges, including high recalcitrance, heterogeneity, and post-thaw viability issues. Despite advancements in cryopreservation techniques, standardized protocols remain scarce, and species-specific optimization is often necessary. The loss of regenerative potential and metabolic stability post-thaw further complicates long-term conservation efforts. Epigenetic changes, such as DNA methylation and chromatin modifications, may influence cell behavior and secondary metabolite production, necessitating further investigation into long-term genetic and physiological stability.

Future advancements in cryopreservation will depend on the integration of innovative technologies such as vitrification, nanowarming, and laser warming to improve survival rates and minimize cryoinjury. Omics-based approaches will aid in identifying stress-responsive pathways, refining cryoprotectant formulations, and enhancing post-thaw regeneration through optimized hormonal treatments and bioreactor-based systems. Ensuring genetic fidelity and metabolic stability post-cryopreservation is critical for maintaining culture integrity and reproducibility.

Microscopy plays a crucial role in assessing the effects of cryopreservation on plant cells by evaluating membrane integrity, intracellular structures, and ice formation. Advanced imaging techniques, such as confocal and cryo-electron microscopy, can further enhance our understanding of cellular responses, aiding in the refinement of cryopreservation protocols.

Optimizing epigenetic and physiological changes during protoplast reprogramming enhances regeneration and transformation success. Metabolic states, ploidy levels, oxidative stress responses, and chromatin modifications influence developmental resetting, improving applications in agriculture, forestry, and biotechnology. Despite its potential in plant synthetic biology, protoplast technology remains inefficient for large-scale use. Standardized protocols, reduced variability, and improved enzyme impact control are needed. Improving the understanding of protoplast biology will unlock its full potential in gene editing and synthetic biology.

For cryopreservation to become a reliable strategy for woody species, large-scale cryobanking efforts must focus on developing standardized protocols, implementing digital tracking systems, and utilizing AI-driven monitoring to increase accuracy and efficiency. Addressing these challenges will strengthen the role of cryopreservation as a key tool for the long-term conservation of valuable genetic resources and biotechnological applications.

Competing Interests The authors have no relevant financial or non-financial interests to disclose.

References

Abo M, Weerapana E (2019) Chemical probes for redox signaling and oxidative stress. Antioxid Redox Signal 30(10):1369–1386

Aguilar ME, Engelmann F, Michaux-Ferriere N (1993) Cryopreservation of cell suspensions of *Citrus deliciosa* Tan and histological study. Cryo-Lett 14:217–228

von Arnold SV (2008) Somatic embryogenesis. In: George EF, Hall MA, De Klerk GJ (eds) Plant propagation by tissue culture, vol 1, 3rd edn. Springer Netherlands, pp 335–354

von Arnold S, Clapham D, Abrahamsson M (2019) Embryology in conifers. In: Canovas FM (ed) Molecular physiology and biotechnology of trees, Advances in botanical research, vol 89. Academic Press Ltd-Elsevier Science Ltd, London, pp 157–184

von Arnold S, Zhu T, Larsson E, Uddenberg D, Clapham D (2020) Regulation of somatic embryo development in Norway spruce. Methods Mol Biol 2122. Humana Press Inc:241–255

Arora K, Rai MK, Sharma AK (2022) Tissue culture mediated biotechnological interventions in medicinal trees: recent progress. Plant Cell Tissue Organ Cult 150(2):267–287

Bajaj YPS (1976) Regeneration of plants from cell suspensions frozen at −20, −70 and −196°C. Physiol Plant 37(4):263–268

Bajaj YPS (1989) Cryopreservation of plant protoplasts. In: Bajaj YPS (ed) Plant protoplasts and genetic engineering I. Springer, Berlin, Heidelberg, pp 97–106

Bajerski F, Nagel M, Overmann J (2021) Microbial occurrence in liquid nitrogen storage tanks: a challenge for cryobanking? Appl Microbiol Biotechnol 105(20):7635–7650

Baker CJ, Mock NM (1994) An improved method for monitoring cell death in cell suspension and leaf disc assays using evans blue. Plant Cell Tissue Organ Cult 39:7–12

Ballesteros D, Martínez MT, Sánchez-Romero C, Montalbán IA, Sales E, Moncaleán P, Arrillaga I, Corredoira E (2024) Current status of the cryopreservation of embryogenic material of woody species. Front Plant Sci 14:1337152

Ben-Amar A, Daldoul S, Allel D, Reustle G, Mliki A (2013) Reliable encapsulation-based cryopreservation protocol for safe storage and recovery of grapevine embryogenic cell cultures. Sci Hortic 157:32–38

Ben-Amar A, Allel D, Bouamama-Gzara B (2024) Osmotic priming-induced cryotolerance uncovers rejuvenation of grapevine cell cultures: morphogenetic changes and gene expression pattern highlighting enhanced embryogenic potential. Protoplasma 261(6):1251–1266

Benelli C, De Carlo A, Engelmann F (2013) Recent advances in the cryopreservation of shoot-derived germplasm of economically important fruit trees of *Actinidia*, *Diospyros*, *Malus*, *Olea*, *Prunus*, *Pyrus* and *Vitis*. Biotechnol Adv 31(2):175–185

Benson EE (2008) Cryopreservation of phytodiversity: a critical appraisal of theory practice. Crit Rev Plant Sci 27(3):141–219

Benson E, Betson F, Fuller BJ, Harding K, Kofanova O (2013) Translating cryobiology principles into trans-disciplinary storage guidelines for biorepositories and biobanks: a concept paper. Cryo-Lett 34(3):277–312

Bi WL, Pan C, Hao XY, Cui ZH, Kher MM, Marković Z, Wang QC, Teixeira da Silva JA (2017) Cryopreservation of grapevine (*Vitis* spp.)—a review. In Vitro Cell Dev Biol Plant 53(5):449–460

Bialoskorska M, Rucinska A, Boczkowska M (2024) Molecular mechanisms underlying freezing tolerance in plants: implications for cryopreservation. Int J Mol Sci 25:101110

Binder WD, Zaerr JB (1980) Freeze preservation of suspension cultured cells of a hardwood, poplar. Cryobiology 17:625

Boisson AM, Gout E, Bligny R, Rivasseau C (2012) A simple and efficient method for the long-term preservation of plant cell suspension cultures. Plant Method 8:4

Bonga JM (2017) Can explant choice help resolve recalcitrance problems in in vitro propagation, a problem still acute especially for adult conifers? Trees-Struc Funct 31(3):781–789

Bonga JM, Klimaszewska KK, von Aderkas P (2010) Recalcitrance in clonal propagation, in particular of conifers. Plant Cell Tissue Organ Cult 100(3):241–254

Carra A, Carimi F, Bettoni JC, Pathirana R (2019) Progress and challenges in the application of synthetic seed technology for *ex situ* germplasm conservation in grapevine (*Vitis* spp.). In: Synthetic seeds, pp 439–467

Carra A, Wijerathna-Yapa A, Pathirana R, Carimi F (2024) Development and applications of somatic embryogenesis in grapevine (*Vitis* spp.). Plants (Basel) 13:3131

Cazzolla Gatti R, Reich PB, Gamarra JGP, Crowther T, Hui C, Morera A, Bastin JF, de-Miguel S, Nabuurs GJ, Svenning JC, Serra-Diaz JM, Merow C, Enquist B, Kamenetsky M, Lee J, Zhu J, Fang J, Jacobs DF, Pijanowski B, Banerjee A, Giaquinto RA, Alberti G, Almeyda Zambrano AM, Alvarez-Davila E, Araujo-Murakami A, Avitabile V, Aymard GA, Balazy R, Baraloto C, Barroso JG, Bastian ML, Birnbaum P, Bitariho R, Bogaert J, Bongers F, Bouriaud O, PHS B, Brearley FQ, Broadbent EN, Bussotti F, Castro da Silva W, Cesar RG, Cesljar G, Chama Moscoso V, HYH C, Cienciala E, Clark CJ, Coomes DA, Dayanandan S, Decuyper M, Dee LE, Del Aguila Pasquel J, Derroire G, MNK D, Van Do T, Dolezal J, Dordevic ID, Engel J, Fayle TM, Feldpausch TR, Fridman JK, Harris DJ, Hemp A, Hengeveld G, Herault B, Herold M, Ibanez T, Jagodzinski AM, Jaroszewicz B, Jeffery KJ, Johannsen VK, Jucker T, Kangur A, Karminov VN, Kartawinata K, Kennard DK, Kepfer-Rojas S, Keppel G, Khan ML, Khare PK, Kileen TJ, Kim HS, Korjus H, Kumar A, Kumar A, Laarmann D, Labriere N, Lang M, Lewis SL, Lukina N, Maitner BS, Malhi Y, Marshall AR, Martynenko OV, Monteagudo Mendoza AL, Ontikov PV, Ortiz-Malavasi E, Pallqui Camacho NC, Paquette A, Park M, Parthasarathy N, Peri PL, Petronelli P, Pfautsch S, Phillips OL, Picard N, Piotto D, Poorter L, Poulsen JR, Pretzsch H, Ramirez-Angulo H, Restrepo Correa Z, Rodeghiero M, RDP RG, Rolim SG, Rovero F, Rutishauser E, Saikia P, Salas-Eljatib C, Schepaschenko D, Scherer-Lorenzen M, Seben V, Silveira M, Slik F, Sonke B, Souza AF, Sterenczak KJ, Svoboda M, Taedoumg H, Tchebakova N, Terborgh J, Tikhonova E, Torres-Lezama A, van der Plas F, Vasquez R, Viana H, Vibrans AC, Vilanova E, Vos VA, Wang HF, Westerlund B, LJT W, Wiser SK, Zawila-Niedzwiecki T, Zemagho L, Zhu ZX, Zo-Bi IC, Liang J (2022) The number of tree species on Earth. Proc Natl Acad Sci 119(6):e2115329119

Chabrillange NA-BF, Noirot M, Duval Y, Engelmann F (2000) Cryopreservation of oil-palm embryogenic suspensions. In: Engelmann FT, H. (eds) Cryopreservation of tropical plant germplasm. IPGRI, International Plant Genomic Resources Institute, Rome, pp 341–343

Chang T, Zhao G (2021) Ice inhibition for cryopreservation: materials, strategies, and challenges. Adv Sci (Weinh) 8(6):2002425
Chen P, Wang S, Chen Z, Ren P, Hepfer RG, Greene ED, Campbell LH, Helke KL, Nie X, Jensen JH, Hill C, Wu Y, Brockbank KGM, Yao H (2023) Nanowarming and ice-free cryopreservation of large sized, intact porcine articular cartilage. Commun Biol 6(1):220
Chopra P, Nayak D, Nanda A, Ashe S, Rauta PR, Nayak B (2016) Fabrication of poly(vinyl alcohol)-Carrageenan scaffolds for cryopreservation: effect of composition on cell viability. Carbohydr Polym 147:509–516
Ciniglia CP, Pinto G, Sansone C, Pollio A (2010) Acridine orange/Ethidium bromide double staining test: a simple *in-vitro* assay to detect apoptosis induced by phenolic compounds in plant cells. Allelopath J 26(2):301–308
Corredoira E, Martínez MT, Sanjosé MC, Ballester A (2017) Conservation of hardwood forest species. In: Ahuja MR, Jain SM (eds) Biodiversity and conservation of woody plants. Springer International Publishing, Cham, pp 421–453
De Carlo A, Lambardi M, Ozudogru EA (2011) Cryogenic technologies for the long-term storage of Citrus germplasm. Methods Mol Biol (Clifton, NJ) 710:185–200
Denchev P, Grossnickle S (2019) Somatic embryogenesis for conifer seedling production – the biology of scaling. Reforesta 7:109–137
Donaldson L (2020) Autofluorescence in plants. Molecules 25:2393
Donini M, Marusic C (2019) Current state-of-the-art in plant-based antibody production systems. Biotechnol Lett 41(3):335–346
Dussert S, Mauro MC, Deloire A, Hamon S, Engelmann F (1991) Cryopreservation of grape embryogenic cell-suspensions – 1-influence of pretreatment, freezing and thawing conditions. Cryo-Lett 12(5):287–298
Dussert S, Mauro MC, Deloire A, Hamon S, Engelmann F (1992) Cryopreservation of grape embryogenic cell-suspensions – 1-influence of post-thaw culture conditions and application to different strains. Cryo-Lett 13:15–22
Dutt M, Vasconcellos M, Song KJ, Gmitter FG, Grosser JW (2010) *In vitro* production of autotetraploid Ponkan mandarin (*Citrus reticulata* Blanco) using cell suspension cultures. Euphytica 173(2):235–242
Eibl R, Meier P, Stutz I, Schildberger D, Huhn T, Eibl D (2018) Plant cell culture technology in the cosmetics and food industries: current state and future trends. Appl Microbiol Biotechnol 102(20):8661–8675
Eibl R, Senn Y, Gubser G, Jossen V, Van Den Bos C, Eibl D (2021) Cellular agriculture: opportunities and challenges. Annu Rev Food Sci Technol 12:51–73
Endoh K, Kasuga J, Arakawa K, Ito T, Fujikawa S (2009) Cryo-scanning electron microscopic study on freezing behaviors of tissue cells in dormant buds of larch (*Larix kaempferi*). Cryobiology 59(2):214–222
Engelmann F, Dambier D, Ollitrault P (1994) Cryopreservation of cell-suspensions and embryogenic calluses of *Citrus* using a simplified freezing process. Cryo-Lett 15(1):53–58
Engelmann F, Arnao MTG, Wu Y, Escobar R (2008) Development of encapsulation dehydration. In: Plant cryopreservation: a practical guide. Springer, New York, pp 59–75
Espinosa-Leal CA, Puente-Garza CA, Garcia-Lara S (2018) *In vitro* plant tissue culture: means for production of biological active compounds. Planta 248(1):1–18
Fazan L, Song YG, Kozlowski G (2020) The woody planet: from past triumph to manmade decline. Plan Theory 9:1593
Feher A (2019) Callus, dedifferentiation, totipotency, somatic embryogenesis: what these terms mean in the era of molecular plant biology? Front Plant Sci 10:536
Fernando D, Kowalczyk M, Guindos P, Auer M, Daniel G (2023) Electron tomography unravels new insights into fiber cell wall nanostructure; exploring 3D macromolecular biopolymeric nano-architecture of spruce fiber secondary walls. Sci Rep 13(1):2350
Ferrucci A, Lupo M, Turco S, Pavese V, Marinoni DT, Botta R, Cristofori V, Mazzaglia A, Silvestri C (2023) A roadmap of tissue culture and biotechnology in European hazelnut (*Corylus avellana* L.). Plant Physiol Biochem 205:108167

Filonova LH, Bozhkov PV, Von Arnold S (2000) Developmental pathway of somatic embryogenesis in *Picea abies* as revealed by time-lapse tracking. J Exp Bot 51(343):249–264

Find JI, Floto F, Krogstrup P, Moller JD, Norgaard JV, Kristensen MMH (1993) Cryopreservation of an embryogenic suspension culture of *Picea sitchensis* and subsequent plant regeneration. Scand J For Res 8(2):156–162

Find JI, Kristensen MMH, Norgaard JV, Krogstrup P (1998) Effect of culture period and cell density on regrowth following cryopreservation of embryogenic suspension cultures of Norway spruce and Sitka spruce. Plant Cell Tissue Organ Cult 53(1):27–33

Gallego A, Malik S, Yousefzadi M, Makhzoum A, Tremouillaux-Guiller J, Bonfill M (2017) Taxol from *Corylus avellana*: paving the way for a new source of this anti-cancer drug. Plant Cell Tissue Organ Cult 129(1):1–16

Germanà MA, Aleza P, Grosser JW, Dutt M, Wang N, Cuenca J, Kaur P (2020) Citrus biotechnology. In: The genus *Citrus*. Elsevier Inc., pp 171–192

Gonzalez-Arnao MT, Martinez-Montero ME, Cruz-Cruz CA, Engelmann F (2014) Advances in cryogenic techniques for the long-term preservation of plant biodiversity. In: Biotechnology and biodiversity, Sustainable development and biodiversity. Springer, Cham, pp 129–170

González-Arnao MT, Dolce N, González-Benito ME, Castillo Martínez CR, Cruz-Cruz CA (2017) Approaches for *in vitro* conservation of woody plants germplasm. In: Ahuja MR, Jain SM (eds) Biodiversity and conservation of woody plants. Springer International Publishing, Cham, pp 355–419

González-Benito ME, Martín C, Vidal JR (2009) Cryopreservation of embryogenic cell suspensions of the Spanish grapevine cultivars 'Albariño' and 'Tempranillo'. Vitis J Grapevine Res 48(3):131–136

Grasso G, Colella F, Forciniti S, Onesto V, Iuele H, Siciliano AC, Carnevali F, Chandra A, Gigli G, del Mercato LL (2023) Fluorescent nano- and microparticles for sensing cellular microenvironment: past, present and future applications. Nanoscale Adv 5(17):4311–4336

Grosser JW, Gmitter FG (2011) Protoplast fusion for production of tetraploids and triploids: applications for scion and rootstock breeding in *Citrus*. Plant Cell Tissue Organ Cult 104(3):343–357

Grosser JW, Gmitter FG, Louzada ES, Chandler JL (1992) Production of somatic hybrid and autotetraploid breeding parents for seedless *Citrus* development. HortSci 27(10):1125–1127

Grosser JW, Gmitter FG, Tusa N, Recupero GR, Cucinotta P (1996) Further evidence of a cybridization requirement for plant regeneration from *Citrus* leaf protoplasts following somatic fusion. Plant Cell Rep 15(9):672–676

Grout BW (1995a) Cryopreservation of plant protoplasts. In: JG D, McLellan MR (eds) Cryopreservation and Freeze-Drying Protocols, Methods in molecular biology (Clifton, N.J.), vol 38. Humana Press, Totowa, pp 91–101

Grout BWW (1995b) Cryopreservation of protoplast, suspension and callus cultures. In: Grout B (ed) Genetic preservation of plant cells *in vitro*, Springer lab manual. Springer, Berlin, Heidelberger, pp 29–46

Grout BWW (2007) Cryopreservation of plant cell suspensions. In: Day JG, Stacey GN (eds) Cryopreservation and freeze-drying protocols. Humana Press, Totowa, pp 153–161

Gubser G, Vollenweider S, Eibl D, Eibl R (2021) Food ingredients and food made with plant cell and tissue cultures: state-of-the art and future trends. Eng Life Sci 2(3–4):87–98

Gupta PK, Timmis R (2005) Mass propagation of conifer trees in liquid cultures-progress towards commercialization. Plant Cell Tissue Organ Cult 8(3):339–346

Hao YJ, You CX, Deng XX (2002) Effects of cryopreservation on developmental competency, cytological and molecular stability of *Citrus* callus. Cryo-Lett 23(1):27–35

Hartel O (2003) Gottlieb Haberlandt (1854–1945): a Portrait. In: Plant tissue culture: 100 years since Gottlieb Haberlandt. Springer, Wien, pp 55–66

Heine-Dobbernack E, Kiesecker H, Schumacher HM (2008) Cryopreservation of dedifferentiated cell cultures. In: Plant cryopreservation: a practical guide. Springer, New York, pp 141–176

Ikeuchi M, Sugimoto K, Iwase A (2013) Plant callus: mechanisms of induction and repression. Plant Cell 25(9):3159–3173

Ikeuchi M, Ogawa Y, Iwase A, Sugimoto K (2016) Plant regeneration: cellular origins and molecular mechanisms. Development 143(9):1442–1451

Jahed KR, Saini AK, Sherif SM (2023) Coping with the cold: unveiling cryoprotectants, molecular signaling pathways, and strategies for cold stress resilience. Front Plant Sci 14:1246093

Jiang F, Zhu J, Liu H-L (2013) Protoplasts: a useful research system for plant cell biology, especially dedifferentiation. Protoplasma 250(6):1231–1238

Kadkade PGB, Christopher B, Schnabel-Preikstas B, Yu B (1999) Cryopreservation of plant cells. USA Patent

Kartha KK, Fowke LC, Leung NL, Caswell KL, Hakman I (1988) Induction of somatic embryos and plantlets from cryopreserved cell-cultures of white spruce (*Picea glauca*). J Plant Physiol 132(5):529–539

Kim H-H, Popova E (2023) Unifying principles of cryopreservation protocols for new plant materials based on alternative cryoprotective agents (CPAs) and a systematic approach. Cryo-Lett 44(1):1–12

Kim SI, Choi HK, Son JS, Yun JH, Jang MS, Kim HR, Song JY, Kim JH, Choi HJ, Hong SS (2001) Cryopreservation of *Taxus chinensis* suspension cell cultures. Cryo-Lett 22(1):43–50

Knapp E, Flores R, Scheiblin D, Scheiblin D, Modla S, Czymmek K, Czymmek K, Yusibov V (2012) A cryohistological protocol for preparation of large plant tissue sections for screening intracellular fluorescent protein expression. BioTechniques 52(1):31–37

Kobayashi S, Sakai A, Oiyama I (1990) Cryopreservation in liquid nitrogen of cultured navel organe (*Citrus sinensis* Osb.) nucellar cells and subsequent plant regeneration. Plant Cell Tissue Organ Cult 23(1):15–20

Kobayashi S, Kutsuna N, Tanino KK, Uemura M, Kawamura Y (2014) Confocal cryomicroscopic analysis and cryodynamics of endoplasmic reticulum in herbaceous plant cells. Env Exp Bot 106:44–51

van der Kooi CJ, Ollerton J (2020) The origins of flowering plants and pollinators. Science 368(6497):1306–1308

Kreis W (2019) Exploiting plant cell culture for natural product formation. J Appl Bot Food Qual 92:216–225

Kristensen MMH, Find JI, Floto F, Moller JD, Norgaard JV, Krogstrup P (1994) The origin and development of somatic embryos following cryopreservation of an embryogenic suspension culture of *Picea sitchensis*. Protoplasma 182(1–2):65–70

Laine E, Bade P, David A (1992) Recovery of plants from cryopreserved embryogenic cell suspensions of *Pinus caribaea*. Plant Cell Rep 11:295–298

Lee EK, Jin YW, Park JH, Yoo YM, Hong SM, Amir R, Yan ZJ, Kwon E, Elfick A, Tomlinson S, Halbritter F, Waibel T, Yun BW, Loake GJ (2010) Cultured cambial meristematic cells as a source of plant natural products. Nat Biotechnol 28(11):1213–U1111

Lelu-Walter MA, Thompson D, Harvengt L, Sanchez L, Toribio M, Paques LE (2013) Somatic embryogenesis in forestry with a focus on Europe: state-of-the-art, benefits, challenges and future direction. Tree Genet Genom 9(4):883–899

Li LQ, Li XL, Fu CH, Zhao CF, Yu LJ (2013) Sustainable use of *Taxus media* cell cultures through minimal growth conservation and manipulation of genome methylation. Process Biochem 48(3):525–531

Li JW, Ozudogru EA, Li J, Wang MR, Bi WL, Lambardi M, Wang QC (2018) Cryobiotechnology of forest trees: recent advances and future prospects. Biodivers Conserv 27(4):795–814

Liu S, Li X, Zhu J, Jin Y, Xia C, Zheng B, Silvestri C, Cui F (2024) Modern technologies provide new opportunities for somatic hybridization in the breeding of woody plants. Plants 13:2539

Ma Y, Gao L, Tian Y, Chen P, Yang J, Zhang L (2021) Advanced biomaterials in cell preservation: hypothermic preservation and cryopreservation. Acta Biomater 131:97–116

Malhotra EV, Bansal S, Gupta S (2024) Plant cryopreservation: a molecular perspective. Plant Cell Tissue Organ Cult 158:19

Mamun NHA, Aidun CK, Egertsdotter U (2018) Improved and synchronized maturation of Norway spruce (*Picea abies* (L.) H.Karst.) somatic embryos in temporary immersion bioreactors. In Vitro Cell Dev Biol-Plant 54(6):612–620

Máthé C, Demeter Z, Resetár A, Gonda S, Balázs A, Szoke E, Kiss Z, Simon A, Székely V, Riba M, Garda T, Gere B, Noszály Z, Attila Molnár V, Vasas G (2012) The plant tissue culture collection at the Department of botany, University of Debrecen. Acta Biol Szegediensis 56(2):179–182

McDonald KL, Auer M (2006) High-pressure freezing, cellular tomography, and structural cell biology. BioTechniques 4(2):137, 139

Mohamad Puad NI, Abdullah TA (2018) Monitoring the growth of plant cells in suspension culture. In: Amid A, Sulaiman S, Jimat DN, Azmin NFM (eds) Multifaceted protocol in biotechnology. Springer Singapore, Singapore, pp 203–214

Mojarrad Nanas S, Moieni A, Mokhtassi-Bidgoli A (2023) Pretreatment with abscisic acid accompanied by sucrose improves callus survival after cryopreservation of hazel (*Corylus avellana* L.) by desiccation. Plant Cell Tissue Organ Cult 152(2):267–273

Mukundan NS, Satyamoorthy K, Babu VS (2025) Advancing plant protoplasts: innovative techniques and future prospects. Plant Biotechnol Rep (Online). https://doi.org/10.1007/s11816-025-00957-1

Mursalimov S, Ohno N, Matsumoto M, Bayborodin S, Deineko E (2021) Serial block-face scanning electron microscopy reveals that intercellular nuclear migration occurs in most normal tobacco male meiocytes. Front Plant Sci 12:672642

Murthy HN, Lee EJ, Paek KY (2014) Production of secondary metabolites from cell and organ cultures: strategies and approaches for biomass improvement and metabolite accumulation. Plant Cell Tissue Organ Cult 118(1):1–16

Mustafa NR, de Winter W, van Iren F, Verpoorte R (2011) Initiation, growth and cryopreservation of plant cell suspension cultures. Nat Protoc 6(6):715–742

Nadarajan J, Pritchard HW (2014) Biophysical characteristics of successful oilseed embryo cryoprotection and cryopreservation using vacuum infiltration vitrification: an innovation in plant cell preservation. PLoS One 9(5):e96169

Nag KK, Street HE (1975a) Freeze preservation of cultured plant cells: I. The pretreatment phase. Physiol Plant 34(3):254–260

Nag KK, Street HE (1975b) Freeze preservation of cultured plant cells: II. The freezing and thawing phases. Physiol Plant 34(3):261–265

Nagel M, Pence V, Ballesteros D, Lambardi M, Popova E, Panis B (2024) Plant cryopreservation: principles, applications, and challenges of banking plant diversity at ultralow temperatures. Ann Rev Plant Biol 75:797–824

Narida A, Hsieh WC, Huang CL, Wen ZH, Tsai S, Lin C (2023) Novel long-term cryo-storage using vitrification and laser warming techniques. Biopreserv Biobank 21(4):427–432

Nausch H, Buyel JF (2021) Cryopreservation of plant cell cultures – diverse practices and protocols. New Biotechnol 62:86–95

Neubauer A, Ruaud S, Waller M, Frangedakis E, Li FW, Notzold SI, Wicke S, Bailly A, Szovenyi P (2022) Step-by-step protocol for the isolation and transient transformation of hornwort protoplasts. Appl Plant Sci 10(2):e11456

Nicolas WJ, Fassler F, Dutka P, Schur FKM, Jensen G, Meyerowitz E (2022) Cryo-electron tomography of the onion cell wall shows bimodally oriented cellulose fibers and reticulated homogalacturonan networks. Curr Biol 32(11):2375–2389 e2376

Normah MN, Sulong N, Reed BM (2019) Cryopreservation of shoot tips of recalcitrant and tropical species: advances and strategies. Cryobiology 87:1–14

Ochoa-Villarreal M, Howat S, Jang MO, Kim IS, Jin YW, Lee EK, Loake GJ (2015) Cambial meristematic cells: a platform for the production of plant natural products. New Biotechnol 32(6):581–587

Omar AA, Dekkers MGH, Graham JH, Grosser JW (2008) Estimation of transgene copy number in transformed citrus plants by quantitative multiplex real-time PCR. Biotechnol Prog 24(6):1241–1248

Omar AA, Dutt M, Gmitter FG, Grosser JW (2016) Somatic embryogenesis: still a relevant technique in *Citrus* improvement. Methods Mol Biol (Clifton, NJ) 1359:289–327

Palanyandy SR, Gantait S, Subramaniam S, Sinniah UR (2020) Cryopreservation of oil palm (*Elaeis guineensis Jacq.*) polyembryoids via encapsulation–desiccation. 3 Biotech 10:9

Park YS (2014) Conifer somatic embryogenesis and multi-varietal forestry. In: Fenning T (ed) Challenges and opportunities for the world's forests. Springer, pp 425–439
Park YS, Beaulieu J, Bousquet J (2016) Multi-varietal forestry integrating genomic selection and somatic embryogenesis. In: Park Y-S, Bonga JM, Moon H-K (eds) . Vegetative propagation of forest trees, Seoul, pp 302–322
Pasternak T, Lystvan K, Betekhtin A, Hasterok R (2020) From single cell to plants: mesophyll protoplasts as a versatile system for investigating plant cell reprogramming. Int J Mol Sci 21:4195
Pearce KC, Samuels FMD, Volk GM, Levinger NE (2024) Direct evidence that cryoprotectant mixtures facilitate individual component permeation into living plant cells. Cryobiology 116:104928
Popova E, Shukla M, Kim HH, Saxena PK (2021) Root cryobanking: an important tool in plant cryopreservation. Plant Cell Tissue Organ Cult 144(1):49–66
Potter KM, Jetton RM, Bower A, Jacobs DF, Man G, Hipkins VD, Westwood M (2017) Banking on the future: progress, challenges and opportunities for the genetic conservation of forest trees. New For 48(2):153–180
Pritchard HW, Grout BWW, Short KC (1986a) Osmotic stress as a pregrowth procedure for cryopreservation 2. Water relations and metabolic state of sycamore and soybean cell suspensions. Ann Bot 57(3):371–378
Pritchard HW, Grout BWW, Short KC (1986b) Osmotic stress as a pregrowth procedure for cryopreservation 3. Cryobiology of sycamore and soybean cell suspensions. Ann Bot 57(3):379–387
Pritchard HW, Grout BWW, Short KC (1986c) Osmotic stress as a pregrowth procedure for cryopreservation: 1. Growth and ultrastructure of sycamore and soybean cell suspensions. Ann Bot 57(1):41–48
Rajeckaite V, Jakstys B, Rafanavicius A, Maciulevicius M, Jakutaviciute M, Satkauskas S (2018) Calcein release from cells in vitro via reversible and irreversible electroporation. J Membr Biol 251(1):119–130
Reed BM (2008) Plant cryopreservation: a practical guide. Springer, New York, p 513
Reed BM (2011) Choosing and applying cryopreservation protocols to new plant species or tissues. Acta Horticulturae. International Society for Horticultural Science 908:363–372
Reed KM, Bargmann BOR (2021) Protoplast regeneration and its use in new plant breeding technologies. Front Genome Edit 3:734951
Reed BM, Uchendu E (2008) Controlled rate cooling. In: Plant cryopreservation: a practical guide. Springer, New York, pp 77–92
Reed BM, Normah MN, Kushnarenko SV (2016) Cryopreservation of plant cells, tissues, and organs. In: Plant tissue culture, development, and biotechnology. CRC Press, pp 489–495
Reinhoud PJ, Uragami A, Sakai A, Van Iren F (1995) Vitrification of plant cell suspensions. In: Day JG, Pennington MW (eds) Cryopreservation and freeze-drying protocols. Humana Press, Totowa, pp 113–120
Reinhoud PJ, van Iren F, Kijne JW (2000) Cryopreservation of undifferentiated plant cells. In: Engelmann FT, H. (eds) Cryopreservation of tropical plant germplasm. IPGRI, International Plant Genomic Resources Institute, Rome, pp 91–102
Ren L, Wang MR, Wang QC (2021) ROS-induced oxidative stress in plant cryopreservation: occurrence and alleviation. Planta 254(6):124
Rensing KH, Samuels AL (2004) Cellular changes associated with rest and quiescence in winter-dormant vascular cambium of *Pinus contorta*. Trees 18(4):373–380
Reyna-Llorens I, Ferro-Costa M, Burgess SJ (2023) Plant protoplasts in the age of synthetic biology. J Exp Bot 74(13):3821–3832
Reza M, Engelhardt P, Ruokolainen J (2013) 3-D Structure of cellulose microfibril in wood using electron tomography and single particle reconstruction methods. Microsc Microanal 19:72–73
Rodríguez SM, Ordás RJ, Alvarez JM (2022) Conifer biotechnology: an overview. Forests 13:1061
Rosvall O (2019) Using Norway spruce clones in Swedish forestry: general overview and concepts. Scand J For Res 34(5):336–341
Sahari Moghaddam A, Kaviani B, Mohammadi Torkashvand A, Abdossi V, Eslami AR (2024) *In vitro* propagation, cold preservation and cryopreservation of *Taxus baccata* L., an endangered medicinal and ornamental shrub. Acta Sci Pol Hortorum Cultus 23(1):13–28

Sakai A, Engelmann F (2007) Vitrification, encapsulation-vitrification and droplet-vitrification: a review. Cryo-Lett 28(3):151–172

Sakai A, Kobayashi S, Oiyama I (1990) Cryopreservation of nucellar cells of navel orange (*Citrus sinensis* Osb. var. *brasiliensis* Tanaka) by vitrification. Plant Cell Rep 9(1):30–33

Sakai A, Kobayashi S, Oiyama I (1991a) Cryopreservation of nucellar cells of navel orange (*Citrus sinensis* Osb) by a simple freezing method. Plant Sci 74(2):243–248

Sakai A, Kobayashi S, Oiyama I (1991b) Survival by vitrification of nucellar cells of navel orange (*Citrus sinensis* var. *brasiliensis* Tanaka) cooled to −196 °C. J Plant Physiol 137(4):465–470

Sakai A, Hirai D, Niino T (2008) Development of PVS-based vitrification and encapsulation-vitrification protocols. In: Plant cryopreservation: a practical guide. Springer, New York, pp 33–57

Sala-Cholewa K, Milewska-Hendel A, Pérez-Pérez R, Grzebelus E, Betekhtin A (2024) Reconstruction pattern of the cell wall in Fagopyrum protoplast-derived hybrid cells. Plant Cell Tissue Organ Cult 157(2):26

Samuels FMD, Stich DG, Bonnart R, Volk GM, Levinger NE (2021) Non-uniform distribution of cryoprotecting agents in rice culture cells measured by cars microscopy. Plan Theory 10:589

Samuels FMD, Pearce KC, Soderlund S, Stich DG, Bonnart R, Volk GM, Levinger NE (2023) Direct observation of common cryoprotectant permeation into rice callus by CARS microscopy. Cell Rep Phys Sci 4:101469

San Jose MC, Corredoira E, Oliveira H, Santos C (2015) Cryopreservation of somatic embryos of *Alnus glutinosa* (L.) Gaertn. and confirmation of ploidy stability by flow cytometry. Plant Cell Tissue Organ Cult 123(3):489–499

Sanchez Carrillo IB, Hoffmann PC, Barff T, Beck M, Germain H (2023) Preparing *Arabidopsis thaliana* root protoplasts for cryo electron tomography. Front Plant Sci 14:1261180

Santos RB, Abranches R, Fischer R, Sack M, Holland T (2016) Putting the spotlight back on plant suspension cultures. Front Plant Sci 7:297

Sarkar P, Bosneaga E, Yap EG Jr, Das J, Tsai WT, Cabal A, Neuhaus E, Maji D, Kumar S, Joo M, Yakovlev S, Csencsits R, Yu Z, Bajaj C, Downing KH, Auer M (2014) Electron tomography of cryo-immobilized plant tissue: a novel approach to studying 3D macromolecular architecture of mature plant cell walls in situ. PLoS One 9(9):e106928

Schmale K, Rademacher T, Fischer R, Hellwig S (2006) Towards industrial usefulness – cryo-cell-banking of transgenic BY-2 cell cultures. J Biotechnol 124(1):302–311

Schrijnemakers EW, Van Iren F (1995) A two-step or equilibrium freezing procedure for the cryopreservation of plant cell suspensions. In: Day JG, McLellan MR (eds) Cryopreservation and freeze-drying protocols, vol 38. Methods in molecular biology, Clifton, pp 103–111

Schumacher HM, Westphal M, Heine-Dobbernack E (2021) Cryopreservation of plant cell lines using alginate encapsulation. In: Wolkers WF, Oldenhof H (eds) Cryopreservation and freeze-drying protocols. Springer US, New York, pp 639–645

Skrlep K, Bergant M, De Winter GM, Bohanec B, Zel J, Verpoorte R, Van Iren F, Camloh M (2008) Cryopreservation of cell suspension cultures of *Taxus* x *media* and *Taxus floridana*. Biol Plant 52(2):329–333

Sliwinska E, Loureiro J, Leitch IJ, Smarda P, Bainard J, Bures P, Chumova Z, Horova L, Koutecky P, Lucanova M, Travnicek P, Galbraith DW (2022) Application-based guidelines for best practices in plant flow cytometry. Cytometry A 101(9):749–781

Smertenko A, Bozhkov PV (2014) Somatic embryogenesis: life and death processes during apical-basal patterning. J Exp Bot 65(5):1343–1360

Steponkus PL (1985) Cryobiology of isolated protoplasts: applications to plant cell cryopreservation. In: Kartha K (ed) Cryopreservation of plant cells and organs. CRC Press, pp 49–60

Steponkus PL, Dowgert MF, Gordon-Kamm WJ (1983) Destabilization of the plasma membrane of isolated plant protoplasts during a freeze-thaw cycle: the influence of cold acclimation. Cryobiology 20(4):448–465

Steponkus PL, Dowgert MF, Ferguson JR, Levin RL (1984) Cryomicroscopy of isolated plant protoplasts. Cryobiology 21(2):209–233

Steponkus PL, Uemura M, Webb MS (1995) Freeze-induced destabilization of cellular membranes and lipid bilayers. In: Permeability and stability of lipid bilayers. CRC Press, pp 77–104

Sugawara YS, Sakai A (1974) Survival of suspension-cultured sycamore cells cooled to the temperature of liquid nitrogen. Plant Physiol 54:722–724

Sun H, Aidun CK, Egertsdotter U (2010) Effects from shear stress on morphology and growth of early stages of Norway spruce somatic embryos. Biotechnol Bioeng 105(3):588–599

Suzuki N, Mittler R (2006) Reactive oxygen species and temperature stresses: a delicate balance between signaling and destruction. Physiol Plant 126(1):45–51

Truernit E, Haseloff J (2008) A simple way to identify non-viable cells within living plant tissue using confocal microscopy. Plant Methods 4:15

Tusa N, Grosser JW, Gmitter FG (1990) Plant-regeneration of valencia sweet orange, femminello lemon, and the interspecific somatic hybrid following protoplast fusion. J Am Soc Hortic Sci 115(6):1043–1046

Vasanth K, Vivier MA (2011) Improved cryopreservation procedure for long term storage of synchronised culture of grapevine. Biol Plant 55(2):365–369

Vidot K, Gaillard C, Rivard C, Siret R, Lahaye M (2018) Cryo-laser scanning confocal microscopy of diffusible plant compounds. Plant Methods 14:89

Volk GM (2010) Application of functional genomics and proteomics to plant cryopreservation. Curr Genomics 11(1):24–29

Wang Q, Perl A (2006) Cryopreservation of embryogenic cell suspensions by encapsulation-vitrification. Methods Mol Biol (Clifton, NJ) 318:77–86

Wang Q, Gafny R, Sahar N, Sela I, Mawassi M, Tanne E, Perl A (2002) Cryopreservation of grapevine (*Vitis vinifera* L.) embryogenic cell suspensions by encapsulation-dehydration and subsequent plant regeneration. Plant Sci 162(4):551–558

Wang QC, Mawassi M, Sahar N, Li P, Violeta CT, Gafny R, Sela I, Tanne E, Perl A (2004) Cryopreservation of grapevine (*Vitis* Spp.) embryogenic cell suspensions by encapsulation-vitrification. Plant Cell Tissue Organ Cult 77(3):267–275

Wang Q, Li P, Hanania U, Sahar N, Mawassi M, Gafny R, Sela I, Tanne E, Perl A (2005) Improvement of *Agrobacterium*-mediated transformation efficiency and transgenic plant regeneration of *Vitis vinifera* L. by optimizing selection regimes and utilizing cryopreserved cell suspensions. Plant Sci 168(2):565–571

Wang MR, Bi W, Shukla MR, Ren L, Hamborg Z, Blystad DR, Saxena PK, Wang QC (2021a) Epigenetic and genetic integrity, metabolic stability, and field performance of cryopreserved plants. Plan Theory 10(9):1889

Wang MR, Lambardi M, Engelmann F, Pathirana R, Panis B, Volk GM, Wang QC (2021b) Advances in cryopreservation of *in vitro*-derived propagules: technologies and explant sources. Plant Cell Tissue Organ Cult 144(1):7–20

Wen B, Wang R, Cheng H, Song S (2010) Cytological and physiological changes in orthodox maize embryos during cryopreservation. Protoplasma 239(1–4):57–67

Wendling I, Trueman SJ, Xavier A (2014a) Maturation and related aspects in clonal forestry-part I: concepts, regulation and consequences of phase change. New For 45(4):449–471

Wendling I, Trueman SJ, Xavier A (2014b) Maturation and related aspects in clonal forestry-part II: reinvigoration, rejuvenation and juvenility maintenance. New For 45(4):473–486

Wesley-Smith J, Walters C, Pammenter NW, Berjak P (2015) Why is intracellular ice lethal? A microscopical study showing evidence of programmed cell death in cryo-exposed embryonic axes of recalcitrant seeds of *Acer saccharinum*. Ann Bot 115(6):991–1000

Whaley D, Damyar K, Witek RP, Mendoza A, Alexander M, Lakey JR (2021) Cryopreservation: an overview of principles and cell-specific considerations. Cell Transplant 30:963689721999617

Wilson SA, Roberts SC (2012) Recent advances towards development and commercialization of plant cell culture processes for the synthesis of biomolecules. Plant Biotechnol J 10(3):249–268

Withers LA (1978) The freeze-preservation of synchronously dividing cultured cells of *Acer pseudoplatanus* L. Cryobiology 15(1):87–92

Withers LA (1985) Cryopreservation of cultured plant cells and protoplasts. In: Kartha K (ed) Cryopreservation of plant cells and organs. CRC Press, pp 243–267

Withers LA, Davey MR (1978) A fine-structural study of the freeze-preservation of plant tissue cultures – I. The frozen state. Protoplasma 94(3–4):207–219
Withers LA, King PJ (1980) A simple freezing unit and routine cryopreservation method for plant cell cultures. Cryo-Lett 1:213–220
Withers LA, Street HE (1977) Freeze preservation of cultured plant cells. III The pregrowth phase. Physiol Plant 39(2):171–178
Xu J, PerezSanchez P, Sadravi S (2025) Unlocking the full potential of plant cell-based production for valuable proteins: challenges and innovative strategies. Biotechnol Adv 79:108526
Yamada T, Kuroda K, Jitsuyama Y, Takezawa D, Arakawa K, Fujikawa S (2002) Roles of the plasma membrane and the cell wall in the responses of plant cells to freezing. Planta 215(5):770–778
Yin Z, Chen L, Zhao B, Zhu Y, Wang Q (2012) Cryopreservation of embryogenic cell suspensions by encapsulation–vitrification and encapsulation–dehydration. In: Loyola-Vargas VM, Ochoa-Alejo N (eds) Plant cell culture protocols. Humana Press, Totowa, pp 81–93
Yin JY, Lai M, Yu XY, Su DD, Xiong XY, Li YL (2025) Comprehensive strategies for paclitaxel production: insights from plant cell culture, endophytic microorganisms, and synthetic biology. Hortic Res 12(3):uhae346
Yue JJ, Yuan JL, Wu FH, Yuan YH, Cheng QW, Hsu CT, Lin CS (2021) Protoplasts: from isolation to crispr/cas genome editing application. Front Genome Edit 3:717017
Yuorieva N, Sinetova M, Messineva E, Kulichenko I, Fomenkov A, Vysotskaya O, Osipova E, Baikalova A, Prudnikova O, Titova M, Nosov AV, Popova E (2023) Plants, cells, algae, and cyanobacteria in vitro and cryobank collections at the institute of plant physiology, Russian academy of sciences-a platform for research and production center. Biology 12:838

Conserving the Genetic Resources of Exceptional Plant Species Through Pollen Banking

4

Daniel Ballesteros, Jose L. Canón-Prieto, Anna Nebot, Hugh W. Pritchard, Yu Tu, Hongying Chen, Raquel Folgado, Max Winkeljohn, Megan Philpott, Valerie Pence, Ana María López Peralta, Laura Prieto, and Jeremy Foster

Abstract

Exceptional plants have seeds that cannot be stored in conventional seed banks long term. Alternative methods based on tissue/explant cryobiotechnology are needed; however, the implementation of these high-tech approaches is still very limited. One route to immediately strengthen the ex situ conservation of exceptional plants is to establish pollen banks. Pollen from several exceptional species

D. Ballesteros (✉)
Department of Botany and Geology, University of Valencia, Burjassot, Spain

Botanical Garden of the University of Valencia, Valencia, Spain

Trait Diversity and Function, Royal Botanic Gardens, Kew, Wakehurst, UK
e-mail: daniel.ballesteros@uv.es

J. L. Canón-Prieto
Department of Botany and Geology, University of Valencia, Burjassot, Spain

A. Nebot
Botanical Garden of the University of Valencia, Valencia, Spain

H. W. Pritchard · Y. Tu · H. Chen
Kunming Institute of Botany, Chinese Academy of Sciences, Kunming, China

R. Folgado
The Huntington Library, Art Museum, and Botanical Gardens, San Marino, CA, USA

M. Winkeljohn · M. Philpott · V. Pence
Center for Conservation and Research of Endangered Wildlife, Cincinnati Zoo and Botanical Garden, Cincinnati, OH, USA

A. M. L. Peralta
Empresa de Transformación Agraria, S.A. (TRAGSA), Guadalajara, Spain

S. Gantait, P. Chmielarz (eds.), *Conservation of Plant Genetic Resources*,
https://doi.org/10.1007/978-981-96-8331-4_4

has demonstrated tolerance to desiccation and freezing. If these attributes are shared among most of the exceptional species reported, banks of dry pollen would constitute a low-tech method for the long-term conservation of wide genetic diversity via the safe storage of male gametes. Moreover, pollen banks could be used as a fundamental conservation tool in the restoration of exceptional plants through assisted reproductive technology (ART). We present here an analysis of the storage physiology of the pollen of exceptional plants, showing that important groups producing recalcitrant seeds, such as oaks (*Quercus* spp.), chestnuts (*Castanea* spp.), and horse chestnuts (*Aesculus* spp.), produce desiccation-tolerant "orthodox" pollen. We report three examples of success in the conservation of exceptional plants through pollen preservation: (1) the 'Global Oak Pollen Bank' for *Quercus* sp., (2) the long-term cryogenic storage of *Castanea dentata* pollen leading to the successful recovery of extinct genotypes through ART, and (3) the efforts of the Chicago Botanic Garden to preserve various endangered species via pollen banking.

Keywords

Castanea dentata · Cryopreservation · Desiccation tolerance · Freezing tolerance · *Quercus* · Trees

4.1 Introduction

Seed storage at approximately 15–25% relative humidity and −18 ± 3 °C is the standard and most commonly used method for the conservation of plant genetic resources (FAO 2014; CPC 2019; Breman et al. 2021; Walters and Pence 2021). This relatively low-tech solution is the basis for the conservation of tens of thousands of plant species in more than 1000 conventional seed banks globally (Breman et al. 2021; Walters and Pence 2021). However, not all plant species produce seeds that can be stored in conventional seed banks for the long term. These species are known as "exceptional plants" and are included in this category for diverse reasons or exceptionality factors (EFs) (Pence et al. 2022):

(i) The plants of the species do not produce seeds or are extremely limited in quantity, viability, or accessibility (EF1).
(ii) The seeds produced do not survive standard drying (at c. 15% relative humidity) of traditional seed bank facilities (EF2).

L. Prieto
Centro Nacional de Recursos Genéticos Forestales 'Alaquàs', Subdirección General de Política Forestal y Lucha Contra la Desertificación, Alaquàs, Spain

J. Foster
Chicago Botanic Garden, Glencoe, IL, USA

(iii) Although the seeds survive drying, they do not remain viable over a reasonable length of time in the seed bank, e.g., a P_{50} of $\leq$20 years (EF3).
(iv) The seeds stored do not have the ability to germinate and produce plants upon removal from the seed bank according to standard germination tests, generally because they are deeply dormant and/or have very small embryos (EF4).

Hence, for the ex situ conservation of exceptional plant genetic resources, alternative methods to conventional seed banking are needed, and these methods can be implemented by living plant collections or through approaches based on tissue/explant cryobiotechnology (Ballesteros and Pritchard 2020; Pence et al. 2020; Walters and Pence 2021; Pence and Bruns 2022). However, these alternatives are more costly and time-consuming than conventional seed banking is and often require innovative research to develop effective in vitro culture or preservation protocols (Dulloo et al. 2009; Li and Pritchard 2009; Pence 2011, 2014; Panis et al. 2020; Nagel et al. 2024). Moreover, the actual implementation of these high-tech approaches is still very limited (Ballesteros and Pritchard 2020; Pence and Bruns 2022; Nagel et al. 2024), even though the relative cost of maintaining a stored sample is quite similar for cryobanking and conventional banking (Li and Pritchard 2009; Acker et al. 2017).

One route to immediately strengthen the ex situ conservation of exceptional plants is to establish pollen banks (Ballesteros and Pritchard 2020). Pollen is the major gene dispersal element of many flowering plants, particularly those that are wind pollinated (Jordano 2010). As a dispersal unit, one of the characteristics that makes pollen a valuable germplasm is its potential tolerance to drying and low temperatures after being released from anthers (Franchi et al. 2011; Marks et al. 2014; Nebot et al. 2021). Additionally, pollen is the male gametophyte of flowering plants. In other words, the pollen of a plant species is the source of the sperm that is needed to fertilize the female structures of that plant species and promote the formation of seeds (the next plant generation). For this reason, pollen is a fundamental tool in plant-assisted reproductive technology (ART). These pollen features make this reproductive unit widely useful in agriculture, horticulture, and forestry for hybridization programs. Consequently, pollen preservation is highly desirable (Dinato et al. 2020; Impe et al. 2022), including for the conservation of wild and threatened species (Towill 2004; Nebot et al. 2021; van der Walt et al. 2022; Ferreira et al. 2024). The storage of pollen can preserve wide genetic diversity via the safe storage of male gametes, and those pollen banks can also be used as a fundamental conservation tool in the restoration of exceptional plants through ART.

In this chapter, we use examples from current research to demonstrate a number of possibilities for conserving exceptional plant genetic resources through pollen banking. First, we analyze the potential of pollen from some exceptional plants to be effectively stored (Sect. 4.2). Pollen of some exceptional species, such as oaks (*Quercus* sp.), tolerates drying and freezing (Ballesteros and Pritchard 2020). If these functional attributes are shared among other reported exceptional species, banks of dry pollen could constitute a low-tech method for the long-term conservation of wide genetic diversity via the safe storage of male gametes. We report three examples of the successful development and implementation of pollen cryobanking for the conservation of exceptional plants. The first example is an initiative funded by the International Oak Society for the development of a "Global Oak Pollen Bank" aiming to preserve ex situ a large genetic diversity of many *Quercus* sp. (Sect. 4.3). Pollen banks have great potential not only for the ex situ conservation of endangered species but also for plant production, as the exchange of pollen between institutions and botanical gardens can facilitate high-quality seed/seedling generation (for habitat restoration projects). In this context, living collection-based programs for the production of cycad plants using preserved pollen are underway at The Huntington, USA, and Nong Nooch TBG, Thailand (Calonje et al. 2011, 2017; Griffith et al. 2015; Nadarajan et al. 2018). The final two examples are the assisted "resurrection" of formerly extinct genotypes of American chestnut, *Castanea dentata* (Sect. 4.4), and the efforts led at the Chicago Botanic Garden to preserve the pollen of endangered species, including a Hawaiian species, *Brighamia insignis*, which is extinct in the wild, and *Amorphophallus titanum*, an endangered species that is declining in the wild and has little genetic diversity in ex situ collections. The exchange of their pollen following a "pedigree" approach similar to that used in zoos will help preserve the genetic diversity of these species for proper propagation and conservation (Sect. 4.5; Wood et al. 2020).

4.2 Storage Physiology of Pollen

Pollen long-term storage and conservation in germplasm banks can be achieved with a relatively easy protocol similar to that used for orthodox seeds if the pollen can withstand drying to RH between 20% and 50% (Volk 2011; Nebot et al. 2021). Once dried, storage can be performed at the standard temperatures of seed banks (ca. −20 °C) or, as recommended—owing to the generally short longevity of pollen— at cryogenic temperatures of liquid nitrogen, i.e., < -180 °C (Hoekstra 1995; Volk 2011; Nebot et al. 2021; Rajasekharan and Rohini 2023).

Pollen from a wide range of plant species is potentially tolerant to drying and low temperatures after being released from the anthers, surviving ambient conditions from a few days to several months, depending on the species (Hoekstra 1995, 2002). Pollen that tolerates drying (or desiccation tolerance (DT)) is often released from anthers at low moisture contents (<30%). This trait is the opposite of pollen that is released at high moisture contents, usually >30%, and has been categorized as "partially hydrated," which may be desiccation sensitive (DS) or recalcitrant (following the seed literature) (Franchi et al. 2011). Longevity is also an important trait, as the longer the pollen remains viable, the greater the potential distance it can travel and fertilize flowers (Hoekstra 2002). Pollen is relatively short-lived compared with other plant dispersules (e.g., seeds of many species or spores; Hoekstra 2002), with binucleate pollen tending to be longer-lived than trinucleate pollen (Hoekstra 2002; Marks et al. 2014). However, such natural longevity can be extended through the control of humidity and the lowering of temperature.

Table 4.1 Desiccation tolerance (DT) and sensitivity (DS) of pollen of temperate trees presenting seeds potentially sensitive to drying (EF2) or with drying tolerant seeds that have a short life span at −20 °C (EF3)

Family (-aceae)	Species	EF[a]	Pollen DT/DS	Reference
Laur-	*Laurus nobilis*	2	DS	Franchi et al. (2011)
Jugland-	*Carya oliviformis*	3	DS	Franchi et al. (2011)
Jugland-	*Corylus avellana*	3	DS	Franchi et al. (2011)
Jugland-	*Juglans regia*	3	DS	Franchi et al. (2011)
Salic-	*Populus tremula*	3	DS	Franchi et al. (2011)
Salic-	*Populus* sp.	3	DS	Franchi et al. (2011)
Ulm-	*Ulmus campestris*	3	DS	Franchi et al. (2011)
Sapind-	*Acer pseudoplatanus*	2	DT	Franchi et al. (2011)
Sapind-	*Acer saccharinum*	2	DT	Franchi et al. (2011)
Sapind-	*Aesculus chinensis*	2	DT	Xu et al. (2014)
Sapind-	*Aesculus hippocastanum*	2	DT	Franchi et al. (2011)
Fag-	*Castanea dentata*	2	DT	Riendeau (2023)
Fag-	*Castanea sativa*	2	DT	Franchi et al. (2011)
Cycad-	*Cycas* sp.	2	DT	Osborne et al. (1992); Nadarajan et al. (2018)
Zami-	*Encephalartos* sp.	2	DT	Osborne et al. (1992)
Fag-	*Quercus* spp.	2	DT	Franchi et al. (2011)
Arec-	*Euterpe oleracea*	3	DT	Oliveira et al. (2001)
Fag-	*Fagus sylvatica*	3	DT	Franchi et al. (2011)
Salic-	*Populus balsamifera*	3	DT	Mercier (1995)
Salic-	*Salix caprea*	3	DT	Franchi et al. (2011)

Information on pollen and seed DT and DS was obtained from a literature review made by the authors using peer-reviewed papers, book chapters, and databases (such as Seed Information Database-SID: https://ser-sid.org/, or the Global Working List of Exceptional Plants: https://cincinnatizoo.org/exceptional-plant-conservation-network/global-working-list-of-exceptional-plants/)
[a]Based on seed response/traits

In relation to exceptional species, we have found that DT pollen is quite common among tree species from temperate regions of the world that also have DS seeds (e.g., *Quercus* sp., *Aesculus* sp., *Castanea* sp.; EF2) or seeds that are freezing-sensitive with a short life span in the dry state (e.g., *Populus balsamifera*, *Salix caprea*, *Acer* sp.; EF3) (Table 4.1). Moreover, it is also common in some of the most endangered plants, cycads (Table 4.1), which are also considered exceptional, with varying challenges in handling their seeds (Calonje et al. 2017; Seed Information Database 2023). The evolution of the desiccation tolerance trait in the pollen of these species, which are keystone species in forest ecosystems (e.g., *Quercus* sp.), facilitates the creation of pollen banks.

However, DS pollen is also observed in exceptional tree species that have DS seeds (e.g., *Laurus nobilis*, EF2) or seeds that are freezing sensitive with a short life span in the dry state (e.g., some *Populus* sp., *Juglans regia*, *Carya oliviformis*,

Table 4.2 Desiccation sensitive (DS) pollen of temperate trees that have DT orthodox seeds

Family (-aceae)	Species	Pollen (DT/DS)	Reference
Betul-	*Betula alba = B. pendula*	DS	Franchi et al. (2011)
Betul-	*Betula alnus = Alnus glutinosa*	DS	Franchi et al. (2011)
Coryl-	*Ostrya carpinifolia*	DS	Franchi et al. (2011)
Hamamelid-	*Liquidambar styraciflua*	DS	Franchi et al. (2011)
Ulm-	*Celtis laevigata*	DS	Franchi et al. (2011)
Anacardi-	*Pistacia vera*	DS	Zamani Bahramabadi et al. (2018)

Information on pollen and seed DT and DS was obtained by assessing peer-reviewed papers, book chapters, and databases (such as SID). Some of the information gathered on pollen and seed DT and DS was obtained from data published by Franchi et al. (2011)

Table 4.3 New empirical determinations of desiccation tolerance (DT) and sensitivity (DS) of pollen of temperate trees

Family (-aceae)	Species	Pollen (DT/DS)	Seed (DT/DS)	Pollen survives LN
Sapind-	*Aesculus assamica*	DT	DS	Yes
Sapind-	*Aesculus glabra*	DT	DS	Yes
Sapind-	*Aesculus hippocastanum*	DT	DS	Yes
Sapind-	*Aesculus indica*	DT	DS	Yes
Sapind-	*Aesculus parviflora*	DT	DS	Yes
Sapind-	*Aesculus wilsonii*	DT	DS	Yes
Aquifoli-	*Ilex aquifolium*	DT	DT	Yes
Betul-	*Corylus avellana*	DT	DT	Yes
Betul-	*Corylus colurna*	DT	DT	Yes
Arali-	*Hedera helix*	DT	DT	N/A
Tax-	*Taxus baccata*	DT	DT	Yes

Pollen was considered DT when it tolerated desiccation <50% RH for 1 to 3 days (reaching water content <0.15 g H_2O/g dry weight). DT pollen remained alive for several weeks when stored at such dry conditions and at 20 °C, showing greater longevity at RH ranging between 11 and 33%. Information on seed DT and DS was obtained from SER, INSR, RBGK, and Seed Information Database (SID) (2023). N/A, exposure to LN was not investigated. Some of the results obtained from the authors are in contradiction with previous information published (Table 4.1), which sometimes was based on 'personal observations' rather than empirical data

Corylus avellana, EF3) (Table 4.1). This observation highlights the need to explore further the mechanisms of DT in pollen and to search for cryobiotechnology solutions for the conservation of these species. DS pollen is also found in diverse tree species with known DT orthodox seeds (Table 4.2). Although seeds are amenable for conventional seed banking, research is needed to develop successful protocols for pollen banks if the conservation of male genotypes is needed for breeding and ART purposes (Impe et al. 2022).

Table 4.4 Exceptional plant species whose pollen has been cryopreserved

Family (-aceae)	Species	Pollen binucleate/ trinucleate	Desiccation-tolerant pollen	Reference
Sapind-	*Aesculus chinensis*	Binucleate	Yes	Xu et al. (2014)
Mor-	*Artocarpus heterophyllus*	NA	Yes	Rajasekharan and Rohini (2023)
Caric-	*Carica papaya*	Binucleate	Yes	Dinato et al. (2020)
Rut-	*Citrus* sp.	Binucleate	Yes	Zhang et al. (2017)
Arec-	*Cocos nucifera*	Trinucleate	Yes	Karun et al. (2014)
Arec-	*Elaeis guineensis*	Binucleate	Yes	Tandon et al. (2007)
Arec-	*Euterpe oleracea*	NA	Yes	Oliveira et al. (2001)
Sapind-	*Litchi chinensis*	Binucleate	Yes	Dinato et al. (2020)
Anacardi-	*Mangifera indica*	Binucleate	Yes	Dutta et al. (2013)
Passiflor-	*Passiflora quadrangularis*	NA		Rajasekharan and Rohini (2023)
Fag-	*Quercus robur*	Binucleate	Yes	Rajasekharan and Rohini (2023)
Myrt-	*Syzygium cumini*	Binucleate	Yes	Rajasekharan and Rohini (2023)

Species have been considered exceptional if the seeds were categorized as recalcitrant or intermediate in the literature. Pollen was considered DT when it tolerated desiccation <50% RH for 1 to 3 days (reaching water content <0.15 g H_2O/g dry weight). Information on number of pollen nuclei to consider it binucleated (B) or trinucleated (T) as well as seed DT and DS was obtained from a literature review made by the authors using peer-reviewed papers, book chapters, and databases (such as Kew SID). N/A, data not available

Finally, detailed research on pollen storage biology is lacking for the vast majority of species, and empirical studies are needed to precisely determine the level of pollen and seed DT and DS (e.g., compare data between Tables 4.2 and 4.3 for *Corylus* sp.), especially for tropical species. This research needs to be extended further to explore the long-term cryogenic storage capability of genetic resources through pollen banking for plant species characterized as having EF2 and EF3 seeds but with DT pollen that is known to survive cryopreservation, e.g., *Quercus* spp., *Aesculus* spp., *Castanea* spp., *Corylus* spp., *Citrus* spp., and many tropical crops (Tables 4.3 and 4.4).

4.3 Case Study 1: The "Global Oak Pollen Bank" (GOPB)

4.3.1 Oaks (*Quercus* sp.) and the Conservation of Genetic Resources of an Important Group of Exceptional Plants

The conservation of oak (*Quercus*) genetic resources ex situ through germplasm storage is challenging, as their seeds are considered "recalcitrant," meaning that they do not tolerate the drying and low temperatures required for conventional seed storage (Ballesteros and Pritchard 2020; Pence et al. 2020). Oak seed storage today is typically short-term (one/two seasons for temperate species) and primarily for plant production purposes (Valladares et al. 2017; Ballesteros and Pritchard 2020; Pence et al. 2020). This generally involves storing acorns at cool temperatures (4 °C) without drying, i.e., at their harvest moisture content. Hence, most oak ex situ collections are based on living tree collections in botanical gardens and arboretas (Carrero et al. 2020). These collections, although very important tools for the conservation of these exceptional species, have several limitations: (1) they are directly exposed to biotic (e.g., pest) and abiotic (e.g., climatic) threats; (2) they are a relatively costly option requiring high maintenance and space; and (3) the genetic diversity conserved between and within species is low, i.e., few individuals (Li and Pritchard 2009; Pence 2011; Carrero et al. 2020; Fernández et al. 2023). Consequently, long-term collections of a wide range of genetic diversity are critical to avoid seriously jeopardizing future restoration actions. The solution for the generation of long-term germplasm banks of oaks currently relies on the cryopreservation of cells/tissues, such as shoot meristems, calli, somatic embryos, and embryonic axes (Ballesteros and Pritchard 2020; Pence et al. 2020). These high-tech approaches have positively impacted the ex situ conservation of some tree species (Corredoira et al. 2017; Li et al. 2018; Benelli et al. 2024). However, its development is still very limited for most oaks (Ballesteros and Pritchard 2020) and is generally focused on a single type of clonal tissue, largely for the conservation of elite varieties resulting from genetic improvement (Corredoira et al. 2017; Martínez et al. 2019; Vieitez et al. 2012). Furthermore, strategies such as shoot tip cryopreservation face their own challenges, ranging from difficulty in establishing in vitro tissue culture lines using material from mature/wild individuals to differences in survival between species following cryopreservation (Pence and Chaiken 2021; Winkeljohn et al. 2022). These techniques will therefore require additional research and development before they can be implemented more broadly for long-term conservation of diverse oak species.

To promote the development of oak germplasm banks and achieve long-term conservation of the greatest possible genetic diversity of each species, we suggest the integration of various cryobiotechnological approaches (Ballesteros and Pritchard 2020). This strategy implies the creation of cryobanks where the conservation of different types of germplasms of clonal origin (somatic embryos, shoot tips) and nonclonal origin (zygotic embryos, pollen) are combined (Pritchard et al. 2014; Ballesteros and Pritchard 2020). In these banks, the genetic diversity of the species is represented by a certain number of "maternal genotypes" (somatic embryos, shoot axes, and zygotic embryos) and by a large "pool" of "paternal" genotypes (i.e., pollen). Preserving many different types of germplasm in this way will help ensure the effective capture of a species' diversity and facilitate various breeding and restoration efforts in the future.

In Sect. 4.2, we explore the importance of desiccation tolerance for the establishment of pollen banks. At least four oak species produce desiccation-tolerant pollen that can be successfully stored at low temperatures, which opens up the possibility of its widespread use in oak conservation (Ballesteros and Pritchard 2020). Indeed, protocols for the collection and cryopreservation of oak and other tree pollen have been published in recent years, although these protocols are typically limited to only a few economically important species (Batos and Miljković 2023). However, empirical data on pollen tolerance to desiccation and low temperatures are limited (1% of known *Quercus* species), and how these traits might vary with species' habitat and natural history is unknown. To date, all investigated species are from temperate forests, and there are no available data on any tropical oak species (Ballesteros and Pritchard 2020). This lack of data hinders the large-scale implementation of oak pollen storage for the conservation of the genus's vast biodiversity. To our knowledge, there are currently no long-term oak pollen banks for biodiversity conservation purposes. However, if the high tolerance of oak pollen to drying and low temperatures is widespread across species and the range of habitats they occupy, pollen banks could become crucial tools for their ex situ conservation. Importantly, except for trees growing in living collections (Carrero et al. 2020), there are currently practically no germplasm collections of wild *Quercus* stored for the long term (Ballesteros and Pritchard 2020). Thus, there is no "backup" of any species of this fundamental, iconic, useful, and threatened genus. Addressing this gap is therefore the primary objective of the Global Oak Pollen Bank initiative.

4.3.2 The "Global Oak Pollen Bank" (GOPB) Initiative

Based on the conservation challenges described above, we developed the "Global Oak Pollen Bank" project, funded by the International Oak Society through the Oak Conservation and Research Funds. This initiative has two main objectives:

(i) To conduct fundamental cryobiotechnology research on the drying and freezing tolerance, as well as longevity, of oak pollen from diverse species across territories, habitats, and the oak evolutionary tree.
(ii) To create the first "Global Oak Pollen Bank" (GOPB), which will preserve ex situ the genetic diversity of a variety of endangered and key oak species in Spain and three of the most important oak hotspots (China, Mexico, and the USA).

We hypothesize that because these species are reputed to have binucleate pollen, the pollen tends to desiccate and freeze, possibly across the whole genus. However, the presumed universality of these traits, which facilitate the creation and implementation of pollen banks, is under investigation. Pollen from oaks from dry and warm (temperate) environments may be more tolerant to drying than that from oaks from more humid (tropical) environments. Species that produce recalcitrant seeds are generally concentrated in hotter, tropical climates (Pritchard et al. 2022), and it is currently unclear whether this pattern extends to pollen as well. These uncertainties and potential differences frame the hypothesis testing in the project and further underscore the urgent need for fundamental cryobiotechnology research. A practical outcome of the project will be an understanding of whether oak pollen from many species can be banked via a generalized protocol or if different species will require different specific techniques or storage conditions.

4.3.3 The GOPB Partnership

The GOPB was born as a partnership between different research and conservation institutions: the University of Valencia in Spain (coordination); the Kunming Institute of Botany of the Chinese Academy of Sciences in China; the Huntington Library, Art Museum, and Botanical Gardens in the USA; and the Center for Conservation and Research of Endangered Wildlife at the Cincinnati Zoo and Botanical Garden in the USA. The purpose of the partnership was to create a decentralized pollen bank covering a wide range of oak distributions, from the USA in the west to China in the east, with Europe located in the middle. Moreover, two of the main oak hotspots are present: the USA and China (Carrero et al. 2020). The locations of the partners could also facilitate oak pollen collection in neighboring countries that are also oak hotspots, such as Mexico or Vietnam (Carrero et al. 2020).

Since the constitution of the GOPB, diverse institutions have joined this initiative, expanding the collection and storage capacity of the GOBP. Some examples are the Pollen Bank at the Chicago Botanical Garden and the National Germplasm Bank of the Government of Spain. Currently, there is an initiative led by the Global Conservation Consortium for Oaks (GCCO) to make available through a database the extent and diversity of the oak pollen collections held at the GOBP (https://mortonarb.org/science/projects/global-conservation-consortium-for-oak-gcco/). In building this network, the GOPB fulfills one of its primary objectives of being a proof-of-concept for the widescale implementation of oak pollen banking.

4.3.4 Preliminary Results from the GOPB

We collected, investigated and banked pollen from 20 different taxa across the original GOPB Partnership. The taxa are native to diverse climates and environments, including evergreen *Sclerophyllous* species from both Mediterranean (e.g., *Q. coccifera*, *Q. ilex*, and *Q. engelmannii*) and Himalayan subalpine (e.g., *Q. longispica*) climates, deciduous (e.g., *Q. pyrenaica*, *Q. serrata*, and *Q. variabilis*) and evergreen (e.g., *Q. aliena and Q. franchetii*) species from temperate-like climates, evergreen species from subtropical biomes (e.g., *Q. myrsinifolia and Q. virginiana*), and deciduous species inhabiting dry and sandy landscapes (e.g., *Q. havardii*). An important preliminary aspect to determine was the proper maturity stage of the male flowers. Only those extended catkins with full and hanging anthers that were yellowish in color had mature pollen (Fig. 4.1a–d). When the plants were harvested at this stage (Fig. 4.1e) and placed in a dry environment (30–50% RH), the anthers naturally dehisced, and the pollen was released within 8–24 h (Fig. 4.1f).

Fig. 4.1 Conserving the genetic resources of exceptional plant species through pollen banking. (**a**–**g**) Oak pollen maturity stage and harvest: (**a**) overripen (empty) oak anthers; (**b**) green (immature) oak anthers, (**c**) mature oak anthers, (**d**) pollen released overnight from mature oak anthers; (**e**) harvest of oak catkins in the field; (**f**) drying of oak catkins in the lab for pollen harvest; (**g**) pollen harvested, sieved, dried, and stored in tubes at 4 and −20 °C. (**h**) female *Castanea dentata* flowers being pollinated with 20-year cryopreserved pollen during ART experiments. (**i**–**j**) *Amorphophallus titanium* pollen collection at the Chicago Botanic Garden. (Unpublished photographs/diagrams of authors)

Table 4.5 Oak (*Quercus* sp.) pollen collected, investigated for desiccation and freezing tolerance, and stored in the GOPB partnership

		Pollen survives[a]		
Origin	Taxon	Drying at RH:	−20 and −80 °C storage	Liquid nitrogen storage
Spain	*Q. canariensis**	15–30%	Yes	NT
	*Q. cerrioides**	15–30%	Yes	NT
	*Q. coccifera**	15–30%	Yes	NT
	*Q. faginea**	15–30%	Yes	NT
	*Q. ilex**	15–30%	Yes	NT
	Q. macrocarpa#	15–30%	Yes	NT
	*Q. pyrenaica**	15–30%	Yes	NT
	Q. polymorpha#	15–30%	Yes	NT
	*Q. suber**	15–30%	Yes	NT
	Q. virginiana#	15–30%	Yes	NT
USA	*Q. agrifolia*	15–30%	Yes	Yes
	Q. alba	30–40%	Yes	Yes
	*Q. engelmannii**	15–30%	Yes	Yes
	Q. georgiana	30–40%	Yes	Yes
	*Q. havardii**	30–40%	NT	Yes
	Q. suber#	15–30%	Yes	Yes
	Q. myrsinifolia#	15–30%	Yes	Yes
	Q. muehlenbergii	15–30%	Yes	Yes
	Q. virginiana	15–30%	Yes	Yes
China	*Q. aliena*	15–30%	Yes	Yes
	Q. franchetii	15–30%	Yes	Yes
	Q. longispica	15–30%	Yes	Yes
	Q. serrata	15–30%	Yes	Yes
	Q. variabilis	15–30%	Yes	Yes

Taxa indicated by an * are considered threatened by IUCN or are protected by national or regional regulations. Taxa indicated by a # are not native to the continents from where the pollen was harvested and were collected from individuals growing in botanical gardens

[a]Viability rates after drying (for 2–7 days) or low-temperature storage (7–365 days) similar to those of freshly harvested pollen. NT: not tested

Pollen from all the taxa studied in the GOPB project tolerated drying to <40% RH. When dried, oak pollen tolerated storage at −20 and −80 °C for up to 1 year (Table 4.5, Fig. 4.1g). In addition, pollen from the species harvested in the USA and China was also stored in liquid nitrogen (−196 °C), tolerating initial exposure (1 h to 2 days) and up to 3 months of storage. Long-term storage experiments at all temperatures tested are ongoing and will be used for comparative longevity studies. In all the cases, pollen germination after drying and subzero storage was similar to that of freshly harvested pollen.

4.3.5 Pollen Collection at the Spanish National Germplasm Bank

Oak forests in Spain are highly important in terms of size, ecological dynamics, and ecosystem services. Approximately 41% of the wooded areas are composed of *Quercus* spp. forests or pastures, with *Quercus ilex* being the species with the greatest extent (according to the "4th National Forest Inventory" of the Spanish government; MITECO 2008). In addition, there is high richness of oak species present in the national territory. Specifically, there are currently 22 accepted taxa, including species and subspecies. Therefore, the conservation of oak populations in Spain, both in situ and ex situ, is essential for the conservation of Spanish biodiversity.

The Royal Decree 159/2022, of March 1, on the conservation of forest genetic resources and wild flora, establishes the basic rules for the conservation and sustainable use of forest genetic resources of national interest and, specifically, the planning, coordination, and collaboration instruments for their in situ and ex situ conservation. Specifically, Chapter III of the Royal Decree establishes the basic rules for the ex situ conservation of forest genetic resources. For this purpose, two tools have been established to aid ex situ conservation: the National Bank of Forest and Wild Flora Germplasm (hereinafter, the National Bank) and the Network of Forest and Wild Flora Germplasm Bank (hereinafter, the Bank Network).

The Spanish National Bank is located in the Net of National Centers for Forest Genetic Resources (CNRGF Net) of the Ministry for Ecological Transition and the Demographic Challenge (MITECO, acronym for "Ministerio para la Transición Ecológica y el Reto Demográfico") and carries out technical support activities related to the management and conservation of genetic materials for the development of the National Plan for the Conservation of Forest Genetic Resources (Article 3 of the Royal Decree). The Spanish National Bank is also supported by other centers that are part of the Bank Network.

On the other hand, Chapter II of the Royal Decree159/2022 establishes the basic rules regarding the in situ conservation of forest genetic resources and, specifically, regulates the in situ genetic conservation units (UCGs, as per its Spanish acronym) and their corresponding nets. The in situ UCG Net is composed of a set of populations of the main forest taxa to enable the conservation of the maximum genetic diversity of each taxon. Currently, 327 populations of 20 priority species have been proposed that will form the preliminary net and that will be gradually expanded to complete the 62 species established by the Royal Decree. Ex situ conservation must be conceived as a complement to this in situ conservation of natural populations. Therefore, these populations should be considered a priority for ex situ conservation.

Within the current UCG Net, there are 115 populations of *Quercus* spp. of six different species: *Quercus petraea*, *Quercus ilex*, *Quercus faginea*, *Quercus suber*, *Quercus pyrenaica*, and *Quercus robur*. Additionally, there are 17 populations of *Castanea sativa*. Owing to the recalcitrant nature of the seeds of these exceptional taxa, long-term germplasm storage via conventional seed storage methods is very difficult. However, these 132 populations, which represent ca. 40% of the UCGs Net, must also be conserved ex situ, which represents a challenge for the Spanish

National Bank. Hence, there is a focus on the development and application of cryobiotechnology (Popova et al. 2016; Pritchard 2018).

Fortunately, as indicated in the above sections, there are viable alternatives for long-term germplasm conservation for these exceptional taxa, such as pollen storage. This makes the creation of a pollen bank an essential tool for the development and success of the National Bank and for advancing fundamental research in the field of germplasm conservation. Therefore, the National Bank has set up a laboratory with all the necessary facilities for pollen conservation within the CNRGF Net, in the center located in Alaquàs, Valencia.

Currently, ex situ collections of the genera *Quercus* and *Castanea* in Spain are conserved for genetic improvement or commercial seed production purposes but are not focused on genetic diversity conservation. In fact, among the collections of the 15 Spanish germplasm banks dedicated to conservation, which are currently members of the Bank Network, there are no accessions of *Quercus* spp. or *Castanea sativa* germplasm conserved ex situ for the long term. To overcome this limitation, MITECO has taken several actions:

(i) The National Bank launched the pollen bank after initial training provided by the GOPB and began collecting genetic material from *Quercus* spp. during the spring of 2024. Pollen from three species (*Q. ilex*, *Q. fagínea*, and *Q. suber*) was collected from five populations within the UCG Net and from two populations outside of the net (Table 4.6). During the spring of 2025, new collections will be made in priority UCGs throughout the national territory for both *Quercus* spp. and *Castanea sativa*.

(ii) To increase pollen storage among members of the Bank in the Network and other Spanish banks by sharing protocols and promoting workshops and courses. The creation of pollen banks in different centers will facilitate the conservation of a larger number of genotypes and populations throughout the Spanish territory, store duplicates of the accessions in different germplasm banks, and ensure the conservation of oak genetic diversity.

The expectations of the pollen bank as a tool for the conservation of exceptional species with recalcitrant seeds are high. However, owing to the recent creation of the pollen bank of the National Bank, less than a year ago, long-term results concerning the success of the conservation of pollen accessions are pending. In 2025, new pollen germination tests of the old and new accessions will be carried out to reveal progress and verify the appropriate conservation of the accessions.

4.4 Case Study 2: *Castanea dentata* Pollen Cryostored Long-Term and Used to Sire Seeds

In this case study, we discuss an example of the successful long-term storage of pollen from an exceptional and endangered plant species that produces recalcitrant seeds, *Castanea dentata*. Importantly, success was measured as the production of

Table 4.6 Accessions conserved in the pollen bank of Spanish National Germplasm Bank

ID accession	Taxa	Location (city and province)	Genotypes	ID UCG
BNA_00001_7479_P_2024	*Q. faginea*	Oropesa (Toledo)	19	ESP00119
BNA_00002_7479_P_2024	*Q. faginea*	Alcudia de Veo (Castellón)	2	–
BNA_00003_7479_P_2024	*Q. faginea*	Villatobas (Toledo)	18	ESP00107
BNA_00004_7483_P_2024	*Q. ilex*	Castejón (Cuenca)	24	ESP00140
BNA_00005_7228_P_2024	*Q. suber*	Alcorcón (Madrid)	22	–
BNA_00006_7228_P_2024	*Q. suber*	Hoyo de Manzanares (Madrid)	16	ESP00157
BNA_00007_7483_P_2024	*Q. ilex*	Alcoi (Alicante)	12	ESP00312

viable pollen after 21 years of cryogenic storage, the use of pollen through ART to produce seeds and seedlings, and the recovery of the genes of an extinct genotype from those plants.

Castanea dentata (American chestnut) is a deciduous tree that was once common and dominant in the oak-chestnut forests of the central and southern Appalachian Mountain range. However, during the early twentieth century, American chestnut trees were devastated by a fungal disease, chestnut blight, caused by the ascomycete *Cryphonectria parasitica*. By the 1950s, most large trees were killed throughout their entire range, and they are currently considered critically endangered by the IUCN (Stritch 2018).

Many initiatives have been undertaken to recover this species. Some have relied on pollen for intercrossing among American chestnuts selected for native resistance to blight (led by the American Chestnut Cooperators' Foundation) (Griffin et al. 2005), and others have relied on the backcrossing of blight-resistant Chinese chestnuts into American chestnut trees to recover American growth characteristics and genetic makeup (the American Chestnut Foundation) (Diskin et al. 2006). The latter initiative also seeks intercrossing advanced backcross generations to eliminate genes for susceptibility to blight (Diskin et al. 2006). Clearly, any genotype of wild *C. dentata* is valuable not only for its potential resistance to chestnut blight but also for providing genetic diversity to crosses.

4.4.1 "The Frozen Garden" of the Lindner Center for Conservation and Research of Endangered Wildlife's (CREW) CryoBioBank®

The Frozen Garden of CREW's CryoBioBank was initiated in 1988 to store plant tissues for long-term ex situ conservation. During the following decades, samples of pollen, short-lived seeds, zygotic embryos, somatic embryos, shoot tips, spores, and gametophytes of a wide variety of species were banked. Currently, over 280 species are represented in the collection, some with multiple genotypes, and many have been banked more than 20 years ago. Among the banked species, 25% are exceptional or are suspected to be exceptional, along with a number of ferns and bryophytes, represented by spores, gametophytes, and sporophytic tissues.

4.4.2 *Castanea dentata* Pollen Storage and Germination

Pollen samples from *Castanea dentata*, *C. pumila*, and *C. mollissima* trees were collected in June 1993 and 1994 from trees growing in the American Chestnut Foundation Meadowview Research Farms, Virginia, or from a single wild tree growing in Florence, Kentucky (Table 4.5). After being transported by road to CREW (1 day), the pollen was stored in a refrigerator for 6 days before initial viability testing. The germination of pollen ranged between 0 and 33%. Four days later,

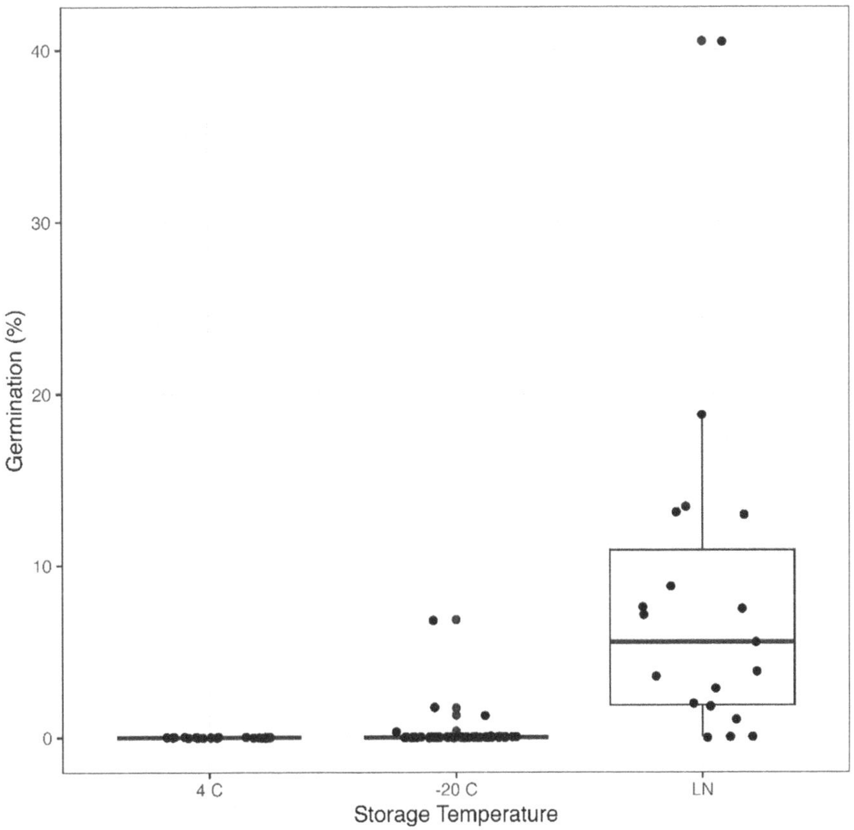

Fig. 4.2 Germination of *Castanea dentata* pollen stored at 4 °C and −20 °C and in liquid nitrogen (LN) at −196 °C for 16 years. (Unpublished photographs/diagrams of authors)

the pollen was dried for 18 hours over silica gel, placed in 2 mL cryovials, and stored at 4 °C or −20 °C or immersed in LN (−196 °C).

In 2009, 16 years after banking, 99 of these samples were evaluated for viability via the hanging drop method. Briefly, the wells on a depression slide were ringed with petroleum jelly, a known quantity of chestnut anthers was shaken in a fixed volume of 0.5% (w/v) sucrose, and a drop was pipetted onto a square coverslip. The coverslip was inverted onto the prepared depression slide, ensuring that the petroleum jelly formed a complete seal. The slides were incubated overnight at 31 °C, and a microscope was used to count germinated and ungerminated pollen grains to assess viability. The results revealed that storage temperature had a significant effect on pollen germination (ANOVA, $F2,96 = 26.39$, $p < 0.001$), and germination in the LN-stored samples was significantly greater than that in the samples stored at 4 °C or −20 °C ($p < 0.001$). Indeed, germination following storage at 4 °C or −20 °C was almost nil (Fig. 4.2).

4.4.3 ART (Pollination) Experiments and Results

In June 2014, 21 years after banking, we removed several cryovials containing *Castanea* sp. pollen from the cryobank. Pollen was transported in a LN-cooled dry shipper to TACF Research Farm at Meadowview, Virginia, and 24 hours after removal, viability was tested by germination in 0.5% sucrose via the hanging drop method. As in 2009, several samples showed germination, with rates ranging between 1% and 47% (Table 4.7).

Six of the viable pollen samples recovered from cryopreservation were used to pollinate four genotypes of *C. dentata* trees in the American Chestnut (Fig. 4.1h). At harvest in the following fall, 22 of the 24 genetic crosses produced burs, with 15 of them producing nuts. The unpollinated bagged control flowers developed 21 burs, but none of these produced nuts (Table 4.7). In total, 70 nuts were produced from the crosses, and 50 seedlings were successfully produced from those nuts (Table 4.7).

From pollen of the extinct Florence wild tree harvested in 1993 and 1994, 15 nuts in 6 out of 8 crosses were produced, but, ultimately, only 11 plants from three of the crosses survived, all from pollen harvested in 1993 (Table 4.7). These seedlings produced through ART represent the recovery of the genes of an extinct genotype of the species, as the original tree from which pollen was harvested succumbed to blight in the early 2000s.

4.4.4 Genetic Identity of the Seedlings (Paternity Tests)

DNA extraction was performed on leaf tissue from mother trees and seedlings via a modified CTAB protocol (Doyle and Doyle 1987). Ten primers developed by Kubisiak et al. (2013) for *C. mollissima* were used for the analysis (Table 4.8). Descriptive statistics were generated via GenAlEx (Peakall and Smouse 2006). Paternal parentage was inferred via Colony 2.0 (Jones and Wang 2010) by calculating the probability that two seedlings produced from pollination by the same paternal donor are full siblings when the maternal parent is known via full-pedigree likelihood methods. Among the purported full-sibling dyads, 72% (55/76) were correctly identified by the analysis, indicating a likely shared paternal donor of the cryopreserved pollen. This finding indicates that a majority of the seedlings produced from the crosses are likely the result of successful pollination via cryostored pollen. Banking pollen long-term in liquid nitrogen appears to be a viable conservation strategy for the long-term maintenance of *C. dentata.*

4.5 Case Study 3: Chicago Botanic Garden's Pollen Bank

Founded in 2023, the Chicago Botanic Garden Pollen Bank is dedicated to the conservation and recovery of rare and endangered plant species worldwide. The primary efforts since its inception have focused on collecting, storing, and preserving pollen from three key plant groups present in their collection or region, most of

Table 4.7 Pollen germination, chestnut production, and seedling growth of cryopreserved pollen and crosses between cryopreserved *Castanea* sp. pollen and *Castanea dentata* trees

			Seed parent														
	Pollen genotype/	% pollen	A1400			A1437			AD144			AD146			Total		
Taxon	code (origin)	germination[a]	B	N	S	B	N	S	B	N	S	B	N	S	B	N	S
C. dentata	Wise 1 (Florence, Kentucky, 1994)	10.8 ± 10.1	8	9	0	3	1	0	1	2	0	3	1	0	15	13	0
C. dentata	Wise 2 (Florence, Kentucky, 1993)		3	1	7	6	0	1	5	1	2	4	0	0	18	2	10
C. dentata	Test A 89-10 (TACF Meadowview farm)	1.0 ± 0.4	5	8	1	6	0	0	5	3	1	8	6	0	24	17	2
C. dentata	KY, Iowa 89-33 (B1 (W × A)) (TACF Meadowview farm)	47.6 ± 22.8	12	22	2	6	5	0	3	6	3	0	0	6	21	33	11
C. pumila	Graves 361 (TACF Meadowview farm)	13.5 ± 8.6	6	2	16	6	0	3	4	1	5	0	0	0	16	3	24
C. mollissima var. *nanking*	Clappers 90-156 (TACF Meadowview farm)	2.2 ± 4.4	4	0	2	5	0	0	5	2	1	3	0	0	17	2	3
	Control (no pollen)		12	0	0	1	0	0	4	0	0	4	0	0	21	0	0
Total															**132**	**70**	**50**

Collection information is indicated for each pollen source. All trees used in crosses were growing at the American Chestnut Foundation Research Farm at Meadowview. Pollen was germinated using the hanging drop method

[a]Pollen germination was measured in 4 different drops per genotype and the average ± SD in shown in the table; *B* burs, *N* nuts, *S* seedlings

Table 4.8 Microsatellite markers used to identify the seedlings

Marker	NA	HO
CmSI0614	4	0.31
CmSI0922	7	0.12
CmSI0028	9	0.73
CmSI0064	5	0.16
CmSI0930	5	0.04
CmSI0644	3	0.02
SmSI0293	3	0.18
CmSI0035	14	0.87
CmSI0921	8	0.55
CmSI0658	3	0.28

N_A refers to number of alleles, H_O refers to observed heterozygosity. All markers were developed by Kubisiak et al. (2013)

which are considered "exceptional species": *Amorphophallus* (EF2), *Quercus* (EF2), and orchids (seed status varies depending on species, with some being orthodox and producing thousands of seeds per capsule and others being EF1 due to the lack of pollinators or very difficult to reach owing to their rarity, EF3, or not studied). The pollen bank operates with two primary objectives: (1) safeguarding the genetic diversity of species native to the Great Lakes region and preserving exceptional species on a global scale and (2) implementing a pedigree approach to botanical conservation similar to that used in zoos (see Wood et al. 2020 for details on this approach). Our current strategies for banking pollen have focused on understanding the general tolerance of pollen to drying and freezing, where most of our germplasm is stored in a deep freezer at −80 °C.

4.5.1 Pollen Banking to Save Rare and Endangered Species

The Chicago Pollen Bank started because of the need to use a pedigree approach for ex situ conservation (Wood et al. 2020). Garden scientists realized that the zoo approach to breeding for conservation via pedigrees could be applied equally to threatened plants. Thus, knowing the maternal and paternal lines allows scientists to perform informed crosses that enable increases in the genetic diversity present in collections by introducing genes that might have been lost owing to inbreeding or simply owing to the lack of founders from certain collections compared with others. Adopting this method for botanic gardens allows the prevention or reduction of inbreeding by exchanging gametes with genes not present in the current collection. The use of pedigrees allows the tracking of the genetic makeup of the plants, an approach that is much more affordable and sustainable than performing DNA analyses to determine the genetics of every plant in collections. Breeding plants living at different institutions has various challenges. When the pollen arrives at an institution, there is no guarantee that there will be a flower available with a receptive pistil.

Banking pollen allows newly arrived pollen to be held until needed, thereby ensuring the success of breeding plants held at different institutions. Pedigree initiatives have focused on two species: *Brighamia insignis* (EF3) and *Amorphophallus titanum* (EF2).

The Chicago Botanic Garden has been storing and exchanging pollen of *Amorphophallus titanum* with other botanical institutions for conservation for several years, prior to developing a formal pollen bank (Fig. 4.1i–j). *A. titanum* has been listed on the IUCN Red List since 1998 and classified as endangered since its last assessment in 2018, with a declining population. Many botanic gardens have this species, but the genetic diversity present in collections is low, as most plants ex situ are from few collection events (Murrell et al. in press). As an exceptional "EF2" species, *A. titanum* seeds cannot be stored. Owing to the scarcity of their flowers, preserving their germplasm to ensure the availability of pollen when plants bloom is important (Fig. 4.1i–j).

One of the earliest species studied more extensively for pollen banking at the Chicago Botanic Garden, in partnership with Dr. Seana Walsh at the National Tropical Botanical Garden in Kauai'i, is *Brighamia insignis* or Alula. This species is a Hawaiian lobelioid endemic to Kauai'i and Ni′ihau and an exceptional species that is short-lived during storage, i.e., 73% of seed viability is lost after 10 years of dry storage (Chau et al. 2019) (EF3). This status makes it a great candidate for pollen banking, as it needs extra assistance to preserve its genetic diversity. The Alula was declared extinct in the wild in 2023 through the IUCN Red List of Threatened Species after the last observed specimen was lost due to a hurricane. This makes the species an important candidate for a breeding program and pollen storage efforts. Our work published in 2021 (outlined in Foster et al. 2024) compared the progeny of various informed crosses between one maternal line and four pollen donors with varied levels of inbreeding. The purpose of this study was to understand whether parental relatedness and inbreeding affect the health of seeds and plant progeny. We found that "Progeny from the wider crosses had increased genetic diversity and were larger and more fertile, while self-pollinated accessions were smaller and less fertile." The results of this study suggest that implementing pedigree-based management strategies in ex situ plant conservation can help enhance genetic diversity and overall plant health, demonstrating the value of pollen banking for ex situ collections.

4.5.2 Building and Sharing Pollen Knowledge

As the urgency to preserve biodiversity and genetic diversity grows, pollen banking remains an underutilized yet vital tool. Currently, most pollen banking efforts are concentrated in wealthier nations with well-funded conservation programs, whereas many resource-limited countries, often home to the highest levels of plant diversity, lack the infrastructure, funding, and technical expertise needed to implement these efforts effectively. Without access to pollen banking, many rare and endangered plant species remain at risk of genetic erosion, habitat loss, and climate change-driven threats.

To bridge this gap, the Chicago Botanic Garden has been actively collaborating with institutions worldwide to expand pollen banking knowledge, research, and capacity. These collaborations include hands-on training workshops, knowledge exchange programs, and direct support for developing and equipping pollen storage facilities. Through these partnerships, some organizations have successfully built their own pollen banks, enabling them to preserve and utilize genetic material for conservation and restoration projects.

4.6 Conclusion

Moving forward, scientists in botanic gardens and plant science institutions are committed to fostering global collaboration, sharing expertise, and strengthening the network needed to safeguard plant genetic diversity (Foster et al. 2024). By expanding pollen banking efforts in biodiversity hotspots and regions with limited conservation resources, this initiative will help protect species that may otherwise be lost. Additionally, increasing access to pollen banking will enhance breeding programs for rare and endangered plants in ex situ collections, ensuring their long-term survival and adaptability in the face of environmental change.

Acknowledgements We thank Fred V. Hebard, Laura L. Georgi, Leslie Leverone, and Josh Wagner from the American Chestnut Foundation Meadowview Research Farm and Theresa Culley's lab at the University of Cincinnati and The Walder Foundation of Illinois for their support.

Competing Interests The authors have no relevant financial or non-financial interests to disclose.

References

Acker JP, Adkins S, Alves A, Horna D, Toll J (2017) Feasibility study for a safety back-up cryopreservation facility. Independent expert report: July 2017. Bioversity International, Rome (Italy), p 100

Ballesteros D, Pritchard HW (2020) The cryobiotechnology of oaks: an integration of approaches for the long-term ex situ conservation of Quercus species. Forests 11(12):1281

Batos B, Miljković D (2023) Pollen cryopreservation of coniferous Serbian spruce (*Picea omorika*/Panč./Purkyne) and deciduous Pedunculate oak (*Quercus robur* L.) species. In: Pollen cryopreservation protocols. Humana, New York, NY, pp 561–574

Benelli C, Tarraf W, İzgü T, Anichini M, Faraloni C, Salvatici MC, Jouini N, Germanà MA, Danti R, Lambardi M (2024) Long-term conservation for the safeguard of *Abies nebrodensis*: an endemic and endangered species of Sicily. Plan Theory 13(12):1682

Breman E, Ballesteros D, Castillo-Lorenzo E, Cockel C, Dickie J, Faruk A, O'Donnell K, Offord CA, Pironon S, Sharrock S, Ulian T (2021) Plant diversity conservation challenges and prospects—the perspective of botanic gardens and the millennium seed Bank. Plan Theory 10(11):2371

Calonje M, Kay J, Griffith MP (2011) Propagation of cycad collections from seed: applied reproductive biology for conservation. Sibbaldia 9:79–96

Calonje M, Calonje C, Barber G, Xaba P, Lindstrom A, Agoo EM (2017) Optimizing the long-term storage and viability testing of cycad pollen. CYCADS Off Newslett IUCN SSC Cycad Specialist Group 2(1):16–18. ISSN 2473-442X

Carrero C, Jerome D, Beckman E, Byrne A, Coombes AJ, Deng M, González-Rodríguez A, Hoang VS, Khoo E, Nguyen N, Robiansyah I, Rodríguez-Correa H, Sang J, Song Y-G, Strijk JS, Sugau J, Sun WB, Valencia-Ávalos S, Westwood M (2020) The red list of oaks 2020. The Morton Arboretum, Lisle

Center for Plant Conservation (CPC) (2019) Best plant conservation practices to support species survival in the wild. Center for Plant Conservation, Escondido

Chau MM, Chambers T, Weisenberger L, Keir M, Kroessig TI, Wolkis D, Kam R, Yoshinaga AY (2019) Seed freeze sensitivity and ex situ longevity of 295 species in the native Hawaiian flora. American Journal of Botany 106: 1248-70

Corredoira E, Martínez MT, Sanjosé MC, Ballester A (2017) Conservation of hardwood forest species. In: Ahuja MR, Mohan Jain S (eds) Biodiversity and conservation of Woody plants. Springer, pp 421–453

Dinato NB, Santos IRI, Vigna BBZ, Ferreira de Paula A, Favero AP (2020) Pollen cryopreservation for plant breeding and genetic resources conservation. Cryo-Lett 41(3):115–127

Diskin M, Steiner KC, Hebard FV (2006) Recovery of American chestnut characteristics following hybridization and backcross breeding to restore blight-ravaged *Castanea dentata*. For Ecol Manag 223(1–3):439–447

Doyle JJ, Doyle JL (1987) A rapid DNA isolation procedure for small quantities of fresh leaf tissue. Phytochem Bull 19:11–15

Dulloo M, Ebert A, Dussert S, Gotor E, Astorga C, Vasquez N, Rakotomalala JJ, Rabemiafara A, Eira MTS da,Bellachew B, Omondi C, Engelmann F, Anthony F, Watts J, Quamar Z, Snook L (2009) Cost efficiency ofcryopreservation as a long-term conservation method for coffee genetic resources. Crop Science 49:2123–2138

Dutta SK, Srivastav M, Chaudhary R, Lal K, Patil P, Singh SK, Singh AK (2013) Low temperature storage of mango (*Mangifera indica* L.) pollen. Sci Hortic 161:193–197

FAO (2014) Genebank standards for plant genetic resources for food and agriculture. FAO, Rome, Itlay. ISBN 978-92-5-108262-1

Fernández A, León-Lobos P, Contreras S, Ovalle JF, Sershen, van der Walt K, Ballesteros D (2023) The potential impacts of climate change on ex situ conservation options for recalcitrant-seeded species. Front For Glob Change 6:1110431

Ferreira MS, Costa EMR, Cruz CRP, de Jesus ON, Souza FVD (2024) Strategy for establishing a pollen cryobank of wild species of *Passiflora* L. Genet Resour Crop Evol 71:4779–4792

Foster J, Ballesteros D, Fant J, Lothian R, Meyer A, Pence V, Pritchard HW, Walsh S, Walters C, Wolkis D, Havens K (2024) Pollen banking: a critical tool for conserving plant species. https://www.chicagobotanic.org/sites/default/files/2024-08/Pollen-Banking-Critical-Tool-073124.pdf. Accessed 28 Mar 2025

Franchi GG, Piotto B, Nepi M, Baskin CC, Baskin JM, Pacini E (2011) Pollen and seed desiccation tolerance in relation to degree of developmental arrest, dispersal, and survival. J Exp Bot 62(15):5267–5281

Griffin GJ, Elkins JR, McCurdy D, Griffin SL (2005) Integrated use of resistance, hypovirulence, and forest management to control blight on American chestnut. In: Steiner KC, Carlson JE (eds) In: Proc. of conf. on restoration of American chestnut to forest lands

Griffith MP, Calonje M, Meerow AW, Tut F, Kramer AT, Hird A, Magellan TM, Husby CE (2015) Can a botanic garden cycad collection capture the genetic diversity in a wild population? Int J Plant Sci 176(1):1–10

Hoekstra FA (1995) Collecting pollen for genetic resources. In: Guarino L (ed) Collecting plant genetic diversity. CABI Publishing, Wallingford, UK, pp 527–550

Hoekstra FA (2002) Pollen and spores: desiccation tolerance in pollen and the spores of lower plants and fungi. In: Black M, Pritchard HW (eds) Desiccation and survival in plants: drying without dying. CABI Publishing, Wallingford, UK, pp 185–205

Impe D, Ballesteros D, Nagel M (2022) Impact of drying and cooling rate on the survival of the desiccation-sensitive wheat pollen. Plant Cell Rep 41(2):447–461

Jones O, Wang J (2010) COLONY: a program for parentage and sibship inference from multilocus genotype data. Mol Ecol Resour 10(3):551–555

Jordano P (2010) Pollen, seeds and genes: the movement ecology of plants. Heredity 105(4):329–330

Karun A, Sajini KK, Niral V, Amarnath CH, Remya P, Rajesh MK, Samsudeen K, Jerard BA, Engelmann F (2014) Coconut (*Cocos nucifera* L.) pollen cryopreservation. Cryo-Lett 35(5):407–417

Kubisiak TL, Nelson CD, Staton ME, Zhebentyayeva T, Smith C, Olukolu BA, Fang GC, Hebard FV, Anagnostakis S, Wheeler N, Sisco PH, Abbott AG, Sederoff RR (2013) A transcriptome-based genetic map of Chinese chestnut (*Castanea mollissima*) and identification of regions of segmental homology with peach (*Prunus persica*). Tree Genet Genome 9:557–571

Li DZ, Pritchard HW (2009) The science and economics of ex situ plant conservation. Trends Plant Sci 14(11):614–621

Li JW, Ozudogru EA, Li J, Wang M-R, Bi W-L, Lambardi M, Wang Q-C (2018) Cryobiotechnology of forest trees: recent advances and future prospects. Biodivers Conserv 27(4):795–814

Marks TR, Seaton PT, Pritchard HW (2014) Desiccation tolerance, longevity and seed-siring ability of entomophilous pollen from UK native orchid species. Ann Bot 114(3):561–569

Martínez MT, San-José MDC, Arrillaga I, Cano V, Morcillo M, Cernadas MJ, Corredoira E (2019) Holm oak somatic embryogenesis: current status and future perspectives. Front Plant Sci 10:239

Mercier S (1995) The role of a pollen bank in the tree genetic improvement program in Québec (Canada). Grana 34(6):367–370

MITECO (2008) Cuarto Inventario Forestal Nacional. https://www.miteco.gob.es/es/biodiversidad/temas/inventarios-nacionales/inventario-forestal-nacional/cuarto_inventario.html. Accessed 13 Mar 2025

Nadarajan J, Benson EE, Xaba P, Harding K, Lindstrom A, Donaldson J, Seal CE, Kamoga D, Agoo EMG, Li N, King E (2018) Comparative biology of cycad pollen, seed and tissue-a plant conservation perspective. Bot Rev 84:295–314

Nagel M, Pence V, Ballesteros D, Lambardi M, Popova E, Panis B (2024) Plant cryopreservation: principles, applications, and challenges of banking plant diversity at ultralow temperatures. Ann Rev Plant Biol 75:797–824

Nebot A, Philpott VJ, Pajdo A, Ballesteros D (2021) Cryopreservation of fern spores and pollen. In: Wolkers WF, Oldenhof H (eds) Cryopreservation and freeze-drying protocols. Humana, New York, pp 623–637

Oliveira MDSPD, Maués MM, Kalume MADA (2001) Viabilidade de pólen in vivo e in vitro em genótipos de açaizeiro. Acta Bot Bras 15:27–33

Osborne R, Robbertse PJ, Claassen MI (1992) The longevity of cycad pollen in storage. South African Journal of Botany 58:250–4

Panis B, Nagel M, van Den Houwe I (2020) Challenges and prospects for the conservation of crop genetic resources in field genebanks, in in vitro collections and/or in liquid nitrogen. Plan Theory 9(12):1634

Peakall RO, Smouse PE (2006) GENALEX 6: genetic analysis in Excel. Population genetic software for teaching and research. Molecular ecology notes 6:288–95

Pence VC (2011) Evaluating costs for the in vitro propagation and preservation of endangered plants. In Vitro Cell Dev Biol Plant 47:176–187

Pence VC (2014) Tissue cryopreservation for plant conservation: potential and challenges. Int J Plant Sci 175(1):40–45

Pence VC, Bruns EB (2022) The tip of the iceberg: cryopreservation needs for meeting the challenge of exceptional plant conservation. Plan Theory 11(12):1528

Pence VC, Chaiken MF (2021) Shoot tip cryopreservation as a conservation tool for species of Quercus: effects of species and environment on recovery. Cryo-Lett 42(3):159–167

Pence VC, Ballesteros D, Walters C, Reed BM, Philpott M, Dixon KW, Pritchard HW, Culley TM, Vanhove AC (2020) Cryobiotechnologies: tools for expanding long-term ex situ conservation to all plant species. Biol Conserv 250:108736

Pence VC, Meyer A, Linsky J, Gratzfeld J, Pritchard HW, Westwood M, Bruns EB (2022) Defining exceptional species—a conceptual framework to expand and advance ex situ conservation of plant diversity beyond conventional seed banking. Biol Conserv 266:109440

Popova E, Kim HH, Saxena PK, Engelmann F, Pritchard HW (2016) Frozen beauty: The cryobiotechnology of orchid diversity. Biotechnology advances 34:380–403

Pritchard HW (2018) The rise of plant cryobiotechnology and demise of plant cryopreservation? Cryobiology 85:160–161

Pritchard HW, Moat JF, Ferraz JB, Marks TR, Camargo JLC, Nadarajan J, Ferraz ID (2014) Innovative approaches to the preservation of forest trees. Forest Ecol Manag 333:88–98

Pritchard HW, Sershen, Tsan FY, Wen B, Jaganathan GK, Calvi G, Pence VC, Mattana E, Ferraz IDK, Seal CE (2022) Chapter 19—Regeneration in recalcitrant-seeded species and risks from climate change. In: Baskin CC, Baskin JM (eds) Plant regeneration from seeds. Academic Press, pp 259–273

Rajasekharan PE, Rohini MR (eds) (2023) Pollen cryopreservation protocols. Humana/Springer, New York, p 625

Riendeau T (2023) Advancing American chestnut (*Castanea dentata*) restoration through science, GIS and partnerships. All Theses and Dissertations 444. https://dune.une.edu/theses/444. Accessed 25 Mar 2025

Society for Ecological Restoration (SER), International Network for Seed Based Restoration (INSR) and Royal Botanic Gardens Kew (REBGK). (2023) Seed Information Database (SID). https://ser-sid.org/. Accessed 26 Mar 2025

Stritch L (2018) *Castanea dentata*. The IUCN red list of threatened species 2018: e.T62004455A62004469. https://doi.org/10.2305/IUCN.UK.2018-1.RLTS.T62004455A62004469.en. Accessed 27 Feb 2025

Tandon R, Chaudhury R, Shivanna KR (2007) Cryopreservation of oil palm pollen. Curr Sci 92(2):182–183

Towill LE (2004) Pollen storage as a conservation tool. In Gerrant et al (Eds) Ex situ plant conservation: supporting species survival in the wild. Island Press, Washington, DC, pp.180–188

Valladares F, Gil P, Forner A (coord.) (2017) Bases científico-técnicas para la Estrategia estatal de infraestructura verde y de la conectividad y restauración ecológicas. Ministerio de Agricultura y Pesca, Alimentación y Medio Ambiente. Madrid, 357 pp

van der Walt K, Alderton-Moss J, Lehnebach CA (2022) Cross-pollination and pollen storage to assist conservation of *Metrosideros bartlettii* (Myrtaceae), a critically endangered tree from Aotearoa New Zealand. Pacific Conserv Biol 29(2):141–152

Vieitez AM, Corredoira E, Martínez MT, San-José MC, Sánchez C, Valladares S, Vidal N, Ballester A (2012) Application of biotechnological tools to Quercus improvement. Europ J Forest Res 131(3):519–539

Volk GM (2011) Collecting pollen for genetic resources conservation. In. In: Guarino L, Ramanatha VR, Goldberg E (eds) Collecting plantgenetic diversity: technical guidelines. Biovers Int, Rome, pp 1–10

Walters C, Pence VC (2021) The unique role of seed banking and cryobiotechnologies in plant conservation. Plants People Planet 3(1):83–91

Winkeljohn M, Pence VC, Culley TM (2022) Improving culture initiation of mature oak shoots through use of silver thiosulfate. Appl Plant Sci 10(5):e11497

Wood J, Ballou JD, Callicrate T, Fant JB, Griffith MP, Kramer AT, Lacy RC, Meyer A, Sullivan S, Traylor-Holzer K, Walsh SK (2020) Applying the zoo model to conservation of threatened exceptional plant species. Conserv Biol 34(6):1416–1425

Xu J, Li B, Liu Q, Shi Y, Peng J, Jia M, Liu Y (2014) Wide-scale pollen banking of ornamental plants through cryopreservation. Cryo-Lett 35(4):312–319

Zamani Bahramabadi E, Jonoubi P, Rezanejad F (2018) Some cytological and physiochemical features relating to non-storability of pistachio (*Pistacia vera* L.) pollen. Grana 57(6):456–463

Zhang JM, Lu XX, Xin X, Yin GK, He JJ, Huang B, Jiang D, Chen XL (2017) Cryopreservation of citrus anthers in the national crop genebank of China. In Vitro Cell Dev Biol-Plant 53:318–327

Pollen Cryobank as an Additional Strategy for Conservation of Genetic Resource: A Case Study on Papaya

5

Josimare Q. Conceição, Everton Hilo de Souza, Eva M. R. Costa, Hellen C. P. Moura, Diego F. M. Cortes, Ronilze L. Silva, Juliana L. S. Pereira, Onildo N. Jesus, Carlos A. S. Ledo, and Fernanda V. D. Souza

Abstract

Pollen conservation plays a crucial role in maintaining plant genetic resources, improving the efficiency of breeding programs, and facilitating germplasm conservation and exchange. However, the viability of papaya pollen gradually decreases during prolonged storage, underscoring the need for long-term preservation strategies, such as cryopreservation. This approach has gained increasing recognition because of its reliability and ability to ensure the integrity of preserved material. The maintenance of alleles crucial for crop improvement, the facilitation of germplasm exchange and institutional collaboration, and their application in cases of floral asynchrony are essential aspects of both conservation and breeding efforts. This chapter presents a case study on the cryopreservation of papaya pollen grains from accessions conserved in the Papaya Active Germplasm Bank at Embrapa Cassava and Fruits, located in Bahia, Brazil. This study details the methodological steps undertaken to establish a protocol capable of supporting the development of a cryobank for papaya pollen grains. The results obtained were robust, providing a solid foundation for the implementation of a cryobank to preserve pollen grains from different *Carica papaya* accessions.

J. Q. Conceição · E. H. de Souza · E. M. R. Costa · H. C. P. Moura · D. F. M. Cortes · R. L. Silva · J. L. S. Pereira
Universidade Federal do Recôncavo da Bahia, Rua Rui Barbosa, s/n, Campus de Cruz das Almas, Cruz das Almas, BA, Brazil

O. N. Jesus · C. A. S. Ledo · F. V. D. Souza (✉)
Embrapa Mandioca e Fruticultura, Rua Embrapa, s/n, Chapadinha, Cruz das Almas, BA, Brazil

Fundação Getúlio Vargas, FGV Conhecimento, Rua Jornalista Orlando Dantas, Botafogo, Rio de Janeiro, RJ, Brazil
e-mail: fernanda.souza@embrapa.br

S. Gantait, P. Chmielarz (eds.), *Conservation of Plant Genetic Resources*,
https://doi.org/10.1007/978-981-96-8331-4_5

Keywords

Carica papaya L. · Conservation · Cryopreservation · Dehydration · Liquid nitrogen · Genebank

5.1 Introduction

The narrow genetic base and susceptibility to pests and diseases are challenges that hinder the expansion of papaya farming (Santos et al. 2017; Novanda et al. 2021), requiring actions to minimize the impacts caused by these factors. The conservation of germplasms with genetic variability representative of the *Carica* genus is crucial for ensuring the sustainability of the crop and developing adaptive solutions over time, particularly in the context of climate change.

Among the preservation strategies used, cryopreservation, which involves the storage of biological material in liquid nitrogen (−196 °C), is particularly promising. It is widely used for long-term preservation of plant genetic resources (Souza et al. 2016; Wang et al. 2018; Ren et al. 2019; Guerra et al. 2020, 2021; Oliveira et al. 2021; Silva et al. 2021; Gowthami et al. 2024). In papaya research, explants such as seeds, shoot tips (Ashmore et al. 2001; Tsai et al. 2009), and other structures, as well as pollen grains, have been successfully cryopreserved (Ganeshan 1986; Shashikumar et al. 2007; Ashmore et al. 2007, 2011; Rajasekharana and Ganeshan 2019; Rajasekharan et al. 2023).

Embrapa Cassava and Fruits have field collections and a seed bank structured to conserve papaya germplasm. The Papaya Active Germplasm Bank (AGP-Papaya) has 145 accessions of four species: *Carica papaya* L., *Vasconcellea quercifolia* A.St.-Hil., *V. monoica* A.DC., and *Jaracatia spinosa* A.DC. that are preserved under field conditions and by seeds (classified as orthodox) in a cold chamber at 10 °C (Ferreira et al. 2022).

The conservation of papaya pollen grains is an alternative considered a complementary strategy for the preservation of the germplasm, since pollen grains are a useful source of several alleles of the genetic constitution of the species and are thus able to conserve genetic diversity (Volk 2011). In addition, frozen pollen grains can be used at any time of year, which can be interesting for genetic improvement regardless of seasonal climatic conditions (Souza et al. 2015; Jia et al. 2017; Silva et al. 2017; Ren et al. 2019).

The cryopreservation of pollen grains has been applied to various species with economic importance, such as ornamental plants, pineapple (Silva et al. 2017), Serbian spruce (Batos and Miljkovic 2019), bromeliads (Souza et al. 2015), apple (Wang et al. 2018), olive cultivars (Elhomosany and Sayed 2015), and litchi (Wang et al. 2015). In the case of papaya, the first report of the cryopreservation of pollen grains was by Ganeshan (1986), who preserved pollen grains of *C. papaya* and *C. cauliflora* L. This demonstrated the possibility of using freezing to preserve pollen grains of these species for long periods. Although promising results were obtained, this study was carried out with a low number of varieties and therefore does not guarantee successful application on a larger scale. In addition, the

Table 5.1 Accessions studied from the Papaya Active Germplasm Bank (*Carica papaya* L.) of Embrapa Cassava and Fruits

Accessions	Common name	Origin	Group
BGCA-003	DCG423-5	Taiwan	Formosa
BGCA-004	DCG424-4	Hawaii	Formosa
BGCA-005	Solo Linha IX	Hawaii	Solo
BGCA-008	DCG593-10	Malaysia	Formosa
BGCA-020	DCG424-4 × 439	–	Formosa
BGCA-028	DCG439	Costa Rica	Solo
BGCA-050	S7	Brazil	Solo
BGCA-060	Sunrise cross 2	Hawaii	Solo
BGCA-076	Manga-Mourão	–	Formosa
BGCA-082	Hortus gold	South Africa	Solo
BGCA-087	Waimanalo	–	Solo
BGCA-114	SEED1216	South Africa	Solo
BGCA-115	SEED1250	South Africa	Formosa
BGCA-123	Vermelho Thai	Thailand	Formosa
BGCA-129	–	Brazil	Formosa
BGCA-188	FRF1436	Brazil	Formosa
BGCA-235	JTA	Brazil	Solo

methodology for cryopreserving the authors used involves pollen grain encapsulation, which is another step in the process.

This case study presents the methodology and results obtained from the analysis of 17 accessions from the Papaya Active Germplasm Bank at Embrapa Cassava and Fruits (Embrapa Mandioca e Fruticultura) (Table 5.1). Additionally, it proposes the establishment of a cryobank on the basis of these results. Importantly, the cryopreservation of pollen grains should not replace other existing conservation methods, but it can be employed as an alternative strategy to preserve the characteristics of species of economic interest.

5.2 Experimental Procedures

To develop an efficient protocol for the cryopreservation of pollen grains, the water content is a fundamental factor. It is necessary to reduce the content of water in pollen grains to maintain their integrity during freezing to avoid the formation of ice crystals and rupture of the exine (Souza et al. 2015; Silva et al. 2017; Ren et al. 2019).

5.2.1 Pollen Grain Dehydration

To evaluate pollen grain tolerance to dehydration, anthers with pollen grains were collected from hermaphrodite flowers during anthesis. The moisture level of the anthers with pollen was determined according to the method described by Pixt(1966):

$M = [(w - d)/(w - t)] \times 100$ where M = moisture content of the anther (%), w = anther wet weight, d = anther weight after desiccation, and t = tare weight. The complete anthers were dehydrated with silica gel in a closed compartment for three time intervals (1 h, 2 h, and 4 h). Pollen viability was measured by the in vitro germination percentage of pollen grains and the length of the pollen tube. Fresh pollen grains were used as an absolute control.

5.2.2 Cryopreservation of Pollen Grains

For cryopreservation, anthers dehydrated three times, including the control, were wrapped in aluminum foil sheets, placed in cryogenic tubes (3 mL) and immersed in liquid nitrogen (LN) at −196 °C for 24 h. After removing the LN, the samples were evaluated for postfreezing pollen viability on the basis of in vitro germination and pollen tube length methods. The anthers were thawed at room temperature for a few minutes before conducting the in vitro test.

5.2.3 In Vitro Pollen Grain Germination

To evaluate pollen grain germination, the anthers with pollen grains were distributed in Petri dishes with BK culture medium (Brewbaker and Kwack 1963) with 10% sucrose and kept in a climatic chamber at a temperature of 27 ± 1 °C for 24 h. After this period, the germinated pollen grains were counted, and the pollen tube length (mm) was measured via a Leica EZ4D stereomicroscope and the ImageJ 1.47r program (Rasband 1997–2012). The germination percentage was established by counting the pollen grains that germinated and nongerminated total grains in each Petri dish. The average pollen tube length was calculated from a random sample of 20 tubes per dish, for a total of 60 tubes of each accession evaluated. Pollen grains were considered germinated when the pollen tube length was greater than or equal to the pollen grain length (Souza et al. 2015).

5.2.4 In Vivo Pollen Grain Viability and Fertility Tests

Three pollinations were performed for each papaya accession to test the in vivo viability of the cryopreserved pollen grains for 24 h after dehydration. The efficiency of the cryopreserved pollen grain method was evaluated through in vivo pollination to determine the fertilization capacity by quantifying fruit and seed formation.

The in vivo viability test was conducted using pollen grains from hermaphroditic plants as the male parent and female plants of the same accession as the female recipient (female parent), along with a cross between the accessions BGCA-082 (male) and BGCA-128 (female). One day before anthesis, floral buds from the female plants were covered with parchment paper bags. On the day of flower opening, pollinations were carried out via cryopreserved pollen grains. The anthers were thawed naturally, considering their removal from the LN, deposition in Petri dishes, and transport to the papaya active germplasm bank in the field. Tweezers were used to distribute the anthers and pollen grains on the surface of the stigma during anthesis. After pollination, the flowers were once again bagged to prevent pollination by insects, guaranteeing the formation of fruits from the crosses. During the fruit ripening period (4 to 5 months), seed germination tests were carried out, and the seeds were placed in Gerbox® boxes on Germitest® paper moistened with distilled water at a proportion of 2.5 times the weight of the dry leaf.

5.2.5 Design of the Cryobank

On the basis of these results, it was possible to outline the assembly of a papaya pollen cryobank with better guarantees of viable pollen grain conservation. The number of samples to be deposited in the cryobank must be calculated on the basis of the predicted number of viable pollen grains (Volk 2011) but also on the basis of the characteristics of the species, such as the number of pollen grains per anther.

For the papaya accessions, the number of recommended anthers (NRAs) was 60 per accession on the basis of the average pollen grain production of the various accessions. The number of anthers with viable pollen grains (AVPG) must be calculated on the basis of the pollen viability (PV), presented in Table 5.2, at a dehydration time of 2 h via the following equation:

$$\mathrm{AVPG} = \mathrm{NRA}\left(\frac{\mathrm{PV}}{100}\right)$$

Moreover, if the number of AVPGs is lower than the number recommended for cryopreservation, how many additional anthers (AddA) will be needed to reach the recommended number must be calculated. The calculation is performed as follows: AddA = NRA—AVPG.

However, to reach the recommended number, that is, the number of anthers to be cryopreserved, it is necessary to consider pollen grain viability to adjust the final value of additional anthers.

$$\text{Anthers for cryopreservation} = \left[\frac{(\mathrm{AddA} \times 100)}{\mathrm{PV}}\right] + \mathrm{NRA}$$

It is essential to consider not only the samples intended for the cryobank but also those required for viability monitoring over a 24-month period. This monitoring must be carried out at 12 and 24 months, considering that the papaya cycle can vary from 18 to 24 months (Ferreira et al. 2022). The viability test performed 24 hours after freezing determines whether a sample should be retained in the cryobank (samples with viability below 40% are excluded). The cryotubes are stored in a cryogenic box with dimensions of 52 mm in height, 133 mm in width, and 133 mm in length, which accommodates 2 mL cryovials. For viability tests, 10 anthers were

Table 5.2 In vitro germination percentage and pollen tube length of cryopreserved pollen grains of different papaya accessions (*Carica papaya* L.) after different dehydration times

Accessions	Time		
	1 h	2 h	4 h
In vitro germination (%)			
BGCA-003	17.60 aB	77.08 aA	14.73 aB
BGCA-004	16.32 aB	75.92 aA	14.76 aB
BGCA-005	11.38 aB	52.71 bA	15.37 aB
BGCA-008	12.31 aB	70.29 aA	15.29 aB
BGCA-020	10.95 aB	89.24 aA	16.40 aB
BGCA-028	15.83 aB	59.80 bA	22.72 aB
BGCA-050	11.56 aB	78.68 aA	17.65 aB
BGCA-060	13.37 aB	72.73 aA	21.05 aB
BGCA-076	14.91 aB	65.00 aA	21.90 aB
BGCA-082	10.20 aA	40.81 bA	25.60 aA
BGCA-087	12.03 aC	46.79 bA	28.14 aB
BGCA-114	15.72 aB	73.94 aA	15.25 aB
BGCA-115	12.51 aB	67.98 aA	20.65 aB
BGCA-123	11.59 aB	81.42 aA	25.49 aB
BGCA-129	13.43 aB	47.30 bA	20.43 aB
BGCA-188	13.66 aA	26.31 cA	17.37 aA
BGCA-235	12.08 aB	52.69 bA	18.77 aB
CV (%)	18.46		
Pollen tube length (mm)			
BGCA-003	0.22 bB	0.54 eA	0.12 aC
BGCA-004	0.20 cB	0.37 hA	0.13 aC
BGCA-005	0.18 cB	0.33 iA	0.08 aC
BGCA-008	0.22 bB	0.45 fA	0.08 aC
BGCA-020	0.27 aB	0.65 cA	0.09 aC
BGCA-028	0.17 cB	0.47 fA	0.08 aC
BGCA-050	0.25 aB	0.85 aA	0.07 aC
BGCA-060	0.23 bB	0.63 cA	0.08 aC
BGCA-076	0.24 aB	0.82 aA	0.08 aC
BGCA-082	0.17 cB	0.44 fA	0.08 aC
BGCA-087	0.26 aB	0.58 dA	0.11 aC
BGCA-114	0.10 dB	0.41 gA	0.10 aB
BGCA-115	0.10 dB	0.31 iA	0.09 aB
BGCA-123	0.17 cB	0.44 fA	0.13 aB
BGCA-129	0.21 bB	0.76 bA	0.09 aC
BGCA-188	0.19 cB	0.40 gA	0.10 aC
BGCA-235	0.09 dB	0.24 jA	0.10 aB
CV (%)	24.69		

used at each evaluation period, totaling 20 anthers, which were stored in working cylinders. The number of cryovials required per accession depends on the final number of anthers after necessary adjustments. We stored the anthers at 20 per cryovial to facilitate handling and increase the likelihood of hybridization at different times.

5.3 The Best Conditions for Cryopreservation of *Carica* Accessions

5.3.1 Dehydration

The dehydration for 2 hours in silica provided the best germination rates for all accessions after 24 hours in culture medium, although for some accessions, 1 hour was also efficient (Table 5.3). The germination percentage varied from 34.83% to 93.85%, with more than 70% of the accessions evaluated from this treatment having germination percentages greater than 70%, with values that varied from 34.83% to 93.85%. Exposure for 4 hours resulted in very low water contents that influenced germination rates, with significantly lower percentages (Table 5.3).

With respect to pollen tube length, among the three dehydration intervals, the highest values were reached after dehydration for 2 h. In this dehydration interval, accessions BGCA-050 (1.03 mm) and BGCA-076 (1.11 mm) presented the greatest pollen tube lengths in contrast to accession BGCA-114 (0.12 mm) after 1 h in silica (Table 5.3). The results of this variable confirmed that the time of 4 h had a negative influence on pollen tube growth, as was the case for germination, but very low values were also obtained in the other treatments, depending on the accession, indicating the influence of genotype on the response to the treatments.

Some researchers have reported that the water content of pollen grains for cryopreservation should be approximately 30%, which is similar to what we found after 2 h of dehydration. This water content is also comparable to that obtained for bromeliads (Souza et al. 2015), pineapple plants (Silva et al. 2017), and other ornamental species, such as *Rosa xanthina* Lindl., *Xanthoceras sorbifolium* Bunge, and *Wisteria sinensis* (Sims) Sweet. In these species, the water content decreases to 20.56%, 28.34%, and 26.55%, respectively (Jia et al. 2017).

The removal of water from a tissue to be cryopreserved is one of the steps that must be evaluated, so tests of dehydration and the subsequent viability of pollen grains are necessary (Alba et al. 2011; Souza et al. 2015; Zhang et al. 2017). Knowing how much water can be removed without impairing the function of the tissue after cryopreservation is fundamental for the success of the method. The great majority of species have a large quantity of intracellular water and are highly sensitive to damage from freezing (Kulus and Zalewska 2014). The formation of ice crystals due to excess water in the tissue causes cell death due to physical and mechanical injuries (Benson 2008). Souza et al. (2015), via microscopic analyses, reported that dehydration is an essential step in obtaining successful results with pollen grains. In the pollen of *Aechmea bicolor* L.B.Sm. (Bromeliaceae), those

Table 5.3 Water content, in vitro germination percentage, and pollen tube length of different accessions of *Carica papaya* L. from the Papaya Active Germplasm Bank of Embrapa Cassava and Fruits

Accessions	Times		
	1 h	2 h	4 h
Water content (%)			
BGCA-003	83.27 cA	27.49 bB	11.82 bC
BGCA-004	81.46 dA	24.54 cB	11.21 bC
BGCA-005	80.62 dA	24.53 cB	12.77 aC
BGCA-008	85.40 bA	24.26 cB	12.82 aC
BGCA-020	79.82 dA	26.82 bB	11.92 bC
BGCA-028	86.32 bA	28.37 aB	12.65 aC
BGCA-050	82.91 cA	26.87 bB	13.19 aC
BGCA-060	82.42 cA	24.23 cB	11.03 bC
BGCA-076	81.05 dA	25.92 bB	11.76 bC
BGCA-082	85.48 bA	30.76 aB	12.93 aC
BGCA-087	86.18 bA	30.47 aB	12.93 aC
BGCA-114	87.82 aA	29.91 aB	12.92 aC
BGCA-115	85.38 bA	30.55 aB	13.72 aC
BGCA-123	80.44 dA	31.95 aB	11.77 bC
BGCA-129	82.47 cA	30.69 aB	11.07 bC
BGCA-188	–	–	–
BGCA-235	–	–	–
CV(%)	2.36		
In vitro germination (%)			
BGCA-003	79.89 aA	81.15 bA	25.69 aB
BGCA-004	48.24 cA	60.48 cA	19.51 aB
BGCA-005	65.56 bA	73.46 bA	20.46 aB
BGCA-008	61.53 bB	74.62 bA	15.67 aC
BGCA-020	84.28 aB	93.85 aA	18.24 aC
BGCA-028	71.23 bA	66.83 bA	21.18 aB
BGCA-050	69.55 bB	92.44 aA	18.50 aC
BGCA-060	71.78 bA	82.32 bA	18.94 aB
BGCA-076	68.91 bB	84.59 aA	21.81 aC
BGCA-082	10.70 dB	34.83 dA	16.10 aB
BGCA-087	36.85 cA	45.36 dA	11.16 aB
BGCA-114	55.59 bA	71.19 bA	17.27 aB
BGCA-115	72.90 bA	71.42 bA	11.39 aB
BGCA-123	83.60 aA	90.67 aA	17.83 aB
BGCA-129	70.00 bA	76.97 bA	17.09 aB
BGCA-188	37.33 cB	51.63 cA	17.92 aC
BGCA-235	61.57 bA	72.79 bA	11.69 aB
CV(%)	9.85		
Pollen tube length (mm)			

(continued)

Table 5.3 (continued)

	Times		
Accessions	1 h	2 h	4 h
BGCA-003	0.47 dA	0.50 eA	0.53 bA
BGCA-004	0.23 gB	0.30 gB	0.40 cA
BGCA-005	0.85 aA	0.64 dC	0.74 aB
BGCA-008	0.51 dA	0.39 fB	0.57 bA
BGCA-020	0.80 aA	0.64 dB	0.70 aB
BGCA-028	0.35 eA	0.32 gA	0.19 eB
BGCA-050	0.25 fB	1.03 aA	0.18 eB
BGCA-060	0.68 bB	0.78 bA	0.64 aB
BGCA-076	0.76 bB	1.11 aA	0.67 aB
BGCA-082	0.28 fB	0.36 fA	0.23 dB
BGCA-087	0.55 cA	0.65 dA	0.61 aA
BGCA-114	0.12 gC	0.48 eA	0.27 dB
BGCA-115	0.18 gB	0.31 gA	0.26 dA
BGCA-123	0.27 fB	0.52 eA	0.28 dB
BGCA-129	0.59 cB	0.77 cA	0.48 bC
BGCA-188	0.37 eA	0.29 gA	0.27 dA
BGCA-235	0.19 gC	0.32 gB	0.65 aA
CV (%)	32.12		

Means followed by the same lowercase letters in the columns and uppercase in the rows do not differ by the Scott-Knott test ($p \leq 0.05$) and Tukey test ($p \leq 0.05$), respectively

authors reported that without the dehydration step, storage in liquid nitrogen caused the formation of ice crystals and consequent degradation of the cytoplasmic content and rupture of the exine layer.

The ideal dehydration time varies according to the method, species, and structure of the tissue or organ. For the cryopreservation of pineapple pollen grains, Silva et al. (2017) reported that 6 h on silica gel was necessary to reach ideal conditions. In turn, for *A. bicolor*, 3 h on silica gel was sufficient (Souza et al. 2015). These marked differences between species of the same family, as in the examples above, may be related to the size and morphology of the pollen (Souza et al. 2015). In papaya, Ferreira (2018) did not find significant variations in the size or quantity of pollen grains per anther among the 10 accessions evaluated.

5.3.2 Freezing and In Vitro Viability

The cryopreservation results revealed that the pollen grains of the 17 papaya accessions, under the established conditions, tolerated freezing if adequately dehydrated. In vitro germination percentages were obtained after 24 hours in liquid nitrogen (−196 °C) (Table 5.3; Fig. 5.1a–d) following a 2-hour dehydration period, confirming the results of the previous assay.

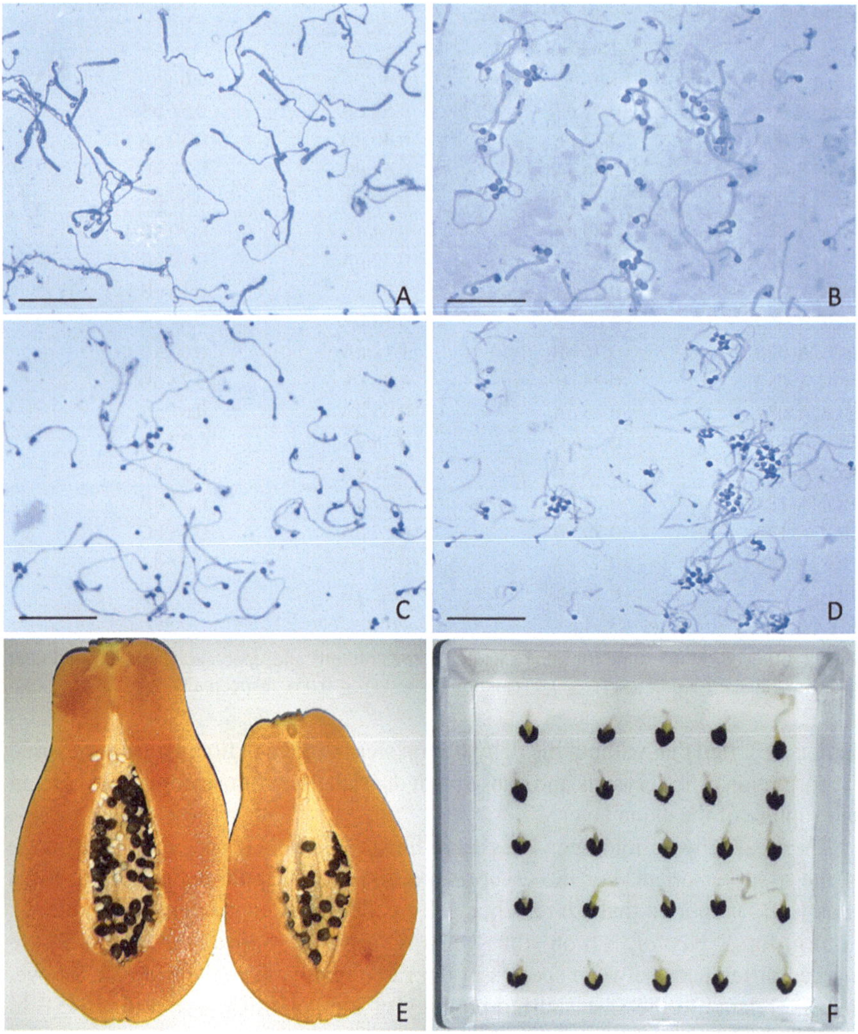

Fig. 5.1 In vitro germination of pollen grains in BK culture medium (Brewbaker and Kwack 1963) with 10% sucrose of accessions BGCA-060 (**a**), BGCA-115 (**b**), BGCA-003 (**c**), and BGCA-114 (**d**) belonging to the Papaya Active Germplasm Bank of Embrapa Cassava and Fruits. (**e**) Fruits with seeds from pollinations with cryopreserved pollen grains of accessions BGCA-082 (♂) x BGCA-128 (♀). (**f**) Germination of seeds from crossing with cryopreserved pollen grains of accessions BGCA-082 (♂) × BGCA-128 (♀). (Source: Authors)

The values observed in the treatments for 1 h and 4 h revealed that both excess water (dehydration for 1 h) and a lack of water in the tissues (dehydration for 4 h) had significant negative effects on the pollen grain viability of all the accessions evaluated. In both treatments, the low germination values obtained for all the accessions clearly indicated that these two dehydration times are not advisable prior to freezing. A particularly noteworthy difference occurred in the treatment for 1 h, which had values far below those observed in the dehydration test (Table 5.2), indicating how strongly the presence of water can damage the tissue via the formation of ice crystals. On the other hand, excess dehydration can hamper the physiological activity of pollen grains by damaging the cells and reducing their viability (Benson 2008), as was the case after treatment for 4 h.

Pollen tube length is a complementary parameter for assessing the behavior of pollen grains after a specific treatment. The results for this variable were similar to those for the three dehydration times. Compared with those of the other two dehydration intervals, the pollen grains of all the accessions dehydrated for 2 h produced significantly longer tubes (Table 5.3). However, great variation in the behavior of the accessions was observed among the pollen grains treated for 2 h, with nine groups separated by significant differences, revealing a strong effect on the accessions in this respect. The longest pollen tubes were produced by accessions BGCA-050 (0.85 mm) and BGCA-076 (0.82 mm), whereas the shortest length was produced by BGCA-235 (0.24 mm). Ren et al. (2019) reported that this difference between accessions may be the result of variations in the water content of each material after dehydration.

In vitro pollen viability has often been used as a measure to support various types of research, ranging from conservation of pollen grains (Soares et al. 2015; Souza et al. 2015; Ahmed et al. 2017; Silva et al. 2017) to investigations of the phenology of flowering and genetic improvement (Salvatierra-Gozález and Jana-Ayala 2016; Aseef et al. 2021). These studies indicate that this viability depends on a series of factors, such as the components of the culture medium, the conditions of the donor plant, and the edaphoclimatic conditions for field cultivation. In studies of the cryopreservation of pollen grains, it is necessary to go beyond this metric, since without evidence of in vivo fertilization, the results obtained may not be conclusive or can even lead to mistaken conclusions, hampering the determination of the best conditions for conservation.

Researchers have successfully cryopreserved pollen grains from various economically important species. Silva et al. (2017) reported that after 4 months, pineapple pollen grains maintained viability greater than 40% for most of the accessions analyzed. Souza et al. (2015) reported that 87% of the pollen grains of *A. bicolor* remained viable for one year. Batos and Miljkovic (2019) reported that after conservation in LN for 14 years, 40% of the pollen grains of *Picea omorika* (Pančić) Purk. remained viable. Elhomosany and Sayed (2015) cryopreserved the pollen of *Olea europaea* L. and obtained a viability of 91.87% after storage for 3 months. Araújo et al. (2020) reported a similar result in species of date palm, where the pollen grains remained stable, with satisfactory results, for 9 months.

The pollen tube acts as a conduit to carry the male gamete cells from the pollen grain to the ovules contained in the ovary, indicating the importance of its length for fecundation efficiency. While the in vitro behavior of this structure alone does not permit conclusive inference about the success of hybridization, it can act as an indicator of how much the pollen grain is affected by freezing in LN.

In vitro germination, although occurring in a controlled experimental setting, does not completely reproduce the growth of the pollen tube in vivo because there are interactions between the composition of the culture medium and the different plant tissues. Nevertheless, according to Soares et al. (2008), the in vitro germination technique produces results that can approximate growth during in vivo germination, thus demonstrating the importance of specifying adequate conditions for each species.

5.4 In Vivo Viability and Fertility Test

Pollinations carried out in vivo with cryopreserved pollen grains after a 2-h dehydration period were successful in all eight combinations, resulting in the production of fruits with viable seeds. Approximately 80% of the seeds germinated in the different combinations (Fig. 5.1e, f). Seed formation and viability are critical factors in assessing the efficiency of a protocol and its applicability for genetic breeding. These findings confirm that the conservation strategy preserved the viability and integrity of the pollen grains.

Similar results were reported by Souza et al. (2015) in bromeliad species, with fruit production in 100% of the pollinated flowers and 88% of the seeds germinating. Silva et al. (2017), studying pineapple species, reported a 75% success rate in pollinating with viable seeds. Vendrame et al. (2008), investigating orchid species, reported an 80% success rate in pollinations resulting in viable seeds. Dinato et al. (2018) conducted pollinations with *Paspalum notatum*, achieving successful hybridizations and confirming that this method is a reliable approach for assessing viability while also addressing challenges such as asynchronous flowering in genetic improvement programs.

5.5 Cryobank Design

The results of this study were obtained through the evaluation of 17 accessions of the *Carica* genus maintained in a germplasm bank associated with a genetic breeding program. Using the data presented in Table 5.3 and considering 60 anthers per accession for cryopreservation, the number of anthers with viable pollen grains was calculated, and the need for any additional anthers was determined when necessary. Figure 5.2 illustrates the steps for establishing the cryobank, considering the accessions studied in this work. Researchers can use this model to establish a comprehensive bank, and it applies to other species as well. The minimum viability rate required for introducing an accession into cryopreservation largely depends on the species and even the variety; however, this study did not consider viability rates below 40%.

To determine the appropriate number of anthers for cryopreservation, it is important to consider several factors, including pollen viability, the number of pollen grains per flower, the number of flowers per plant, the occurrence of floral

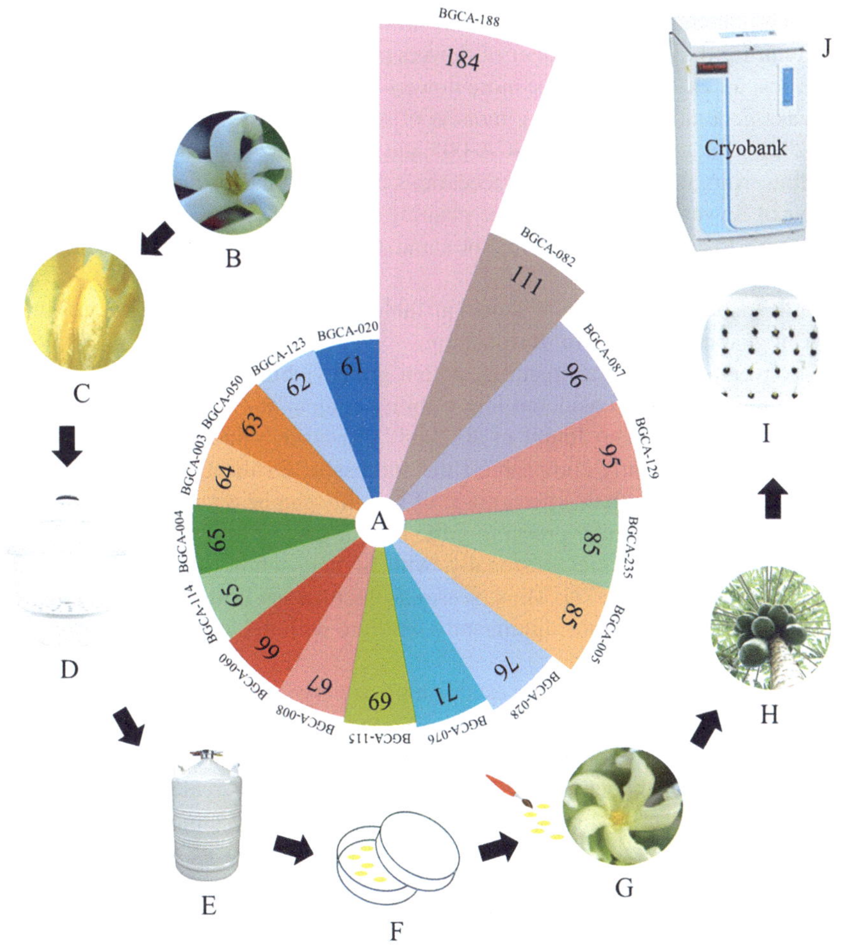

Fig. 5.2 Sequential scheme to validate the viability of cryopreserved pollen grains from *Carica papaya* L. and the establishment of a cryobank as a backup. (**a**) Number of anthers required for cryopreservation in different accessions. (**b**) Hermaphrodite flowers in anthesis. (**c**) Collection of pollen grain. (**d**) Desiccation of pollen grains in activated silica gel for 2 h; (**e**) Immersion of desiccated anthers in liquid nitrogen; (**f**) In vitro viability test of pollen grains; (**g**) In vivo viability test of pollen grains; (**h**) Fruit formation after pollination with cryopreserved pollen grains; (**i**) Germination of seeds from fruits generate by the fertilization with pollen cryopreserved. (**j**) Establishment of the cryobank. (Diagram prepared by authors)

anomalies, the fruiting of the accessions, and the infrastructure available for cryopreservation (Ferreira et al. 2022). In this study, 60 anthers, a number derived from both the average number of anthers per flower and the quantity generally used in cross-pollinations, were used. On average, a hermaphroditic flower produces 10 anthers, which are deposited on the stigmas of female flowers for hybrid seed production. However, it is important to pollinate multiple flowers to ensure successful seed production. It is also crucial to avoid collecting flowers with anomalies and to account for the average number of flowers produced per genotype. Some papaya genotypes, for example, produce more flowers than other genotypes do. The flowering peaks generally occur during months of relatively high precipitation (Ferreira et al. 2022). For example, the BGCA-005 and BGCA-082 accessions exhibit moderate flowering, whereas the other accessions evaluated in this study produce a low number of flowers throughout all the planting months (Ferreira et al. 2022). If it is not possible to collect 60 flowers on the same day, the collection can be divided into batches.

For the papaya accessions, as shown in Table 5.4, achieving a viability rate above 40% can guarantee successful fertilization.

In conventional papaya breeding, pollen grain viability is considered high at 80%, with 50% being considered low viability (Bajpai and Singh 2006) and 70% intermediate (Damasceno Junior et al. 2008). However, values between 50% and 80% do not significantly compromise the success of fertilization, and the pollen grains of genotypes with this percentage of viability can be cryopreserved for conservation purposes, thereby preventing the loss of alleles. Importantly, in papaya, pollen grain viability varies seasonally, with higher viability in summer than in winter (Ferreira et al. 2022). Thus, seasonal fluctuations in the gametic viability percentage of papaya do not significantly interfere with the fertilization process of cryopreserved pollen grains.

The availability of flowers for establishing the cryobank follows the flowering dynamics of the papaya active germplasm bank, which has already been evaluated for the 119 accessions conserved in the field (Ferreira et al. 2022). A map of the seasonal flowering frequency of the papaya accessions was created (variation and distribution of flowering in a growing cycle), and the total number of flowers in anthesis during 1 year of evaluation was recorded. These data are essential for synchronization with the establishment of the cryobank. In addition to being important for the allelic conservation of papaya, pollen cryopreservation also opens up the possibility of its use in genetic breeding programs aimed at obtaining intra- or intergeneric hybrids, especially if the genotypes are asynchronous in flowering.

Table 5.4 Number of pollinated flowers, fruits formed, and average number of seeds generated by *Carica papaya* L. with cryopreserved pollen grains after dehydration for 2 h

Accessions	Number of pollinated flowers	Fruits formed	Average number of seeds per fruit
BGCA-005	5	3	854
BGCA-008	4	4	786
BGCA-020	5	4	948
BGCA-028	5	3	342
BGCA-050	2	2	231
BGCA-060	5	3	230
BGCA-188	4	4	210
BGCA-082 × BGCA-128	1	1	489

5.6 Conclusions and Final Considerations

The cryopreservation of pollen grains can be a promising complementary strategy for the conservation of alleles of papaya germplasm. The ideal dehydration time for adjusting the water content of the pollen grains was 2 h. The in vivo pollinations successfully germinated the pollen grains, indicating that the protocol can be adopted to form a papaya cryobank. This technique is important for genetic breeding, especially for overcoming drawbacks such as asynchronous flowering.

This protocol is currently being implemented to establish the *Carica* pollen cryobank at Embrapa Cassava and Fruits, which conserves 145 accessions in the field through seed preservation. The increasing variability in climate conditions has increased the risks associated with field conservation, potentially compromising both germplasm preservation and genetic improvement efforts for the crop.

Acknowledgments The authors would like to thank Conselho Nacional de Desenvolvimento Científico e Tecnológico (FVDS: 307128/2021-0), Coordenação de Aperfeiçoamento de Pessoal de Nível Superior (CAPES—Finance Code 001), Fundação de Amparo à Pesquisa do Estado da Bahia (FAPESB), Embrapa Mandioca e Fruticultura, and Universidade Federal do Recôncavo da Bahia.

Competing Interests The authors have no relevant financial or non-financial interests to disclose.

References

Ahmed S, Rattanpal HS, Ahmad E, Singh G (2017) Influence of storage duration and storage temperature on in vitro pollen germination of *Citrus* species. Int J Curr Microbiol Appl Sci 6:892–902

Alba V, Bisignano V, Alba E, Stradis A, Polignano GB (2011) Effects of cryopreservation on germinability of olive (*Olea europaea* L.) pollen. Genet Resour Crop Evol 58:977–982

Araújo AC, Lédo AS, Polek M, Krueger R, Shepherd A, Volk GM (2020) Optimization of in vitro germination and cryopreservation conditions for preserving data palm pollen in the USDA national plant germplasm system. Plant Cell Tissue Organ Cult 144:223–232

Aseef RM, Soorianathasundaram K, Paramaguru P, Muthulakshmi P, Joel AJ, Chandrasekhar CN (2021) An analysis of pollen parameters, fruit set and fruit development during attempts on intergeneric hybridization of papaya (*Carica papaya* L.). Curr J Appl Sci Technol 40:74–86

Ashmore SE, Azimi M, Drew RA (2001) Cryopreservation trials *in Carica papaya*. Acta Hort 560:117–120

Ashmore SE, Drew RA, Azimi M (2007) Vitrification-based shoot tip cryopreservation of *Carica papaya* and a wild relative Vasconcellea pubescens. Aust J Bot 55:541–547

Ashmore SE, Drew RA, Kaity A (2011) Storage stability using cryopreservation: a case study in *Papaya*. Acta Hort 918:125–130

Bajpai A, Singh AK (2006) Meiotic behavior of *Carica papaya* L.; spontaneous chromosome instability and elimination in important cvs. In North Indian conditions. Cytologia 71:131–136

Batos B, Miljkovic D (2019) The vitality of the serbian spruce (*Picea omorika*) pollen during the long-term cryopreservation. Grana 58:433–446

Benson EE (2008) Cryopreservation theory. In: Reed BM (ed) Plant cryopreservation. A practical guide. Springer, New York, pp 15–32

Brewbaker JL, Kwack BH (1963) The essential role of calcium ion pollen germination and pollen tube growth. Am J Bot 50:859–865

Damasceno Junior PC, Pereira TNS, Pereira MG, Silva FF (2008) Conservação de grão de pólen de mamoeiro. Rev Ceres 55:433–438

Dinato NB, Santos IRI, Leonardecz E, Burson BL, Quarin CL, Paula F, Favero AP (2018) Storage of Bahiagrass pollen at diferent temperatures. Crop Sci 58:2391–2398

Elhomosany AA, Sayed HAE (2015) Effect of low temperature and cryopreservation on in vitro pollen germination of some Olive cultivars. Am Eurasian J Agric Environ Sci 15:1803–1808

Ferreira JAB (2018) Aspectos da biologia floral e reprodutiva de diferentes acessos de mamoeiro. Dissertation. Universidade Federal do Recôncavo da Bahia

Ferreira JAB, Souza FVD, Moura HCP, Souza EH, Cortes DFM, Conceicao JQ, Lêdo CAS (2022) Flowering map of papaya germplasm: support for genetic breeding and conservation programs. Sci Hortic 293:110699

Ganeshan S (1986) Criogenic preservation of papaya pollen. Sci Hortic 28:65–70

Gowthami R, Martinez-Montero ME, Rajasekharan PE (2024) Cryoconservation of Antiallergic medicinal plants: techniques and applications. In: Gantait S, Majumder J, Sharangi AB (eds) Biotechnology of medicinal plants with Antiallergy properties. Springer, Singapore, pp 611–634

Guerra PA, Souza EH, Andrade EC, Max DAS, Oliveira RS, Souza FVD (2020) Comparison of shoot tip culture and cryotherapy for eradication of ampeloviruses associated with *Pineapple mealybug wilt* in wild varieties. In Vitro Cell Dev Biol Plant 56:903–910

Guerra PA, Souza EH, Max DAS, Rossi ML, Ledo CAS, Villalobos-Olivera A, Martinez-Montero ME, Souza FVD (2021) Morphoanatomical aspects of the starting material for the improvement of pineapple cryopreservation by the droplet-vitrification technique. An Acad Bras Ciênc 93:e20190555

Jia MX, Shi Y, Di W, Jiang X, Xu J, Liu Y (2017) ROS-induced oxidative stress is closely related to pollen deterioration following cryopreservation. In Vitro Cell Dev Biol Plant 53:433–439

Kulus D, Zalewska M (2014) Cryopreservation as a tool used in long-term storage of ornamental species—a review. Sci Hortic 168:88–107

Novanda RR, Sutrawati M, Ganefianti DW (2021) Analysis of the risk of profit loss in papaya farming affected by yellow mosaic disease. J La Lifesci 2:1–7

Oliveira RS, Souza FVD, Souza SO, Aona LYS, Souza EH (2021) Cryopreservation and low-temperature storage of seeds of *Tillandsia* species (Bromeliaceae) with ornamental potential. 3 Biotech 186:1–16

Pixton SW (1966) Moisture content-its significance and measurement in stored products. J Stored Prod Res 3:35–47

Rajasekharan PE, Ganeshan S, Vishwakarma PK, Vasugi C (2023) Cryopreservation of pollen grains of *Carica papaya* and different *Vasconcellea* species. In: Rajasekharan P, Rohini M (eds) Pollen cryopreservation protocols, Springer protocols handbooks. Humana, New York, pp 85–98

Rajasekharana PE, Ganeshan S (2019) Current perspectives on pollen cryopreservation in horticultural species. Acta Hortic 1234:47–56

Rasband WS (1997–2012) Image J. US National Institutes of Health, Bethesda. http://imagej.nih.gov/ij/

Ren R, Li Z, Li B, Xu J, Jiang X, Liu Y, Zhang K (2019) Changes of pollen viability of ornamental plants after long-term preservation in a cryopreservation pollen bank. Cryobiology 89:14–20

Salvatierra-Gozález MA, Jana-Ayala C (2016) Floral expression and pollen germination ability in productive mountain papaya (*Vasconcellea pubescens* A.DC.) orchads. Chil J Agric Res 76:136–142

Santos PHD, Silveira SF, Vivas M, Carvalho BM, Pereira MG (2017) Alternative control to black spot disease in papaya genotypes. Summa Phytopathol 43:60–62

Shashikumar S, Ganeshan S, Tejavathi DH (2007) Fertilizing ability of cryopreserved pollen in inter-intra specific crosses in *Carica papaya* L. Acta Hort 740:73–78

Silva RL, Souza EH, Vieira LJ, Pelacani CR, Souza FVD (2017) Cryopreservation of pollen of wild pineapple acessions. Sci Hortic 219:326–334

Silva SSS, Souza EH, Souza FVD, Max D, Rossi ML, Costa MAPC (2021) Post-seminal development and cryopreservation of endemic and endangered bromeliads. An Acad Bras Ciênc 93:e20191133

Soares TL, Silva SO, Costa MAPC, Santos-Serejo JA, Souza AS, Lino LSM, Souza EH, Jesus ON (2008) Germinação *in vitro* e viabilidade de grãos de pólen de diplóides de bananeira. Crop Breed Appl Biotechnol 8:111–118

Soares TL, Souza EH, Carvalho MAP, Oliveira S, Santos Serejo JA (2015) Viability of pollen grains of tetraploide banana. Bragantia 75:145–151

Souza EH, Souza FVD, Rossi ML, Brancalleão N, Ledo CAS, Maertinelli AP (2015) Viability, storage and ultrastructure analysis of *Aechmea bicolor* (Bromeliaceae) pollen grains, an endemic species to the Atlantic forest. Euphytica 204:13–28

Souza FVD, Kaya E, Vieira LJ, Souza EH, Amorim VBO, Skogerboe D, Matsumoto T, Alves AAC, Ledo CAS, Jenderek MM (2016) Droplet-vitrification and morphohistological studies of cryopreserved shoot tips of cultivated and wild pineapple genotypes. Plant Cell Tissue Organ Cult 124:351–360

Tsai S-F, Yeh S-D, Chan C-F, Liaw S-I (2009) High-efficiency vitrification protocols for cryopreservation of in vitro grown shoot tips of transgenic papaya lines. Plant Cell Tissue Organ Cult 98:157–164

Vendrame W, Carvalho VS, Dias JMM, Maguire I (2008) Pollination of dendrobium hybrids using cryopreserved pollen. HortSciense 43:264–267

Volk GM (2011) Collecting pollen for genetic resources conservation. In: Guarino L, Ramanatha VR, Goldberg E (eds) Collecting plant genetic diversity: technical guidelines. Biovers Int, Rome, pp 1–10

Wang L, Wu J, Chen J, Fu D, Zhang C, Cai C, Ou L (2015) A simple pollen collection, dehydration, and long-term storage method for litchi (*Litchi chinensis* Sonn.). Sci Hortic 188:78–83

Wang MR, Longo C, Silva JAT, Volk GM, Wang QC (2018) Cryobiotechnology of apple (*Malus* spp.): development, progress and future prospects. Plant Cell Rep 37:689–709

Zhang JM, Lu XX, Xin X, Yin GK, He JJ, Huang B, Jiang D, Chen XL (2017) Cryopreservation of citrus anthers in the national crop genebank of China. In Vitro Cell Dev Biol Plant 53:318–327

Cryopreservation of Embryos and Embryogenic Samples to Conserve Plant Genetic Resources

6

Teresa Hazubska-Przybył, Paulina Kosek, Urszula Wasileńczyk, Małgorzata Pałucka, Mikołaj Krzysztof Wawrzyniak, Sakari Välimäki, Saila Varis, and Tuija Aronen

Abstract

Cryopreservation is a practical and effective technique for the long-term (>5 years) storage of plant genetic resources at the ultralow temperature of liquid nitrogen (−196 °C; LN). To date, cryopreservation protocols have been developed for various organs and tissues of many plant species, including those from the earliest stages of plant development, i.e., zygotic embryos, embryogenic axes, embryogenic tissues, and somatic embryos. Since zygotic embryos and somatic tissues originate from different sources (ex vitro vs. in vitro), distinct approaches are needed when cryopreservation protocols are developed, depending on their physiological and anatomical characteristics. Effective procedures have been developed for this type of plant material via a wide range of techniques, such as desiccation, controlled slow cooling, vitrification, encapsulation−vitrification, droplet−vitrification, and the cryo−plate method. Currently, cryopreservation is mainly used for storing germplasms of nonwoody plants and a restricted number of tree species. The main limitations of this technique include the need to modify procedures to adapt them to individual plant species and even specific genotypes, as well as the need for specialized facilities and expertise. Therefore, special efforts should be directed toward developing cryopreservation protocols for both in vivo-sourced explants and in vitro cultures. Advancing climate change is increasing the urgency of securing biodiversity through more

T. Hazubska-Przybył (✉) · M. K. Wawrzyniak
Institute of Dendrology, Polish Academy of Sciences, Kórnik, Poland
e-mail: hazubska@man.poznan.pl

P. Kosek · U. Wasileńczyk · M. Pałucka
Kostrzyca Forest Gene Bank, Miłków, Poland

S. Välimäki · S. Varis · T. Aronen
Natural Resources Institute Finland (LUKE), Savonlinna, Finland

S. Gantait, P. Chmielarz (eds.), *Conservation of Plant Genetic Resources*,
https://doi.org/10.1007/978-981-96-8331-4_6

efficient biotechnological techniques. Moreover, advances in micropropagation techniques, when combined with cryopreservation, are creating new opportunities to protect valuable plant genotypes.

Keywords

Cryostorage · Controlled slow cooling · Embryogenic tissue · Somatic embryos · In vitro culture · Liquid nitrogen · Zygotic embryos · Vitrification

6.1 Long-Term Storage in Liquid Nitrogen

The human species is inextricably linked to the world of plants, which are a source of oxygen, food, medicines, aesthetic sensations, etc. Today, however, owing to factors such as pollution, overexploitation, invasive species intrusion, habitat loss, and fragmentation, plant diversity is in serious decline (Lasushe et al. 2022). For these reasons, as well as the increasing threat posed by climate change, many plants need special protection via available tools. Over the past nearly 80 years, there have been international efforts to protect global plant biodiversity (Murray 2017; Lasushe et al. 2022; Convention on Biological Diversity 1992). Biodiversity conservation relies on two basic strategies: in situ and ex situ conservation. Ex situ conservation involves protecting species or genetic material outside their natural habitat, whereas in situ conservation focuses on protecting ecosystems and natural habitats and maintaining or restoring species populations (Engelmann 2012). To achieve conservation goals, both conventional methods and more advanced biotechnological solutions are used, such as maintaining in vitro cultures through repeated subcultures, slow growth in cold storage, or cryopreservation in liquid nitrogen (LN) (Murray 2017; Libiad et al. 2020; Mikuła et al. 2022).

Cryopreservation is one of the most commonly used biotechnological methods for plant germplasm conservation, alongside in vitro culture techniques (Ruta et al. 2020; Benelli 2021; Mikuła et al. 2022). This method is particularly applicable for preserving the seed viability of forest-forming species, especially biocenotic, irregularly seed-producing, endangered, or declining species (Mikuła et al. 2013). The genetic resources of plants that rarely or do not propagate generatively and are propagated vegetatively can be stored in LN. Cryopreservation is also useful for plants that produce seeds that are intolerant to dehydration (recalcitrant seeds), such as certain trees (chestnut, oak), economically important tropical plants (coffee, rubber), and aquatic plants (Yong et al. 2023; Wasileńczyk et al. 2024). Using cryopreservation, valuable plant varieties can be securely stored (Mikuła et al. 2013; Kulus et al. 2025). Cryopreservation enables the long-term, safe storage of germplasm using minimal space and plant material (Sakai 2004). During cryopreservation, cellular metabolism is halted, preventing plant material (e.g., embryogenic tissue) from aging (Sánchez Fernández et al. 2008; Hazubska-Przybył et al. 2013; Benelli 2021; Mosa et al. 2023). In combination with an in vitro propagation system via somatic embryogenesis, it helps prevent somaclonal variation and the loss of embryogenic potential in tissues maintained on proliferation media (Corredoira et al. 2004; Hazubska-Przybył et al. 2020). The key principle behind cryopreservation is to prevent cell damage by lowering the water content to a safe level and then slowing down cellular metabolic activity to a rate that allows survival at the liquid nitrogen temperature (−196 °C). For cryopreservation to be practical, protocols must be reliable and applicable to a wide range of genotypes and must guarantee the genetic stability of the stored plant material and its ability to regenerate after tissue recovery from LN. Cryopreservation is increasingly the method of choice for ex situ genetic diversity conservation because of its many advantages, such as the following:

Long-term storage and protection against threats Cryopreservation allows plant genetic material to be stored for a very long time, potentially for hundreds of years. This ensures the preservation of valuable plant varieties and species for future generations, even if they become extinct in their natural environment (due to small population sizes, urban development threatens the habitat, or extreme sensitivity to environmental changes).

Preservation of genetic diversity Cryopreservation enables the conservation of a broad genetic spectrum of plants, including both cultivated varieties and wild species. This is crucial for maintaining biodiversity and the gene pool, which can be used in the future to breed new varieties resistant to diseases or extreme weather conditions.

Space efficiency Storing genetic material in cryopreserved form requires significantly less space than traditional plant conservation methods, such as botanical gardens. This allows for the accumulation of more genetic material and reduces storage costs.

Low maintenance costs While the cryopreservation process itself requires specialized equipment and expertise, the long-term storage costs of samples in liquid nitrogen are relatively low. There is no need for watering, fertilization, or pest control—only liquid nitrogen is needed, which is ideally supplied to cryogenic tanks via an automatic pipeline.

Convenient transport Cryopreserved genetic material is easier to transport than live materials. It can be shipped across countries and continents without the risk of damage or loss, significantly facilitating international cooperation in the conservation of plant genetic diversity.

Scientific research Cryopreservation supports scientific research on plant genetic material, including comparative analyses, gene identification for valuable traits, DNA barcoding, and the development of new breeding techniques. This contributes to a better understanding and utilization of plant genetic potential.

Safe storage of genetically modified plant material Transformed plant material can be stored in an unaltered state for a long time, enabling field trials of genetically modified plants at any desired time.

A practical biotechnology tool Cryopreservation facilitates the implementation of new biotechnological solutions in forestry, such as the establishment of commercial multivarietal forestry (MVF) plantations on the basis of plant material produced by somatic embryogenesis.

Finally, in cases where the establishment of a nature reserve or other form of in situ protection is impossible due to factors such as a lack of space, conflicts with local communities, or other constraints, cryopreservation serves as an alternative means of protecting biodiversity. However, certain limitations of cryopreservation must be considered. Not all repositories have the financial resources to establish a cryobank equipped with cryogenic tanks for liquid nitrogen (−196 °C) and nitrogen vapor (approximately −170 °C), connected to an automated supply pipeline. Additionally, not all explants are suitable for cryogenic freezing (e.g., recalcitrant seeds), and for many plant species, their ultimate sensitivity to dehydration or low temperatures remains unknown.

6.2 Cryopreservation Techniques for Plant Material

6.2.1 Overview

The cryopreservation process involves the selection of suitable tissues, dehydration, cooling, storage, thawing, and plant regeneration. Sakai's (1956) initial research highlighted the importance of tissue dehydration and precooling. Currently, a particular challenge is establishing cryopreservation protocols for in vitro-cultured cells and tissues, which are especially susceptible to damage from ice crystal formation because of their high water content. The development of a two-step protocol using vitrification and/or dehydration solutions to remove water before slow cooling has proven effective for disorganized tissues such as calli and suspension cells and has been successfully applied to the storage of cold-tolerant species (Gupta et al. 1987; Lynch et al. 1995; Zeliang et al. 2010).

Cryopreservation procedures include two types of techniques: classical techniques based on slow cooling and those based on vitrification (Engelmann 2011). In classical techniques, slow cooling gradually reduces the water content in plant cells before the rapid immersion of samples in LN. These procedures prevent the formation of undesirable ice crystals inside the cell when the subcooled material is immersed at −196 °C. In vitrification techniques, cells are exposed to highly concentrated cryoprotective media and/or air-desiccated under sterile conditions and then rapidly cooled to prevent ice crystal formation (Engelmann 2011). Treating the samples with a highly concentrated vitrification solution for a strictly defined period results in the transition of water from a liquid to an amorphous glassy state, avoiding crystallization (Chmielarz 1998; Sakai 2004; Engelmann 2011). The optimal cooling rate allows the solution to vitrify at a given time and temperature when water loss from the cell is minimal, preventing damage from excessive dehydration (Chmielarz 1998). On the other hand, the increase in the sap concentration inside the cells prevents ice formation.

Compared with classical techniques, vitrification-based techniques have much broader applications, as they can be used to store plant organs such as shoot tips, embryos (zygotic or somatic), or pollen (San José et al. 2015; Delgado-Aceves et al. 2019; Impe et al. 2022; Ayala et al. 2023; Wasileńczyk et al. 2024), which have more diverse histological compositions and, therefore, specific requirements for dehydration conditions caused by freezing (Engelmann 2011). These techniques are less expensive and easier to use, as they do not require additional equipment, and the developed procedures are less time-consuming (Engelmann 2004; Benelli 2021). Engelmann (2011) identified eight available techniques based on vitrification, including dehydration, vitrification, encapsulation–dehydration, encapsulation–vitrification, pregrowth, pregrowth–dehydration, droplet freezing, and droplet vitrification. Recently, D and V cryo-plates have been developed (Maślanka et al. 2016; Niino et al. 2019; Benelli 2021; Vilardo et al. 2023).

6.2.2 Preconditioning

Before cryopreservation, increasing the viability of selected explants after storage in LN is essential. Preconditioning, which improves osmotic balance and reduces cell damage during freezing, is crucial for the survival of plant tissues after cryopreservation (Reed 1996; Chmielarz 1998; Engelmann 2011).

According to Reed (1996), the pretreatment of explants can be divided into three categories: cultural, chemical, and desiccation. The cultural category includes treating the explant with a growth medium, cold hardening, or selecting specific parts of the plant for cryopreservation (Fig. 6.1). Chemical treatment involves the use of osmotic or penetrating cryoprotectants. These agents, which are incorporated into semisolid or liquid media, include sugars, sugar alcohols, and polyethylene glycol (PEG). Additionally, antioxidants or other antistress agents (e.g., ABA and proline) in the pregrowth medium can improve cell conditions and accelerate regeneration after cryopreservation. The desiccation category includes air drying or the use of desiccants.

For example, in protocols developed for embryogenic tissues, careful dehydration of plant tissue was achieved via osmotic agents in the medium before growth (Kartha et al. 1988; Find et al. 1993; Häggman et al. 1998). The abovementioned strategies have been tested on various plant species with positive results (Ryynänen 1996; Bettoni et al. 2019; Song et al. 2020; Sharma and Mukherjee 2023).

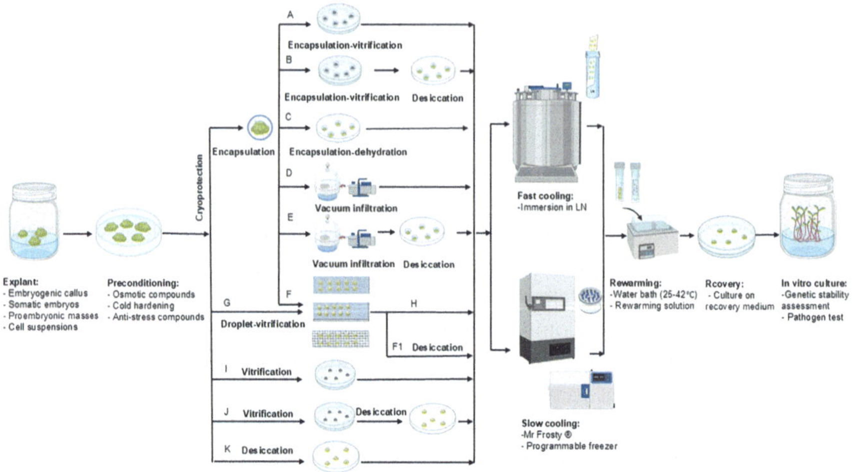

Fig. 6.1 Major steps of explant cryopreservation by encapsulation–vitrification (**a**), encapsulation–vitrification–desiccation (**b**), encapsulation–dehydration (**c**), encapsulation–vacuum infiltration (**d**), encapsulation–vacuum infiltration–desiccation (**e**), encapsulation–droplet vitrification (**f**), encapsulation–droplet vitrification–desiccation (**f**1), droplet vitrification (**g**), droplet vitrification–desiccation (**h**), vitrification (**i**) and vitrification–desiccation (**j**), desiccation (**k**). (Source: prepared by authors)

6.2.3 Slow Cooling

The slow-cooling technique is one of the fundamental methods for the cryopreservation of plant tissue and was developed by Sakai (1960). Sakai (1966) applied a gradual freezing protocol based on his experiments with freezing white mulberry (*Morus alba* L.) cells, poplar (*Populus sieboldii* Miq.) twigs, and willow (*Salix koriyanagi* Kimura ex Goerz) twigs. In the 1970s, Latta (1971) reported that carrot (*Daucus carota* L.) cell cultures precultured with different concentrations of sucrose survived storage at −40 °C in LN. These experiments led to the development of classical (or slow) cooling methods, often referred to as two-step cooling (Kartha 1985; Ballesteros et al. 2024). These methods rely on two fundamental principles: (a) freezing extracellular water, leading to partial, nonlethal dehydration of the intracellular contents, and (b) vitrification of the intracellular contents into a glassy state. According to Engelmann (2011), these methods involve several successive steps: pregrowth of samples, cryoprotection, slow cooling (up to 2 °C/min) to approximately −40 °C, rapid cooling at hundreds of °C per second to −196 °C by immersion in LN, storage of samples in LN, rapid rewarming, and recovery.

Before slow freezing, dimethyl sulfoxide (DMSO) is commonly used as a cryoprotectant and is often added to cryovials in a liquid medium. PEG, sugars, and/or glycerol may also be incorporated into cryovials to dehydrate tissue and prevent ice crystallization. Additionally, the air-drying of explants, such as the embryonic axes of *Acer*, has been successfully applied in conjunction with slow cooling (Pukacki and Juszczyk 2015). Similarly, Kim et al. (2006) obtained positive results after air-drying somatic embryos of *P. lactiflora* Pall. before freezing in LN. Research on the potential for storing desiccated somatic embryos of certain coniferous species was already undertaken in the early twenty-first century by Bomal and Tremblay (2000) and Percy et al. (2001).

A standard slow-cooling rate of 1 °C/min is commonly achieved when devices such as Mr. Frosty® are placed in a −80 °C freezer. With programmable freezers, it is possible to tailor slow-freezing protocols to meet the specific requirements of different plant species or genotypes. However, the high cost of programmable freezers poses a challenge to the widespread adoption of these methods. Despite these limitations, traditional slow-cooling methods are still widely utilized for the large-scale cryopreservation of various plant genetic resources, as shown in Tables 6.1 and 6.2.

6.2.4 Vitrification

Vitrification is the direct transition of water to a glassy state without ice crystal formation and is achieved by rapid cooling from the liquid to the glassy phase. This process prevents ice crystal formation both inside and outside the cell, minimizing callus or cell damage caused by ice crystals. Vitrification-based cryopreservation involves the pretreatment of material with a highly concentrated vitrification solution, followed by loading and dehydration (Fig. 6.1). When exposed to LN, the high concentration of the vitrification solution ensures that the material rapidly transitions to a vitrified state, maintaining its structural integrity (Zhang et al. 2024). In nature, few plants have the ability to vitrify, either because they lack vitrification-inducing substances or because they contain high water contents. The cryoprotectants used in vitrification must possess three key properties: high glass-forming ability, the ability to dehydrate plant cells to induce a vitrified state, and a concentration that does not cause excessive toxicity to the plants.

Plant vitrification solution (PVS), a widely used cryoprotectant, can substitute for cellular water, modify freezing behavior, and prevent excessive water loss. This makes it a valuable tool for preserving plant material and maintaining tissue viability at cryogenic temperatures (Ren et al. 2020; Ballesteros et al. 2024).

The vitrification-based procedure is a fast and efficient method widely used in plant germplasm conservation because of its adaptability, high freezing rate, and ease of use. In a typical procedure, explants are precultured on sucrose-enriched media, followed by exposure to a loading solution containing 2 M glycerol and 0.4 M sucrose. After rapid rewarming following LN immersion, explants are treated with a 1.2 M sucrose solution to remove cryoprotectants before being transferred to in vitro culture conditions.

The choice of method and duration of PVS exposure are critical factors that must be determined empirically for each species and explant type to ensure successful cryopreservation (Ballesteros et al. 2024). Vitrification has been successfully applied to cryopreserve various plant species, including suspension cells, axillary buds, and shoots (see Tables 6.1 and 6.2).

Table 6.1 Cryopreservation of zygotic embryos and their parts, somatic embryos, and embryogenic callus/cultures of selected woody plant species

Plant species	Frozen material	Survive (%)	Method	References	Application
Abies alba	Embryogenic tissue	92–100	Slow cooling	Salaj et al. (2022)	Pulpwood, paper manufacture
Alnus glutinosa	Clumps of globular/heart shaped somatic embryos	90	Sucrose preculture, PVS2	San José et al. (2015)	Threatened by *Phytophthora alni*
Castanea sativa	Embryonic axes; clumps of globular and heart-shaped somatic embryo	93–100; 33 and 68	Air-drying; sucrose pre-culture, air-drying sucrose pre-culture, PVS2	Corredoira et al. (2011) and Gaidamashvili et al. (2021)	Timber, fruit tree
Coffea arabica	Seeds	>80	Rapid drying	Coelho et al. (2017)	Crop
C. arabica	Zygotic embryos	79–83; 100	Encapsulation–vitrification; droplet vitrification	Valdés et al. (2021)	Crop
Coffea spp.	Seeds	30–70	Desiccation; slow cooling	Dussert et al. (2007)	Crop
Coryllus spp.	Embryogenic axes	45–100	Air-drying	Reed and Hummer (2001)	Crop
Cryptomeria japonica	Embryogenic tissue	75	Slow cooling	Taniguchi et al. (2020)	Forestation
Elaeis guineensis	Zygotic embryos; embryogenic tissue	96.67; 68.33	Air-drying; sucrose pre-culture, PVS2	Wei et al. (2023)	Crop, oil
Fraxinus americana and *F. pennsylvanica*	Embryogenic tissues	91; 39	Vitrification; slow cooling	Richins et al. (2024)	Wood; threatened by *Agrilus planipennis*
F. excelsior	Seeds	78	Desiccation	Chmielarz (2009)	Forestry tree
Melia azedarach	Mature somatic embryos	36; 30	Encapsulation–dehydration; pregrowth–dehydration	Scocchi et al. (2007)	Forestry tree

(continued)

Table 6.1 (continued)

Plant species	Frozen material	Survive (%)	Method	References	Application
Olea europaea	Somatic embryos	Not stated	Droplet–vitrification	Bradaï and Sánchez-Romero (2021a)	Fruit tree crop
Persea americana	Somatic embryos	Up to 100	Droplet vitrification	O'Brien et al. (2016)	Crop
Picea abies	Embryogenic tissue	Up to 100	Slow cooling	Gupta et al. (1987) and Varis et al. (2022)	Forest tree
P. koraiensis	Embryogenic tissue	100	Slow cooling	Gao et al. (2021)	Ornamental, timber
Pinus koraiensis	Embryogenic tissue	Up to 77.83	Slow cooling	Peng et al. (2021)	Forest tree
Pinus sylvestris	Embryogenic tissue	59–93	Slow cooling	Latutrie and Aronen (2013)	Forest tree
Pinus sp.	Seeds	24–100	Direct immersion in LN	Pita et al. (1998)	Forestry trees
Quercus ilex	Somatic embryos	17–63	Sucrose preculture, PVS2	Martínez et al. (2022)	Firewood
Q. petraea	Plumules	Up to 83	Cryo-plates	Wasileńczyk et al. (2024)	Forest tree
Q. suber	Globular secondary embryos	88–93	Vitrification	Valladares et al. (2004)	Wine bottling, threatened
Theobroma cacao	Somatic embryos	Not stated	Vitrification	Fang et al. (2009)	Crop
Syzygium maire	Zygotic embryos axes	0	Desiccation, slow cooling	van der Walt et al. (2022)	Threatened, endemic
Vitis spp.	Cell cultures	48–81	Encapsulation/dehydration	Ben-Amar et al. (2013)	Fruit crop
V. vinifera	Zygotic embryos	2–60	Vitrification PVS2	Esquivel-Figueroa et al. (2023)	Fruit crops
V. vinifera cv.	Zygotic embryos	63	Vitrification	García-Coronado et al. (2016)	Fruit crops
Cocos nucifera	Zygotic embryos	13–17	Vitrification, desiccation, directly immersion in LN	N'Nan et al. (2012)	Fruit crops

(continued)

Table 6.1 (continued)

Plant species	Frozen material	Survive (%)	Method	References	Application
Syzygium maire	Zygotic embryos	19	Droplet vacuum infiltration vitrification (DVIV) method and a novel metal-mesh vacuum infiltration vitrification (MVIV) method using plant vitrification solution 2 (PVS2)	van der Walt et al. (2023)	Threatened ornamental tree, the fruit are edible
Cocos nucifera	Plumules	30	Droplet–plumules	da Silva Lédo et al. (2020)	Fruit crops
Quercus robur	Plumules	5–18	Vitrification, desiccation, directly immersion in LN	Chmielarz et al. (2011)	Forest tree
Pistacia vera	Embryonic axe	85–100	Desiccation, directly immersion in LN	Benmahioul et al. (2015)	Crop

Droplet vitrification Droplet vitrification is a cryopreservation method first described by Panis et al. (2005). This method follows a conventional vitrification protocol, except for the cooling process, in which explants are placed on aluminum foil in tiny droplets of vitrification solution before being rapidly immersed in liquid nitrogen (LN) (Figs. 6.1 and 6.2). This technique involves several key steps: pretreatment, dehydration, ultrarapid cooling, and LN storage.

First, the plant material is treated with a loading solution (LS) containing osmoprotectants such as glycerol and sucrose, which increase tolerance to dehydration. Next, the material is exposed to plant vitrification solution (PVS2), which contains cryoprotective agents such as glycerol, ethylene glycol, dimethyl sulfoxide (DMSO), and sucrose. This step facilitates water removal and prevents ice formation during freezing. The dehydration process is conducted at low temperatures to increase survival rates. After sufficient exposure to PVS2, the treated samples were placed in small droplets of vitrification solution on sterile aluminum foil strips. These strips are then rapidly immersed in LN, ensuring ultrafast cooling that transitions cellular water into a glass-like state without the formation of damaging ice crystals. The frozen samples are stored in LN until further use.

Table 6.2 Cryopreservation of zygotic embryos and their parts, somatic embryos, and embryogenic callus/cultures of selected nonwoody plant species

Plant species	Frozen material	Shoot regrowth (%)	Method	References	Application
Agapanthus praecox	Embryogenic tissue	53–85	Vitrification PVS2	Ren et al. (2020)	Ornamental plant
Agave tequilana	Somatic embryos	83	V-cryo-plate	Delgado-Aceves et al. (2021)	Crop
Agave tequilana cv	Somatic embryos	90	V-cryo-plate	Delgado-Aceves et al. (2019)	Crop
Banana	Embryogenic tissue	80	Vitrification PVS2	Roostika et al. (2022)	Fruit crop
Catharanthus roseus L. (G.)	Embryogenic cell suspensions	55	Vitrification, slow cooling	Mujib et al. (2022)	Medicinal plant
Coelogyne nervosa	Somatic embryos	70	Droplet–vitrification	Chithra Devi et al. (2021)	Medicinal plant
Gentiana capitata, *Gentiana decumbens*	Embryogenic cell suspensions	Growth rate: 28, 33	Encapsulation–dehydration	Markowski et al. (2024)	Medicinal plant
Musa balbisiana Colla	Zygotic embryos	90	Dehydration	Singh et al. (2021a)	Fruit crops
Narcissus L.	Globular somatic embryos	20–27	V-cryo-plate, PVS2	Maślanka et al. (2016)	Medicinal and ornamental plant
Oryza sativa L	Embryogenic tissue	42–48	Vitrification PVS2, slow cooling	Zeliang and Pattanayak (2024)	Crop
Oryza sativa L	Embryogenic tissue	75–100	Vitrification PVS2, PVS3	Taratima et al. (2022)	Crop
Panax ginseng C. A. Meyer	Embryogenic tissue	68	Vitrification	Lei et al. (2021)	Medicinal plant
Petiveria alliacea L	Somatic embryos	70–85	Vitrification encapsulation–dehydration, D-cryo-plate	de Almeida et al. (2017)	Medicinal plant
Petiveria alliacea L.	Somatic embryos	100	V-cryo-plate	de Pettinelli et al. (2020)	Medicinal plant
Podophyllum hexandrum Royle	Zygotic embryos	90	V-cryo-plate	Parasher et al. (2023)	Medicinal plant

(continued)

Table 6.2 (continued)

Plant species	Frozen material	Shoot regrowth (%)	Method	References	Application
Swertia chirayita	Somatic embryos	93	Encapsulation, dehydration	Gupta and Sood (2022)	Medicinal plant
Musa sp.	Embryogenic cell suspensions	64	Slow freezing	Panis et al. (1990)	Fruit crop
Agapanthus praecox	Embryogenic callus	49.14–86.85	Vitrification	Zhang et al. (2015)	Ornamental plant
Anemarrhena asphodeloides Bunge	Embryogenic callus	60	Vitrification	Sen-Rong and Ming-Hua (2012)	Medicinal plant
Oryza rufipogon	Embryogenic callus	85–95	Vitrification	Zeliang et al. (2010)	Crop
Crocus spp.	Embryogenic callus	67	Encapsulation–vitrification	Baghdadi et al. (2009)	Medicinal and ornamental plant
Gentiana cruciata	Embryogenic cell suspension	80	Encapsulation/dehydration method	Mikuła et al. (2011)	Medicinal and ornamental plant
Saccharum officinarum	Embryogenic callus	34–42, 20–94	Vitrification	Distabanjong et al. (2015) and Martínez-Montero et al. (1998)	Crop
Dioscorea bulbifera	Embryogenic callus	60–70	Vitrification	Hong et al. (2009)	Crop and ornamental plant
Catharanthus roseus L.	Embryogenic cell suspensions	21–78	Vitrification, slow cooling	Fatima et al. (2009)	Medicinal plant
Banana	Zygotic embryos	70–100	Air dehydration, directly immersion in LN	Singh et al. (2021b)	Fruit crop
Arachis hypogaea	Embryonic axes	98	Droplet vitrification	Tacán et al. (2017)	Crop
Phaseolus vulgaris	Embryonic axes	96	Droplet vitrification	Tacán et al. (2017)	Fruit crop
Jatropha curcas	Zygotic embryos	100	Desiccation and rapid cooling, vitrification	Prada et al. (2015)	Medicinal plant

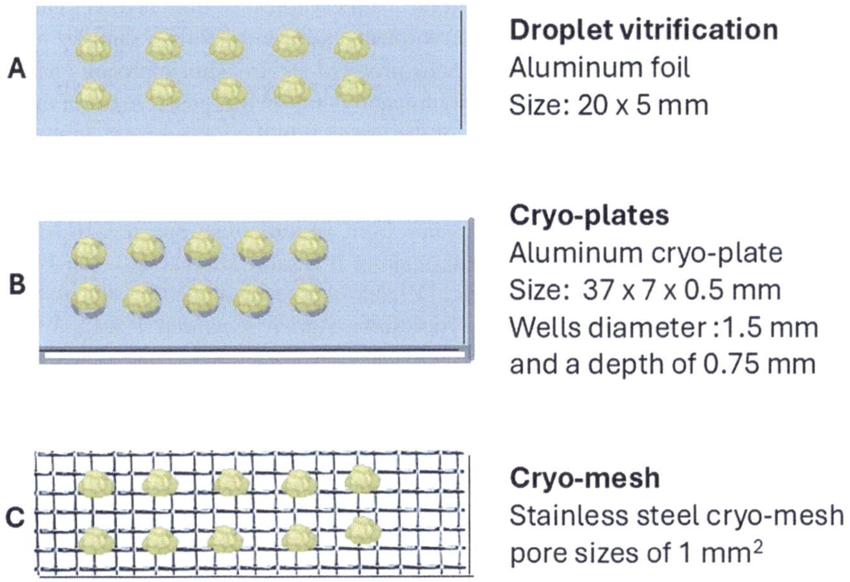

Fig. 6.2 Aluminum plate with explants immersed in PVS drops (**a**), cryo-plate with explants (**b**), and cryo-mesh with explants (**c**) used for cryopreservation. (Source: prepared by authors)

The droplet vitrification method has been widely adopted and modified for various plant species, including potatoes (Yamamoto et al. 2015), hops (Malhotra et al. 2024), coelogyne (Chithra Devi et al. 2021), narcissus (Maślanka et al. 2016), and chrysanthemums (Yamamoto et al. 2011; Teixeira da Silva et al. 2015), to preserve genetic resources effectively. The main advantage of this technique is that explants remain in direct contact with LN during cooling and with the unloading solution during reheating, without direct handling, reducing the risk of mechanical damage and contamination.

V-Cryo-Plate Method Two cryopreservation approaches using cryoplates have been developed: the V-cryo-plate method and the D-cryo-plate method. Both techniques use calcium alginate to attach explants to aluminum cryoplates (7 × 37 × 0.5 mm), which typically contain 10 wells of varying sizes (1.0–1.5 mm in diameter and 0.5–0.75 mm in depth) (Yamamoto et al. 2011). The V-cryo-plate method combines encapsulation vitrification with droplet vitrification, whereas the D-cryo-plate method utilizes calcium alginate encapsulation with dehydration. While both techniques follow similar general procedures, the key difference lies in the treatment of the explants: in the V-cryo-plate method, explants are treated with plant vitrification solution (PVS), whereas in the D-cryo-plate method, explants are air-dried (Yamamoto et al. 2011; Niino et al. 2019).

The V-cryo-plate method, developed for plant cryopreservation, involves encapsulating plant material on aluminum cryo-plates, allowing multiple samples to be processed simultaneously through various protocol stages. This approach enables ultrafast cooling and heating rates, significantly improving postcryopreservation recovery and the overall effectiveness of the process. In the V-cryo-plate approach, samples undergo osmotic dehydration by exposure to PVS, which consists of highly concentrated cryoprotectant mixtures. Among these, PVS2 has been identified as the most effective for dehydrating tissues from various plant species. However, PVS3 has been found to be less toxic, making it a safer alternative when PVS2 induces cytotoxic effects. For example, Delgado-Aceves et al. (2021) successfully used the V-cryo-plate method to cryopreserve *Agave tequilana* F.A.C. Weber somatic embryos.

V-Cryo-Mesh Method The cryo-mesh cryopreservation method (Fig. 6.2) is very similar to the V-cryo-plate method described by Yamamoto et al. (2011) and the droplet vitrification method, which uses a strip of aluminum foil. The key difference between these techniques lies in the material used. The cryo-mesh method involves the use of a stainless steel mesh strip, whereas the V-cryo-plate method involves the use of a steel-aluminum plate with cavities measuring 7 mm × 37 mm × 0.5 mm (length × width × depth; Yamamoto et al. 2011). The droplet vitrification method relies on an aluminum foil strip (Teixeira da Silva et al. 2015). This method has also been applied in vacuum infiltration, where the porous mesh facilitates the deep penetration of the cryoprotectant into the tissue. Van der Walt et al. (van der Walt et al. 2023) successfully used the cryo-mesh method to cryopreserve embryonic axes, achieving a 19% survival rate. As a result, this technique serves as an alternative approach for the cryopreservation of various explants.

Vacuum Methods The vacuum infiltration vitrification (VIV) method uses a vacuum to enhance the penetration of the cryoprotectant into the tissue (Fig. 6.1). Soejima et al. (2012) investigated the effect of vacuum on glycerol penetration into tissue and compared the effectiveness of cryoprotectants injected into tissue under atmospheric pressure and vacuum (20, 40, and 60 cmHg for 10, 20, and 30 minutes) on survival after LN. The efficiency of the conventional vitrification (CV) method was compared with that of the VIV method using plant embryos (*Carica papaya* L., *Passiflora edulis* Sims, and *Laurus nobilis* L.) (Soejima et al. 2012; Nadarajan and Pritchard 2014). The results obtained for *C. papaya* embryos revealed the highest viability after cryopreservation with 30-minute PVS2 treatment at 0 °C, reaching 25%. In contrast, with 5-minute VIV treatment and an approximately tenfold shorter PVS2 exposure time, embryo viability after cryopreservation increased to 55%. Previous studies on the use of VIV for embryo cryopreservation have demonstrated the advantages of this approach over CV by reducing the duration of cryoprotectant treatment to just a few minutes, minimizing temperature dependence during cryoprotectant application, and enhancing postcryopreservation regrowth (Nadarajan and Pritchard 2014; Lukić et al. 2022; van der Walt et al. 2023).

6.2.5 Desiccation-Based Methods

Air-Drying Physical (air) drying is one of the simplest methods for cryoprotection of plant explants. By drying the explants to a moisture content tolerated by the tissue, the intracellular aqueous solution is concentrated, reducing the water fraction susceptible to freezing damage. This process is crucial for the cryopreservation of various plant materials, including recalcitrant seed tissues (embryonic axes), shoot tips, and embryogenic material such as calli and somatic embryos. The optimal moisture content for these explants is generally reduced to approximately 20–30% of the fresh weight (Panis 2019; Ballesteros and Pritchard 2020). However, the precise moisture content can vary depending on the species. After drying, the explants are rapidly cooled to LN temperatures, which minimizes the formation of ice crystals and promotes the vitrification of the cellular contents (Walters and Pence 2021; Ballesteros et al. 2024).

Two primary methods are used in plant cryopreservation, which differ in how the explants are dried. One method involves drying the explants "naked," whereas the other method uses encapsulation–dehydration, where explants are first encapsulated in alginate beads.

D-Cryo-Plate Method Niino et al. (2013) developed the D-cryo-plate (D for desiccation) method, which is similar to the V-cryo-plate technique but has a key difference. In the D-cryo-plate method, plant tissue is encapsulated in alginate on aluminum plates. However, instead of using PVS for cryoprotection, explants undergo desiccation under dry conditions or in dry air. This approach utilizes controlled dehydration for cryoprotection, offering an alternative to vitrification-based methods (Ballesteros et al. 2024). In contrast, in the encapsulation–dehydration procedure, alginate-encapsulated explants are often treated with a high-concentration sucrose solution (0.75–2.0 M) for a period ranging from 12 hours to 7 days, depending on the species and explant type, before desiccation (Panis 2019).

6.2.6 Thawing

Rewarming, also known as "thawing," is a crucial step in all cryopreservation procedures, regardless of the explant type (Bettoni et al. 2021). During rewarming, cryopreserved samples, typically stored below −130 °C, transition from a glassy state to a liquid state (devitrification). This process can lead to the formation or growth of ice crystals, which can be lethal to cells if not properly managed (Benson 2008). To minimize this risk, rapid rewarming is essential. However, determining the optimal rewarming temperature and duration is critical. For example, the temperature of a warming environment typically ranges from 25 °C (room temperature) to 42 °C, depending on the specific protocol. Different approaches are used to warm the explants: some protocols submerge the entire vial containing the explants in a warm water bath (37–42 °C) for 2–4 minutes, whereas others expose the explants (e.g., in alginate beads or aluminum plates) directly to a warming medium at room temperature or in a warm liquid (25–42 °C) in a Petri dish.

After rewarming, in vitrification and encapsulation–vitrification methods, it is common to immerse the material in an "unloading" solution containing 0.3–1.2 M sucrose for a short period to remove cryoprotectants and further aid recovery (Corredoira et al. 2004; San José et al. 2015; O'Brien et al. 2016; Ballesteros et al. 2024). In slow-cooling techniques, postcryopreservation treatments for embryogenic samples typically include removing the contents of the cryovials into a filter-placed medium used for the proliferation of embryogenic tissue or supplemented with a decreasing concentration of sucrose. The cryoprotectants can be removed by placing the filter with explants into a Büchner funnel and adding hormone-free liquid media, which is drained off by suction before placing the filter explant into semisolid media. The proliferation medium is usually refreshed at least once within 24 hours, meaning that the workload is similar to that of the precryo dehydration treatment with increasing osmoticum but now with decreasing osmoticum (Klimaszewska et al. 1992; Find et al. 1998; Hazubska-Przybył et al. 2013; Varis et al. 2022).

6.2.7 Factors Affecting Survival

The most important factor affecting the outcome of cryopreservation is the condition of the starting material. The use of young plant material increases the likelihood of regeneration after thawing. Smaller cells, tissues, or organs have a greater tolerance to freezing, and optimal regrowth occurs in the late lag phase, early exponential phase, or early stationary phase. In contrast, cells in the late lag or early exponential phase have the lowest survival rates (Ishikawa et al. 1996). The age of the callus affects chromosomal stability, with older calluses showing more chromosomal aberrations (Jain and Maluszynski 2004). Moreover, desiccation and freezing tolerance also depend on the physiological and developmental stages of the explants. Therefore, plant material is carefully selected to ensure the highest tolerance to desiccation and cooling (Normah and Makeen 2008).

The success of cryopreservation is determined by the steps taken during the preparation of the material for freezing, including cold hardening, osmotic/chemical treatments, the desiccation rate/method, the final water content, and the choice of cryopreservation method, along with the technical factors associated with a particular procedure. Dehydration tolerance can be induced through cold hardening by exposing explants to low temperatures (Benson 2008). For example, Ryynänen (1996, 1998) exposed *Betula pendula* L. shoot tips in vitro to 5 °C for 3 weeks to improve their regeneration rate after thawing with LN. Similarly, osmotic dehydration, carried out in the presence of osmotic agents such as glucose, sucrose, or polyethylene glycol, or chemicals such as ABA (Edesi et al. 2020), often leads to increased survival of explants after cryopreservation (Suzuki et al. 2006; Hazubska-Przybył et al. 2010, 2013; Varis et al. 2017). Comparable effects are achieved through the use of cryoprotectants that penetrate the cell, such as DMSO, proline, glycerol, and sorbitol (Chmielarz 1998; Nagel et al. 2024). In many cases, cryopreservation success has been improved by desiccating explants under a stream of sterile air or over silica gel (Kim et al. 2006; Hazubska-Przybył et al. 2010; Pukacki and Juszczyk 2015; Panis 2019).

Selecting the appropriate cryopreservation method for a specific type of plant material plays a key role in its success. Important criteria for method selection include survival and regeneration rates, genetic stability, protocol duration, and challenges associated with handling explants (Gallard et al. 2006; Benelli 2021). The ultimate goal of cryopreservation is to recover plant material (Fig. 6.1, step 6). The samples were thawed at room temperature (25 °C) or in a water bath (approximately 40 °C), and the explants were then incubated in regeneration/proliferation media, usually for 24–48 hours. During this period, any toxic compounds formed due to the stress of cryopreservation and thawing are released from the explants into the medium.

The effectiveness of a cryopreservation method is assessed on the basis of key parameters such as viability, survival, growth, regrowth, explant recovery, and shoot/embryogenic tissue formation. Additionally, challenges in technology transfer and protocol validation between laboratories can significantly impact cryopreservation results. The lack of reproducibility in available protocols or difficulties in adapting them may arise from various factors, including differences in laboratory materials, equipment, and technical expertise among cryopreservation laboratories (Benelli 2021).

6.2.8 Genetic Fidelity

Cryopreservation procedures commonly utilize in vitro culture techniques and demand precise control over multiple steps, including explant excision, preculture, osmo- and cryoprotection, dehydration, freeze–thaw cycles, unloading, and postculture for plant recovery. Since metabolic processes are suspended during cryostorage, genetic changes are not expected to occur. However, during the cryoprotection–regeneration process, stresses caused by physical and chemical conditions may lead to cryo-injuries and genetic alterations. Ensuring the genetic fidelity of preserved material is highly important for the ex situ conservation of genetic resources.

Genetic stability at the phenotypic, chromosomal, DNA, and epigenetic levels—and the factors contributing to genetic stability in cryopreserved embryogenic materials of woody plants—has recently been reviewed and discussed by Ballesteros et al. (2024). According to their study, plants regenerated via cryostorage generally do not exhibit morphological variation. At the DNA level, variability in microsatellite markers or methylation after storage in LN has rarely been detected. In contrast, somaclonal variation, such as telomere shortening, has been observed in embryogenic materials after extended subcultures, highlighting the genetic stability provided by cryopreservation (Aronen et al. 2021).

In nonwoody plants, changes in DNA methylation have been detected in plants recovered from cryopreserved shoot tips of *Fragaria gracilis* 'Joho' (Wang et al. 2021). Peredo et al. (2009) assessed epigenetic integrity in *Humulus lupulus* L. plants recovered from shoot tips cryopreserved for 3 years. DNA methylation changes were observed at 36% of the loci. Further analysis revealed that only 2.6–9.8% of these changes were induced by the cryopreservation procedure, whereas the remainder were attributed to in vitro culture processes. In contrast to these findings, no significant alterations in DNA methylation were detected in regenerants recovered from *Gentiana cruciata* L. embryogenic cell suspensions cryopreserved via encapsulation–dehydration (Mikuła et al. 2011).

In vitro tissue culture is now a fundamental aspect of cryopreservation technology, playing a key role in establishing and maintaining stock cultures, as well as supporting postculture recovery of shoot tips, cell suspensions, and embryogenic tissues in various plant species (Us-Camas et al. 2014; Wang et al. 2021). In vitro tissue culture imposes stressful conditions and induces the generation of reactive oxygen species (ROS), leading to ROS-induced oxidative stress (Miguel and Marum 2011; Us-Camas et al. 2014). Zhang et al. (2015) reported changes in cell morphology and ultrastructure and detected dynamic shifts in ROS components (O_2^-, H_2O_2, and OḤ), antioxidant systems, and programmed cell death (PCD) events during embryonic callus cryopreservation of *Agapanthus praecox* Willd.

6.3 Zygotic Embryos and Their Parts as Cryostored Germplasms

According to Wyse and Dickie (2017), dried seed banking is estimated to be an effective method for preserving genetic diversity in up to 92% of seed plant species. However, seed tolerance to desiccation varies, leading Roberts (1972) to classify seeds into two main categories: orthodox and recalcitrant. Orthodox seeds can withstand desiccation to 1–3% moisture content without losing viability, whereas recalcitrant seeds tolerate desiccation only in the 12–40% range (Chmielarz 1998; Bramasto et al. 2020). Some orthodox seeds can be stored at ambient to freezing temperatures (−20 °C) for decades or even centuries under proper conditions. Later, a third category, intermediate seeds, was identified. These seeds can be stored conventionally for a limited time (Ellis et al. 1990) but are less tolerant to desiccation than orthodox seeds are.

Recalcitrant seeds cannot be stored in traditional seed banks because of the formation of lethal ice crystals in their tissues. This category includes species from temperate and tropical forests, plantations, shrubs, and trees, such as *Acer* spp., *Quercus* spp., *Theobroma*, and *Coffea*. Among woody plants, only approximately two-thirds can be stored as dried seeds (Wyse et al. 2018). However, many recalcitrant seeds can be preserved through cryopreservation. Liquid nitrogen provides an excellent medium for storing seeds, zygotic embryos, and embryonic axes, either in LN at −196 °C or in LN vapor, with minimal viability loss (Chmielarz 1998; Engelmann 2011). Numerous examples of both woody and nonwoody plant species whose seeds are successfully stored in LN exist (Stanwood 1985; Kholina and Voronkova 2013; Pammenter and Berjak 2014; Michalak et al. 2015; Song et al. 2024).

Despite its effectiveness, cryopreservation faces limitations when storing whole seeds, particularly large nuts with high oil contents (Reed and Hummer 2001). In such cases, zygotic embryos can be extracted and stored separately. Since cotyledons often have lower cryotolerance, excising the embryonic axes can improve regeneration (Vieitez et al. 2011). For example, the embryonic axes of multiple *Corylus* spp. (hazelnut) can be desiccated and cryopreserved, or whole seeds can be cryopreserved, with the embryonic axes excised and recovered postthawing (Reed and Hummer 2001). In *Quercus* spp., embryonic axes are the preferred explants for cryopreservation because of their relatively simple regrowth procedures and high genetic variability (Wasileńczyk et al. 2024). Recently, Wasileńczyk et al. (2024) reported an 83.7% survival rate of *Q. petraea* (Matt.) Liebl. embryonic axes (plumules) via the cryoplate method. Similarly, Corredoira et al. (2011) reported an almost 100% survival rate for *Castanea sativa* L. embryonic axes after air-drying. However, an attempt to cryopreserve the embryonic axes of the endangered *Syzygium maire* Sykes & Garn.-Jones failed (van der Walt et al. 2022).

For certain conifers, such as *Pinus radiata* D. Don and *Pinus* hybrids, cotyledons of zygotic embryos have been successfully cryopreserved and regenerated via organogenesis (Toivonen and Kartha 1989; Hargreaves et al. 2004; Ayala et al. 2023). This method can capture more genotypes than can somatic embryogenesis (SE), which has low initiation rates in some species, including many pines. Additionally, SE initiation windows are often narrow, whereas mature seed embryos used as explants for organogenesis provide greater flexibility (Ayala et al. 2023). However, in many species, reducing the size of zygotic explants can inhibit shoot growth due to wound-induced oxidative damage (Ballesteros et al. 2014). To mitigate this, explants for cryostorage may include embryonic axes with attached cotyledon segments or entire cotyledons.

Cryopreservation protocols have enabled the long-term storage of desiccation-resistant seeds, such as *Coffea arabica* L. seeds, which remain viable for up to 2 years when stored at 15 °C and 10% humidity (Chmielarz 1998). However, isolated zygotic embryos can be stored for much longer periods. Valdés et al. (2021) demonstrated the effectiveness of three cryopreservation methods—encapsulation, vitrification, and droplet vitrification—for *C. arabica* zygotic embryos, with survival rates ranging from 78% to 100% after thawing. In contrast, Dussert et al. (2007) reported 30–70% survival for *Coffea* spp. seeds after desiccation and slow cooling. Wei et al. (2023) reported a 96% survival rate for *Elaeis guineensis* Jacq. embryos after air-drying, whereas preculturing embryogenic tissues in sucrose followed by PVS2 treatment resulted in 68% survival.

Recently, Bazrafshan et al. (2025) developed a new method to increase *Cocos nucifera* L. embryo dehydration efficiency via a rapid physical dehydration technique with a silica gel desiccator. Compared with previous methods, this device reduced the moisture content of the embryo to 20% in just 6 hours, improving dehydration rates by 2 hours (Sisunandar et al. 2010). In nonwoody plants, a 90%

survival rate was reported for embryos isolated from *Podophyllum hexandrum* seeds via the V-cryo-plate method (Parasher et al. 2023).

The optimal moisture content (MC) is a critical parameter for successful regrowth and is species specific (Normah and Makeen 2008; Ballesteros et al. 2014). The drying rate of recalcitrant seed-derived explants is particularly important for cryopreservation success (Ballesteros et al. 2014). Ballesteros et al. (2014) emphasized that embryo morphology and anatomy are key factors in optimizing explant drying prior to LN immersion, ultimately affecting postcryopreservation survival. Other crucial factors include explant physiological state, desiccation rate manipulation, preculture treatments, type of cryoprotectant used, cooling and rewarming conditions, recovery media, and postthawing recovery optimization (Normah and Makeen 2008).

Zygotic embryos or excised embryonic axes acquire desiccation tolerance through rapid or ultrarapid drying (Pritchard and Manger 1998; Walters et al. 2002; Makeen et al. 2005). Pritchard and Manger (1998) and Normah and Makeen (2008) suggested that desiccation tolerance likely results from minimizing the duration of intermediate water content in metabolically active tissues, reducing the risk of degradation. Desiccation or dehydration of plant material is typically achieved via laminar flow, silica gel (Coelho et al. 2017), or vitrification (Wei et al. 2023). Dehydration-based cryopreservation methods require cryoprotectants such as glycerol, mannitol, sorbitol, DMSO, sucrose, or polyethylene glycol (Chmielarz 1998; O'Brien et al. 2020). For this purpose, tissues and organs are treated with cryoprotectant mixtures such as PVS2, PVS3, or PVS4 (Sakai et al. 1990; Nishizawa et al. 1993; O'Brien et al. 2020), which are routinely used in cryopreservation protocols.

6.4 Embryogenic Cultures and Somatic Embryos as Cryostored Germplasms

Cryopreservation is a key method for preserving the juvenility and regenerability of embryogenic cultures obtained through SE in both nonwoody and woody plants, as shown in Tables 6.1 and 6.2. The first successful protocols for the cryopreservation of embryogenic cultures were developed in the 1990s for *Musa* sp. and *Citrus sinensis* L. (Panis et al. 1990; Sakai et al. 1990). Since then, it has been possible to safely store valuable embryogenic cell lines in LN without the risk of contamination, loss of embryogenic and morphogenic potential, or genetic instability of the cultures (Park and Bonga 2010; Hazubska-Przybył et al. 2013). This solution is particularly important for embryogenic cultures of elite trees, as it facilitates the characterization of the obtained embryogenic lines (e.g., their ability to produce high-quality somatic embryos) and the testing of progeny from particular genotypes under field conditions (Adams et al. 2016). Moreover, it is a crucial tool for the long-term storage of valuable embryogenic tissues or somatic embryos obtained from selected elite trees in breeding programs (Häggman et al. 2000).

The development of effective protocols for the cryopreservation of embryogenic cultures is a complex and multistep process. One of the most difficult challenges is the diverse, genotype-dependent response of embryogenic cultures (Bradaï and Sánchez-Romero 2021a), even from the same donor tree. Other potential problems may arise from the developmental stage of embryogenic cells and their physiological state, the sensitivity of cultures to drying and cryoprotection, the low survival rate of material recovered from LN, and the genetic instability of cryopreserved tissues/embryos (Lambardi et al. 2008). Currently, embryogenic cultures of many plant species are stored in LN via either slow cooling or rapid freezing methods, typically as tissue pieces/clumps or cell suspensions (see Tables 6.1 and 6.2). As shown in Tables 6.1 and 6.2, the traditional slow-cooling method is the most commonly used approach for cryopreserving embryogenic tissues in plants. Rapid cooling methods, on the other hand, have been intensively developed over the past two decades as a simpler alternative to the former. Among these, protocols based on encapsulation–dehydration and vitrification have been developed mainly for embryogenic cultures of both woody and nonwoody plants (Hazubska-Przybył et al. 2010, 2013; Bekheet et al. 2020; Lei et al. 2021; Wei et al. 2023; Tokgöz et al. 2023).

Embryogenic cultures are usually stored in LN in the form of embryogenic tissue, cell suspensions, or somatic embryos at various developmental stages (Valladares et al. 2004; Hazubska-Przybył et al. 2010; Maślanka et al. 2016; Markowski et al. 2024). As a reservoir of somatic embryos in the early stages of development, embryogenic tissue is highly efficient, with a high survival rate and rapid regeneration after thawing (Zeliang et al. 2010; Latutrie and Aronen 2013; Salaj et al. 2022). Therefore, it is successfully used for the long-term storage of genetic resources for various crops, fruit crops, medicinal and ornamental plants, forestry species, and economically important trees and, less frequently, for ex situ conservation of endangered or protected plants. Moreover, embryogenic cell suspensions serve as efficient sources of somatic embryos, protoplasts, and cell aggregates, which are valuable for genetic manipulation, secondary metabolite biosynthesis, and micropropagation (Markowski et al. 2024). For these reasons, suspension cultures can be easily and safely stored in LN, reducing time, maintenance, and labor costs while eliminating the risk of contamination. Positive cryopreservation results have been obtained for suspension cultures of *Pinus nigra* L., *Gentiana* spp., and *Cyclamen persicum* Mill., among others (Winkelmann et al. 2004; Salaj et al. 2007; Markowski et al. 2024).

The cryopreservation of embryogenic tissues on a production scale via cryoprotectants is expensive and labor-intensive. Therefore, in some cases, somatic embryos have been used as explants for long-term storage at −196 °C (Chmielarz 1998; Valladares et al. 2004; San José et al. 2015; Gaidamashvili et al. 2021). On the other hand, clumps of maturing somatic embryos are much more heterogeneous than embryogenic tissues at the proliferation stage, as they consist of embryos at different developmental stages (Engelmann 2011). For this reason, proliferating embryogenic cultures are usually preferred as explants for cryopreservation (Palanyandy et al. 2020; Suranthran et al. 2012) over mature somatic embryos. Despite this, successful cryopreservation of mature somatic embryos has been reported, e.g., for spruces (*Picea mariana* (Mill.) Britton, Sterns & Poggenb., *P. glauca* (Moench)

Voss) and paradise tree (*Melia azedarach* L.). The key factor in these cases was dehydration of the embryos prior to direct immersion in LN without cryoprotectants or slow cooling (Bomal and Tremblay 2000; Scocchi et al. 2007). Additionally, the maturation of embryos at 5 °C has led to the successful cryopreservation of *P. glauca* (Kong and von Aderkas 2011). Positive results have also been obtained after droplet vitrification of *Olea europaea* L. somatic embryos (Bradaï and Sánchez-Romero 2021b), vitrification of *Theobroma cacao* L. embryos (Fang et al. 2009), and sucrose preculture followed by treatment with PVS2 solution in *Quercus ilex* L. somatic embryos (Martínez et al. 2022).

In the case of nonwoody plants, somatic embryos of mainly medicinal plants have been used as explants suitable for storage in LN (Table 6.2). Kim et al. (2006) achieved more than 40% recovery of *Paeonia lactiflora* embryos following air drying. However, the positive response was influenced by the origin of the explants (cotyledonary vs. anther-derived embryos), which may be related to the ploidy level of the somatic embryos. On the other hand, Maślanka et al. (2016) reported a 26.7% cryopreservation success rate for somatic globular embryos of *Narcissus* L. 'Carlton' via the droplet vitrification method. In this case, the success of recovery depended on the size of the explants tested.

6.5 Role of Cryopreservation in Germplasm Conservation

According to Botanic Gardens Conservation International (BGCI), as many as two out of five plant species are currently threatened with extinction due to habitat loss, overexploitation, disease and pests, pollution, climate change, surface fragmentation, large-scale deforestation, the uncontrolled introduction of invasive species, and other factors (https://www.bgci.org/our-work/saving-plants/conservation-prioritization/). Research teams' estimates highlight the exceptionally rapid loss of biodiversity over the past few centuries, indicating that the sixth mass extinction is already underway. In the case of plants, this has resulted in an approximately 50% reduction in plant biomass compared with that in the prehuman era (Ceballos et al. 2015; Díaz et al. 2019; Breman et al. 2021).

The key to protecting genetic resources lies in within-species genetic diversity, which enables populations to adapt to rapidly changing climates and associated challenges, such as emerging diseases, pests, pathogens, extreme weather events, and forest fires (Barrett and Schluter 2008; Hoffmann and Sgrò 2011; Seidl et al. 2014; Zohner et al. 2020; Lee 2023; Dyderski et al. 2025). Given the need to conserve genetic resources while representing the broadest possible gene pool of particular species (or populations), steps have been taken to establish and disseminate standards for creating collections of plant and animal genetic resources. These efforts include international agreements such as the Berne Convention (1979/1997), the CITES Convention (1973), the Convention on Biological Diversity (1992), the Habitats Directive (92/43/EEC, 21.05.1992), and recommendations from organizations such as the European Forest Genetic Resources Programme (EUFORGEN), the European Native Seeds Conservation Network (ENSCONET), the Food and Agriculture Organization of the United Nations (FAO), and the International Seed

Table 6.3 Examples of the large cryobanks for nonwoody genetic resources (by Małgorzata Pałucka)

Cryobank	Country	Type of material deposed	Number of gene resources
Agricultural Genetic Resources Preservation Research, Fort Collins	USA	Seeds, shoot tips	621000 plant accessions representing more than 16800 species
ICAR, National Bureau of Plant Genetic Resources, New Delhi	India	Recalcitrant seeds, intermediate seeds	7997
		Orthodox seeds	3995
		Dormant buds	389
		Pollen	673
		DNA	2194
The International Potato Center (CIP)	Denmark	In vitro	7500 accessions (4900 cultivated potatoes and 2600 related wild potatoes)
Royal Botanic Gardens Kew	United Kingdom	Seeds, shoot tips	2.4 billion seeds representing almost 40000 different species
National Genebank of China	China	In vitro, shoot tips, dormant buds/twigs, and pollen	300 accessions in vitro, belonging to 35 crops
Chinese Academy of Medical Sciences and Peking Union Medical College	China	Seeds	164 medicinal plant

Testing Association (ISTA), among others. As a result, many cryobanks maintaining a vast collection of plant genetic resources have been established (see examples in Table 6.3).

In 1994, the European Forest Genetic Resources Programme (EUFORGEN, FAO) was established to promote, among other objectives, the conservation of forest genetic resources across Europe (Kelleher et al. 2015). EUFORGEN's priority is the dynamic in situ and ex situ conservation of forest genetic resources in the context of climate change and nature conservation. Accordingly, a pan-European network of conservation units for different tree species has been established to preserve the adaptive genetic diversity of tree species growing within European ecosystems. The main objective of in situ conservation efforts is to support the natural regeneration of tree species and populations in their native habitats, ensuring both the quality and quantity of regenerating material. This can be achieved by monitoring and managing in situ conservation units through (a) using specific critical values for seed number and density, (b) reducing regeneration time, (c) regulating competition from other tree species, and (d) controlling invasive species.

High levels of genetic diversity and reduced seedling consanguinity should be the result of effective management (Kelleher et al. 2015). However, with rapidly increasing climate change, this strategy for protecting forest genetic resources may not be sufficient. Hence, ex situ conservation efforts are expected to become increasingly important, potentially supplementing or even substituting in situ conservation in the future (Bradley St Clair and Howe 2007).

The essence of ex situ conservation is breeding forests beyond their original sites (dynamic ex situ conservation). However, these efforts tend to be expensive to establish and maintain. Dynamic conservation occurs, for example, when artificially planted trees (e.g., species, provenances, and families) can regenerate from seeds without significant intervention (Amaral et al. 2004). Another form of ex situ conservation is the static approach, which involves creating plantings to preserve the genetic diversity of valuable populations while protecting endangered or rare species. Tree collections are established as seed orchards, clone banks, or clone archives. Tree genetic resources are also maintained in arboreta or botanical gardens (Kelleher et al. 2015). For species at risk of extinction, that are experiencing population declines, or that do not produce seeds, the only safe conservation techniques are vegetative propagation, micropropagation, or cryopreservation.

Cryopreservation as an ex situ measure is particularly necessary when germplasm needs to be stored for an extended period while maintaining a rejuvenated state, as in the case of forest trees (George et al. 2024). In Finland, this method has already been applied within the national program to preserve the genetic resources of dormant buds from *Ulmus laevis* L. and *U. glabra* L. However, its widespread use remains limited for most species because of the lack of well-developed and efficient procedures, as well as reliable methods tested across a broad range of genotypes. Additional barriers include the high costs, specialized skills, and facilities required for successful cryopreservation (George et al. 2024). According to the literature, it is generally assumed that the efficiency of a procedure should be at least 40% (Reed 2001; Vollmer et al. 2016). If the success rate is lower, protocols must be refined to improve cryopreservation outcomes. However, in some cases, a lower recovery rate is acceptable, such as for *Solanum* spp. at the International Potato Center (CIP) in Peru (Vollmer et al. 2016).

Ongoing research has enabled the cryostorage of embryogenic material from both woody and nonwoody plants in liquid nitrogen (LN), not only for scientific purposes but also for conservation efforts. Currently, suspension culture-derived embryogenic tissues from the *Gentiana* genus are cryostored at the Polish Academy of Sciences Botanical Garden-Center for Biological Diversity Conservation (PAS BG-CBDC) (Mikuła et al. 2022). Efficient propagation protocols based on SE facilitate the use of embryogenic material for the conservation, breeding, and restoration of endangered forest species. For example, ashes (*Fraxinus pennsylvanica* Marsh. and *F. americana* L.) that have survived emerald ash borer infestations in the USA and Canada have been propagated via SE (Merkle et al. 2023) and simultaneously cryostored (Richins et al. 2024). Similarly, chestnuts (*Castanea dentata* (Marshall) Borkh., *C. sativa* L.), which are threatened by fungal pathogens, can be successfully cryostored as either embryogenic cultures (Holliday and Merkle 2000; Corredoira et al. 2011) or embryonic axes (Gaidamashvili et al. 2021).

Among coniferous trees, hemilocks (*Tsuga canadensis* L., *T. caroliniana* Engelm., *T. ulleungensis* Havill.) serve as an example of a species in which cryostored SE materials are used for the propagation and conservation of pest-tolerant genotypes (Merkle et al. 2014; Ahn et al. 2018). More recently, this conservation strategy has been applied to the endangered Brazilian conifer *Podocarpus lambertii* Klotzsch, which is classified as Near Threatened on the IUCN Red List (Farjon and Filer 2013; Guerra et al. 2018). However, while cryostored materials play a valuable role in the conservation and restoration of these species, they represent genotypes selected for specific traits, such as pest resistance, and do not necessarily encompass the full genetic diversity of the species.

Another type of cryobank focuses on storing SE materials for large-scale commercial propagation. A notable example is Finland's collection of *Picea abies* (Karst.) L. embryogenic cell lines, which comprises over 5600 accessions (Aronen et al. 2025). Such collections contain a significant number of genotypes, although selection is primarily based on parental breeding values.

The choice of method for preserving genetic resources depends on multiple factors, including species characteristics, the type of biological material, available resources, and specific conservation goals. A combination of different strategies is often employed to ensure the most effective protection. It is impossible to definitively determine which approach—in situ or ex situ—is superior, as both are essential and complementary. The effectiveness of these methods depends on various conditions and challenges.

In agriculture, the need to integrate both methods for plant biodiversity conservation is particularly emphasized. This dual approach aligns with the implementation of the International Treaty and the Global Plan of Action on Plant Genetic Resources for Food and Agriculture (Rajpurohit and Jhang 2015). Field collections serve as invaluable genetic reservoirs, offering resilience to pathogens and pests, tolerance to climatic stresses, and improved traits while being safeguarded against genetic erosion. Conversely, ex situ conservation does not replicate natural evolutionary processes but serves as a critical backup in the event of unforeseen environmental disasters.

Since 2005, the Institute of Plant Breeding and Acclimatization–National Research Institute (Poland) has been supporting potato breeding research and breeding programs through a collection of wild and cultivated *Solanum* L. species, along with interspecific hybrids. The collection of diploid *Solanum* L. species includes genetic resources resistant to viral, bacterial, and fungal pathogens. Moreover, interspecific hybrids—both diploid and tetraploid hybrids—serve as potential parental forms for the breeding of new potato varieties (Zimnoch-Guzowska and Flis 2021; Zimnoch-Guzowska et al. 2022). Tetraploid hybrids are grown in the field and maintained in in vitro collections, whereas unique diploid forms undergo additional cryopreservation via shoot tip vitrification and PVS2 dehydration (Zimnoch-Guzowska et al. 2022). Another method for safeguarding the genetic diversity of *Solanum* L. involves the cryopreservation of pollen through direct immersion in liquid nitrogen (LN), ensuring its availability for future breeding efforts.

A notable example of the synergy between in situ and ex situ collections—including cryopreservation—is the creation of the European Integrated *Allium* L. Main Collection. This initiative brought together national collections from Italy, France, the Czech Republic, Poland, Germany, and Nordic countries (http://

euralliveg.ipk-gatersleben.de). As part of this project, cryogenic collections were established, incorporating the deposition of duplicates in the cryobank of one of the project partners. A segment of the highest-priority garlic collection was tested for infection by five viruses, and comprehensive strategies were developed for the long-term conservation of endangered plant gene pools, including species from Poland's natural habitats and Natura 2000 sites (Zimnoch-Guzowska et al. 2022).

Cryogenic preservation plays a crucial role in various programs aimed at protecting rare and endangered herbaceous plant species. One such initiative, the FlorNaturLBG project (2009–2012) (http://flornaturlbg.pl/), focused on the ex situ conservation of endangered and protected native herbaceous species in western Poland. As part of this effort, a comprehensive, long-term conservation strategy was developed, and seeds from 58 species collected from 129 sites across Poland were stored in LN. A sister project, FlorNaturOB (2010–2013) (https://ogrod-powsin.pl/), aimed at assembling a genetic resource collection of endangered and protected native plants in central and eastern Poland. This resulted in the cryogenic preservation of seeds from 61 species of vascular plants at risk of extinction, which were collected from 161 natural sites in Poland (Zimnoch-Guzowska et al. 2022). To further advance Poland's conservation efforts, the FlorNatur ROBiA project (2013–2015) (https://ogrod-powsin.pl) focused on assessing populations and ensuring ex situ conservation of selected rare and endangered wild plant species. Consequently, a network of regional seed banks was established, preserving seeds from 33 protected, rare, or endangered plant species collected from 75 sites nationwide.

6.6 Conclusion

Cryopreservation enables the preservation of a plant's genetic resources, primarily in the form of embryos isolated from seeds or their fragments (e.g., embryogenic axes or plumules). With the development of efficient micropropagation techniques, particularly somatic embryogenesis, embryogenic tissues, suspension cultures, and somatic embryos are increasingly being used as excellent materials for cryopreservation in liquid nitrogen (LN) across many plant species. Today, a variety of cryopreservation methods are available, continually improving to streamline procedures and enhance their effectiveness. As a result, the most advanced techniques are already in use at gene banks worldwide, and ongoing progress in this field promises the integration of even more innovative approaches into practical applications. This is especially critical in the face of climate change, as many woody and nonwoody plant species are struggling to adapt rapidly to these environmental shifts. However, owing to advancements in micropropagation and cryopreservation, new biotechnological tools are now available to preserve both a vast array of biodiversity and valuable genotypes—important for both ecological balance and economic sustainability—ensuring that they are available for future generations.

Acknowledgments This chapter is based upon work from COPYTREE—CA21157, supported by COST (European Cooperation in Science and Technology).

Competing Interests The authors have no relevant financial or non-financial interests to disclose.

References

Adams GW, Kunze HA, McCartney A et al (2016) An industrial perspective on the use of advanced reforestation stock technologies. In: Park YS, Bonga JM, Moon H-K (eds) Vegetative propagation of forest trees. National Institute of Forest Science (Nifos), Seoul, pp 323–334

Ahn C-H, Han J-Y, Kim Y-S, Choi Y-E (2018) Propagation and cryopreservation of Ulleungdo hemlock (*Tsuga ulleungensis*) via somatic embryogenesis. Trees 32:1801–1808

Amaral W, Yanchuk A, Kjaer E (2004) Methodologies for *ex situ* conservation. In: FAO, FLD, IPGRI (eds) Plantations and Genebanks (*ex situ*). Forest genetic resources conservation and management. International Plant Genetic Resources Institute, Rome

Aronen T, Virta S, Varis S (2021) Telomere length in Norway spruce during somatic embryogenesis and cryopreservation. Plants 10:1–14

Aronen T, Varis S, Valimaki S et al (2025) Paving the way for commercial somatic embryogenesis: Norway spruce in Finland. In: Proc 6th IUFRO 20902 Conf Might Veg Propag Healthy Product For Face Clim Chall, 3–8 Mar 2024

Ayala LPE, Luna CV, Brugnoli EA et al (2023) Cryopreservation of mature zygotic embryos, shoot bud regeneration, and field establishment of *Pinus elliottii* var. *elliottii x Pinus caribaea var. hondurensis in vitro*-derived plants. Trees 37:417–433

Baghdadi SH, Makhadmeh I, Syouf M et al (2009) Cryopreservation by vitrification of embryogenic callus of wild crocus (*Crocus hyemalis* and *Crocus moabiticus*). In: International symposium on cryopreservation in horticultural species, vol 908, ISHS, Leuven, Belgium, pp 239–246

Ballesteros D, Pritchard HW (2020) The cryobiotechnology of oaks: an integration of approaches for the long-term *ex situ* conservation of Quercus species. Forests 11:1281

Ballesteros D, Sershen, Varghese B et al (2014) Uneven drying of zygotic embryos and embryonic axes of recalcitrant seeds: challenges and considerations for cryopreservation. Cryobiology 69:100–109

Ballesteros D, Martínez MT, Sánchez-Romero C et al (2024) Current status of the cryopreservation of embryogenic material of woody species. Front Plant Sci 14. https://doi.org/10.3389/fpls.2023.1337152

Barrett RDH, Schluter D (2008) Adaptation from standing genetic variation. Trends Ecol Evol 23:38–44

Bazrafshan A, Sudarma S, Kalaipandian S et al (2025) New method for enhancing coconut (*Cocos nucifera* L.) embryo dehydration: an important step towards proficient cryopreservation. Plants 14:600

Bekheet SA, Sota V, El-Shabrawi HM, El-Minisy AM (2020) Cryopreservation of shoot apices and callus cultures of globe artichoke using vitrification method. J Genet Eng Biotechnol 18:1–8

Ben-Amar A, Daldoul S, Allel D et al (2013) Reliable encapsulation-based cryopreservation protocol for safe storage and recovery of grapevine embryogenic cell cultures. Sci Hortic 157:32–38

Benelli C (2021) Plant cryopreservation: a look at the present and the future. Plants 10:2744

Benmahioul B, Daguin F, Kaïd-Harche M (2015) Cryopreservation of pistacia vera embryonic axes – short communication. J For Sci 61:182–187

Benson EE (2008) Cryopreservation of phytodiversity: a critical appraisal of theory & practice. Crit Rev Plant Sci 27:141–219

Bettoni JC, Kretzschmar AA, Bonnart R et al (2019) Cryopreservation of 12 vitis species using apical shoot tips derived from plants grown in vitro. HortScience 54:976–981

Bettoni JC, Bonnart R, Volk GM (2021) Challenges in implementing plant shoot tip cryopreservation technologies. Plant Cell Tiss Organ Cult 144:21–34
Bomal C, Tremblay F-M (2000) Dried cryopreserved somatic embryos of two picea species provide suitable material for direct plantlet regeneration and germplasm storage. Ann Bot 86:177–183
Bradaï F, Sánchez-Romero C (2021a) Effect of cryopreservation on olive (*Olea europaea* L.) plant regeneration via somatic embryogenesis. Plants 10:34
Bradaï F, Sánchez-Romero C (2021b) Effect of cryopreservation on the ex vitro establishment of olive plants regenerated via somatic embryogenesis. Plants 10:396
Bradley St Clair J, Howe GT (2007) Genetic maladaptation of coastal Douglas-fir seedlings to future climates. Glob Chang Biol 13:1441–1454
Bramasto Y, Putri KP, Yuniarti N, Aminah A et al (2020) Seed handling of specific forest tree species: recalcitrant and intermediate seed. Earth Environ Sci 522:012015
Breman E, Ballesteros D, Castillo-Lorenzo E et al (2021) Plant diversity conservation challenges and prospects—the perspective of botanic gardens and the Millennium Seed Bank. Plants 10:2371
Ceballos G, Ehrlich PR, Barnosky AD et al (2015) Accelerated modern human–induced species losses: entering the sixth mass extinction. Sci Adv 1:e1400253
Chithra Devi BS, Sangilimuthu A, Daanya V (2021) Slow-growth conservation- cryopreservation and analysis of genetic stability of in vitro regenerated Coelogyne Nervosa A.Rich A endemic orchid. Int J Res Pharm Sci 12:1695–1698
Chmielarz P (1998) Reakcje roślin na niskie temperatury. Metody kriokonserwacji komórek i organów roślin w ciekłym azocie. Instytut Dendrologii PAN. Wydawnictwo Z. Bartkowiak, Poznań
Chmielarz P (2009) Cryopreservation of dormant European ash (*Fraxinus excelsior*) orthodox seeds. Tree Physiol 29:1279–1285
Chmielarz P, Michalak M, Pałucka M, Wasileńczyk U (2011) Successful cryopreservation of *Quercus robur* plumules. Plant Cell Rep 30:1405–1414
Coelho SVB, Rosa SDVF, Fernandes JS (2017) Cryopreservation of coffee seeds: a simplified method. Seed Sci Technol 45:638–649
Convention on Biological Diversity (1992) Convention on Biological Diversity. United Nations. https://www.cbd.int/doc/legal/cbd-en.pdf
Corredoira E, San-José MC, Ballester A, Vieitez AM (2004) Cryopreservation of zygotic embryo axes and somatic embryos of European chestnut. Cryo Letters 25:33–42
Corredoira E, San José MC, Martínez T et al (2011) Conservación de germoplasma en especies leñosas con técnicas de cultivo *in Vitro* y almacenamiento en frío. Span J Rural Dev:15–24. https://doi.org/10.5261/2011.GEN1.02
da Silva Lédo A, Santana FV, de Oliveira ACA et al (2020) Cryopreservation of Brazilian green dwarf coconut plumules by droplet-vitrification. Cienc Rural 50:1–8
de Almeida PJ, de Oliveira SB, Cantelmo L et al (2017) Cryopreservation of somatic embryos from Petiveria alliacea L. by different techniques based on vitrification. In Vitro Cell Dev Biol Plant 53:339–345
de Pettinelli JA, de Soares BO, Collin M et al (2020) Cryotolerance of somatic embryos of Guinea (Petiveria alliacea) to V-cryoplate technique and histological analysis of their structural integrity. Acta Physiol Plant 42:1–10
Delgado-Aceves L, Gonzalez-Arnao MT, Portillo L, Folgado R (2019) Cryobiotechnology for somatic embryos of Agave Tequilana Cv. Chato: an endangered genetic resource. Cryobiology 91:180–180
Delgado-Aceves L, González-Arnao MT, Santacruz-Ruvalcaba F et al (2021) Indirect somatic embryogenesis and cryopreservation of agave tequilana weber cultivar 'chato'. Plants 10:1–14
Díaz SM, Settele J, Brondízio E et al (2019) The global assessment report on biodiversity and ecosystem services: summary for policy makers. IPBES secretariat, Bonn, Germany, ISBN: 978-3-947851-20-1
Distabanjong K, Distabanjong C, Jang S-W (2015) Developing regeneration system for cryopreservation in sugarcane (*Saccharum officinarum* L.). In VIII International Symposium on In

Vitro Culture and Horticultural Breeding. Acta Hort 1083:427–433. https://doi.org/10.17660/ActaHortic.2015.1083.55

Dussert S, Vasquez N, Salazar K, Anthony F & Engelmann F (2007) VI. Cryopreservation of coffee genetic resources. In: Engelmann F, Dulloo ME, Astorga C, Dussert S, Anthony F (eds.), Conserving coffee genetic resources: complementary strategies for ex situ conservation of coffee (*Coffea arabica* L.) genetic resources: a case study in CATIE, Costa Rica Rome (ITA); Paris : Biodiversity International; IRD, 49-58 ISBN 978-92-9043-752-9

Dyderski MK, Paź-Dyderska S, Jagodziński AM, Puchałka R (2025) Shifts in native tree species distributions in Europe under climate change. J Environ Manag 373:123504

Edesi J, Tolonen J, Ruotsalainen AL et al (2020) Cryopreservation enables long-term conservation of critically endangered species Rubus humulifolius. Biodivers Conserv 29:303–314

Ellis RH, Hong TD, Roberts EH (1990) An intermediate category of seed storage behaviour?: I. Coffee. J Exp Bot 41:1167–1174

Engelmann F (2004) Plant cryopreservation: progress and prospects. In Vitro Cell Dev Biol Plant 40:427–433

Engelmann F (2011) Use of biotechnologies for the conservation of plant biodiversity. In Vitro Cell Dev Biol Plant 47:5–16

Engelmann F (2012) 17 – Germplasm collection, storage, and conservation. In: Altman A, Hasegawa PM (eds) Plant biotechnology and agriculture. Academic Press, San Diego, pp 255–267

Esquivel-Figueroa D, Tiznado-Hernández ME, Islas-Osuna MA et al (2023) Etapas del protocolo de crioconservación y su efecto sobre la viabilidad y regeneración de embriones cigóticos de vid (*Vitis vinífera* L.). Biotecnia 25:170–175

Fang J-Y, Wetten A, Adu-Gyamfi R et al (2009) Use of secondary somatic embryos promotes genetic fidelity in cryopreservation of cocoa (*Theobroma cacao* L.). Agric Food Sci 18:152–159

Farjon A, Filer D (2013) An atlas of the world's conifers: an analysis of their distribution, biogeography, diversity and conservation status. Brill, Leiden

Fatima S, Mujib A, Nasim SA, Siddiqui ZH (2009) Cryopreservation of embryogenic cell suspensions of Catharanthus roseus L. (G) Don. Plant Cell Tissue Organ Cult 98:1–9

Find JI, Floto F, Krogstrup P et al (1993) Cryopreservation of an embryogenic suspension culture of *Picea sitchensis* and subsequent plant regeneration. Scand J For Res 8:156–162

Find JI, Kristensen MMH, Nørgaard JV, Krogstrup P (1998) Effect of culture period and cell density on regrowth following cryopreservation of embryogenic suspension cultures of Norway spruce and Sitka spruce. Plant Cell Tissue Organ Cult 53:27–33

Gaidamashvili M, Khurtsidze E, Kutchava T et al (2021) Efficient protocol for improving the development of cryopreserved embryonic axes of chestnut (*Castanea sativa* Mill.) by encapsulation–vitrification. Plants 10:231

Gallard A, Panis B, Chevalier M et al (2006) 141. Comparative study between encapsulation–dehydration and droplet-vitrification processes for the cryopreservation of pelargonium apices. Cryobiology 53:426–427

Gao F, Chen S, Qin C et al (2021) Optimization of somatic embryogenesis system and cryopreservation of *Picea koraiensis*. J Nanjing For Univ 45:100

García-Coronado H, Báez-Flores M-E, Troncoso-Rojas R et al (2016) Cryopreservation by vitrification of *Vitis vinifera* cv. "Red Globe" zygotic embryos and effect on the expression of DNA methyltransferase genes. J Agric Sci Technol B 6:387–399

George J-P, Yrjänä L, Beuker E, Rusanen M (2024) Ex situ conservation of forest genetic resources in Finland: evaluation and adaptation of the current strategy. http://urn.fi/URN:ISBN:978-952-380-895-9

Guerra MP, do Nascimento Vieira L, de Freitas Fraga HP (2018) Somatic embryogenesis of Brazilian Conifer *Podocarpus lambertii* Klotzsch ex Endl. In: Jain SM, Gupta P (eds) Step wise protocols for somatic embryogenesis of important woody plants, vol I. Springer International Publishing, Cham, pp 83–91

Gupta R, Sood H (2022) Artificial seed production and cryopreservation by encapsulation dehydration for medicinal herb of Himalayan Region, Swertia chirayita. Cryo Letters 43:295–302

Gupta PK, Durzan DJ, Finkle BJ (1987) Somatic polyembryogenesis in embryogenic cell masses of *Picea abies* (Norway spruce) and *Pinus taeda* (loblolly pine) after thawing from liquid nitrogen. Can J For Res 17:1130–1134
Häggman HM, Ryynänen LA, Aronen TS, Krajnakova J (1998) Cryopreservation of embryogenic cultures of Scots pine. Plant Cell Tissue Organ Cult 54:45–53
Häggman HM, Aronen TS, Ryynänen LA (2000) Cryopreservation of embryogenic cultures of conifers. In: Jain SM, Gupta PK, Newton RJ (eds) Somatic embryogenesis in woody plants, vol 6. Springer Netherlands, Dordrecht, pp 707–728
Hargreaves C, Grace L, van der Maas S et al (2004) Cryopreservation of *Pinus radiata* zygotic embryo cotyledons: effect of storage duration on adventitious shoot formation and plant growth after 2 years in the field. Can J For Res 34:600–608
Hazubska-Przybył T, Chmielarz P, Michalak M, Bojarczuk K (2010) Cryopreservation of embryogenic tissues of *Picea omorika* (Serbian spruce). Plant Cell Tissue Organ Cult 102:35–44
Hazubska-Przybył T, Chmielarz P, Michalak M et al (2013) Survival and genetic stability of *Picea abies* embryogenic cultures after cryopreservation using a pregrowth-dehydration method. Plant Cell Tissue Organ Cult 113:303–313
Hazubska-Przybył T, Ratajczak E, Obarska A, Pers-Kamczyc E (2020) Different roles of auxins in somatic embryogenesis efficiency in two *Picea* species. Int J Mol Sci 21:3394
Hoffmann AA, Sgrò CM (2011) Climate change and evolutionary adaptation. Nature 470:479–485
Holliday C, Merkle SA (2000) Preservation of American chestnut germplasm by cryostorage of embryogenic cultures. J Am Chestnut Found 14:46–52
Hong S, Yin M, Shao X, Wang A, Xu W (2009) Cryopreservation of embryogenic cultures of Dioscorea bulbifera l. by encapsulation- dehydration. Cryo Letters 30:440–448
Impe D, Ballesteros D, Nagel M (2022) Impact of drying and cooling rate on the survival of the desiccation-sensitive wheat pollen. Plant Cell Rep 41:447–461
Ishikawa M, Tandon P, Suzuki M, Yamaguishi-Ciampi A (1996) Cryopreservation of bromegrass (*Bromus inermis* Leyss) suspension cultured cells using slow prefreezing and vitrification procedures. Plant Sci 120:81–88
Jain SM, Maluszynski M (2004) Induced mutations and biotechnology on improving crops. In: In vitro applications in crop improvement: Recent Progress. Science Publishers. Publisher: IBH-Oxford, India, CRC Press, pp 169–202
Kartha KK (1985) Meristem culture and germplasm preservation. In: Cryopreservation of plant cells and organs. CRC Press, Boca Raton
Kartha KK, Fowke LC, Leung NL et al (1988) Induction of somatic embryos and plantlets from cryopreserved cell cultures of white spruce (*Picea glauca*). J Plant Physiol 132:529–539
Kelleher CT, Baliuckas V, Bozzano M et al (2015) Approaches to the conservation of forest genetic resources in Europe in the context of climate change. Bioversity International, Rome
Kholina AB, Voronkova NM (2013) Seed cryopreservation of some medicinal legumes. J Bot 2012:186891
Kim HM, Shin JH, Sohn JK (2006) Cryopreservation of somatic embryos of the herbaceous peony (*Paeonia lactiflora* Pall.) by air drying. Cryobiology 53:69–74
Klimaszewska K, Ward C, Cheliak WM (1992) Cryopreservation and plant regeneration from embryogenic cultures of larch (*Larix* × *eurolepis*) and black spruce (*Picea mariana*). J Exp Bot 43:73–79
Kong L, von Aderkas P (2011) A novel method of cryopreservation without a cryoprotectant for immature somatic embryos of conifer. Plant Cell Tissue Organ Cult 106:115–125
Kulus D, Tymoszuk A, Kulpińska A, Osial M (2025) Synergistic effects of iron oxide nanoparticles and indole-3-acetic acid on the germination and development of cold-stored chrysanthemum synthetic seeds. Plant Cell Tissue Organ Cult 160:18
Lambardi M, Ozudogru EA, Benelli C (2008) Cryopreservation of embryogenic cultures. In: Reed BM (ed) Plant cryopreservation: a practical guide. Springer New York, New York, pp 177–210
Lasushe K, Mir AH, Singh PP et al (2022) A comprehensive checklist of threatened plants of Meghalaya, Northeast India. J Asia-Pac Biodivers 15:435–441

Latta R (1971) Preservation of suspension cultures of plant cells by freezing. Can J Bot 49:1253–1254

Latutrie M, Aronen T (2013) Long-term cryopreservation of embryogenic *Pinus sylvestris* cultures. Scand J For Res 28:103–109

Lee H (2023) *Ex situ* conservation efforts for plant diversity protection with a focus on seeds. Open Agric J 17:e187433152307250

Lei X, Wang Q, Yang H et al (2021) Vitrification and proteomic analysis of embryogenic callus of Panax ginseng C. A. Meyer. In Vitro Cell Dev Biol Plant 57:118–127

Libiad M, Khabbach A, El Haissoufi M et al (2020) *Ex-situ* conservation of single-country endemic plants of Tunisia and northern Morocco (Mediterranean coast and Rif region) in seed banks and botanic gardens worldwide. Kew Bull 75:46

Lukić M, Funnekotter B, Considine MJ et al (2022) Evaluation of vacuum infiltration vitrification and cryo-mesh cryopreservation techniques with Arabidopsis thaliana shoot tips. Cryo Letters 43:328–333

Lynch PT, Benson EE, Jones J et al (1995) The embryogenic potential of rice cell suspensions affects their recovery following cryogenic storage. In: Cassells AC, Jones PW (eds) The methodology of plant genetic manipulation: criteria for decision making. Springer Netherlands, Dordrecht, pp 347–349

Makeen MA, Noor NM, Dussert S, Clyde MM (2005) Cryopreservation of whole seeds and excised embryonic axes of Citrus suhuiensis cv. limau langkat in accordance to their desiccation sensitivity. Cryo Letters 26:259–268

Malhotra EV, Mali SC, Sharma S, Bansal S (2024) A droplet vitrification cryopreservation protocol for conservation of hops (*Humulus lupulus*) genetic resources. Cryobiology 115:104887

Markowski M, Czarnomska Z, Tomiczak K et al (2024) The influence of cryopreservation via encapsulation-dehydration on growth kinetics, embryogenic potential and secondary metabolite production of cell suspension cultures of *Gentiana capitata* Buch.-Ham. ex D.Don and Gentiana decumbens L.f. Ind Crop Prod 212:118349

Martínez MT, Suárez S, Moncaleán P, Corredoira E (2022) Cryopreservation of holm oak embryogenic cultures for long-term conservation and assessment of polyploid stability. Plants 11:1266

Martínez-Montero ME, González-Arnao MT, Borroto-Nordelo C et al (1998) Cryopreservation of sugarcane embryogenic callus using a simplified freezing process. Cryo Letters 19:171–176

Maślanka M, Panis B, Malik M (2016) Cryopreservation of Narcissus L. "Carlton" somatic embryos by droplet vitrification. Prop Orn Plants 16(1):28–35

Merkle SA, Montello PM, Reece HM, Kong L (2014) Somatic embryogenesis and cryostorage of eastern hemlock and Carolina hemlock for conservation and restoration. Trees 28:1767–1776

Merkle SA, Koch JL, Tull AR et al (2023) Application of somatic embryogenesis for development of emerald ash borer-resistant white ash and green ash varietals. New For 54:697–720

Michalak M, Plitta BP, Tylkowski T et al (2015) Desiccation tolerance and cryopreservation of seeds of black poplar (*Populus nigra* L.), a disappearing tree species in Europe. Eur J For Res 134:53–60

Miguel C, Marum L (2011) An epigenetic view of plant cells cultured in vitro: somaclonal variation and beyond. J Exp Bot 62:3713–3725

Mikuła A, Tomiczak K, Rybczyński JJ (2011) Cryopreservation enhances embryogenic capacity of *Gentiana cruciata* (L.) suspension culture and maintains (epi)genetic uniformity of regenerants. Plant Cell Rep 30:565–574

Mikuła A, Makowski D, Tomiczak K, Rybczyński JJ (2013) Kultury in vitro i krioprezerwacja w zachowaniu różnorodności roślin–standardy dla banku genów. Pol J Agron 14:3–17

Mikuła A, Chmielarz P, Hazubska-Przybył T et al (2022) Cryopreservation of plant tissues in Poland: research contributions, current status, and applications. Acta Soc Bot Pol 91:9132

Mosa KA, Ahmed AE, Hazem Y et al (2023) Insights into cryopreservation, recovery and genetic stability of medicinal plant tissues. Fitoterapia 169:105555

Mujib A, Fatima S, Malik MQ (2022) Cryo-derived plants through embryogenesis showed same levels of vinblastine and vincristine (anticancer) in *Catharanthus roseus* and had normal genome size. Sci Rep 12:1–13

Murray BG (2017) Plant diversity, conservation and use. In: Thomas B, Murray BG, Murphy DJ (eds) Encyclopedia of applied plant sciences, 2nd edn. Academic Press, Oxford, pp 289–308
N'Nan O, Borges M, Konan JLK et al (2012) A simple protocol for cryopreservation of zygotic embryos of ten accessions of coconut (*Cocos nucifera* L.). In Vitro Cell Dev Biol Plant 48:160–166
Nadarajan J, Pritchard HW (2014) Biophysical characteristics of successful oilseed embryo cryoprotection and cryopreservation using vacuum infiltration vitrification: an innovation in plant cell preservation. PLoS One 9:e96169
Nagel M, Pence V, Ballesteros D et al (2024) Plant cryopreservation: principles, applications, and challenges of banking plant diversity at ultralow temperatures. Annu Rev Plant Biol 75:797–824
Niino T, Yamamoto S, Fukui K et al (2013) Dehydration improves cryopreservation of mat rush (*Juncus decipiens* Nakai) basal stem buds on cryo-plates. Cryo Letters 34:549–560
Niino T, Yamamoto S, Matsumoto T et al (2019) Development of V and D cryo-plate methods as effective protocols for cryobanking. Acta Hortic:249–262. https://doi.org/10.17660/ActaHortic.2019.1234.33
Nishizawa S, Sakai A, Amano Y, Matsuzawa T (1993) Cryopreservation of asparagus (Asparagus officinalis L.) embryogenic suspension cells and subsequent plant regeneration by vitrification. Plant Sci 91:67–73
Normah MN, Makeen AM (2008) Cryopreservation of excised embryos and embryonic axes. In: Reed BM (ed) Plant cryopreservation: a practical guide. Springer New York, New York, pp 211–240
O'Brien C, Constantin M, Walia A et al (2016) Cryopreservation of somatic embryos for avocado germplasm conservation. Sci Hortic 211:328–335
O'Brien C, Hiti-Bandaralage J, Folgado R et al (2020) Cryopreservation for tree species with recalcitrant seeds: the avocado case. https://doi.org/10.20944/preprints202012.0304.v1
Palanyandy SR, Gantait S, Subramaniam S, Sinniah UR (2020) Cryopreservation of oil palm (*Elaeis guineensis Jacq.*) polyembryoids via encapsulation–desiccation. 3 Biotech 10:9
Pammenter NW, Berjak P (2014) Physiology of desiccation-sensitive (recalcitrant) seeds and the implications for cryopreservation. Int J Plant Sci 175:21–28
Panis B (2019) Sixty years of plant cryopreservation: from freezing hardy mulberry twigs to establishing reference crop collections for future generations. Acta Hortic:1–8. https://doi.org/10.17660/ActaHortic.2019.1234.1
Panis B, Withers LA, De Langhe EAL (1990) Cryopreservation of Musa suspension cultures and subsequent regeneration of plants. Cryo Letters 11:337–350
Panis B, Piette B, Swennen R (2005) Droplet vitrification of apical meristems: a cryopreservation protocol applicable to all Musaceae. Plant Sci 168:45–55
Parasher K, Sharma S, Mukherjee P, Qazi PH (2023) Cryopreservation of zygotic embryos of *Podophyllum hexandrum* Royle, an endangered medicinal plant, by vitrification and v cryo-plate techniques. Cryo Letters 44:219–228
Park Y, Bonga JM (2010) Application of somatic embryogenesis in forest management and research. In: Proceedings of the IUFRO Working Party 2.09.02 conference, Suwon, Republic of Korea, pp 3–8
Peng C, Gao F, Wang H et al (2021) Optimization of maturation process for somatic embryo production and cryopreservation of embryogenic tissue in *Pinus koraiensis*. Plant Cell Tissue Organ Cult 144:185–194
Percy REL, Livingston NJ, Moran JA, Von Aderkas P (2001) Desiccation, cryopreservation and water relations parameters of white spruce (*Picea glauca*) and interior spruce (*Picea glauca* × *engelmannii* complex) somatic embryos. Tree Physiol 21:1303–1310
Peredo EL, Arroyo-García R, Reed BM, Revilla MÁ (2009) Genetic stability of in vitro conserved germplasm of *Humulus lupulus* L. Agric Food Sci 18(2):144–151
Pita JM, Sanz V, Escudero A (1998) Seed cryopreservation of seven Spanish native pine species. Silvae Genet 47:220–223
Prada JA, Aguilar ME, Abdelnour-Esquivel A, Engelmann F (2015) Cryopreservation of seeds and embryos of Jatropha curcas L. Am J Plant Sci 06:172–180

Pritchard HW, Manger KR (1998) A calorimetric perspective on desiccation stress during preservation procedures with recalcitrant seeds of *Quercus robur* L. Cryo Letters 19:23–30

Pukacki PM, Juszczyk K (2015) Desiccation sensitivity and cryopreservation of the embryonic axes of the seeds of two Acer species. Trees 29:385–396

Rajpurohit D, Jhang T (2015) In situ and ex situ conservation of plant genetic resources and traditional knowledge. In: Salgotra RK, Gupta BB (eds) Plant genetic resources and traditional knowledge for food security. Springer, Singapore, pp 137–162

Reed BM (1996) Chapter 3. Pretreatment strategies for the cryopreservation of plant tissues. In: Normah MN, Narimah MK, Clyde MM (eds.) In Vitro Conservation of Plant Genetic Resources. Plant Biotechnology laboratory, Faculty of Life Sciences, Universiti Kebangsaan, Malaysia, 73–87

Reed BM (2001) Implementing cryogenic storage of clonally propagated plants. Cryo Letters 22:97–104

Reed BM, Hummer KE (2001) Long-term storage of hazelnut embryonic axes in liquid nitrogen. Acta Hortic 556:177–180

Ren L, Deng S, Chu Y et al (2020) Single-wall carbon nanotubes improve cell survival rate and reduce oxidative injury in cryopreservation of *Agapanthus praecox* embryogenic callus. Plant Methods 16:130

Richins M, Montes C, Merkle S (2024) Conservation of green and white ash germplasm using the cryopreservation of embryogenic cultures. Plants 13:352

Roberts EH (1972) Storage environment and the control of viability. In: Roberts EH (ed) Viability of seeds. Springer Netherlands, Dordrecht, pp 14–58

Roostika I, Damayanti F, Witjaksono (2022) Successful preservation of banana embryogenic callus by minimal growth and cryopreservation technique. J Teknol 84:167–176

Ruta C, Lambardi M, Ozudogru EA (2020) Biobanking of vegetable genetic resources by in vitro conservation and cryopreservation. Biodivers Conserv 29:3495–3532

Ryynänen L (1996) Cold hardening and slow cooling: tools for successful cryopreservation and recovery of in vitro shoot tips of silver birch. Can J For Res 26:2015–2022

Ryynänen L (1998) Effect of abscisic acid, cold hardening, and photoperiod on recovery of cryopreserved in vitro shoot tips of silver birch. Cryobiology 36:32–39

Sakai A (1956) Survival of plant tissue of super-low temperatures. Low Temp Sci B 14:17–23

Sakai A (1960) Survival of the twig of woody plants at −196° C. Nature 185:393–394

Sakai A (1966) Survival of plant tissue at super-low temperatures. IV. Cell survival with rapid cooling and rewarming. Plant Physiol 41:1050–1054

Sakai A (2004) Plant cryopreservation. In: Fuller BJ, Lane N, Benson EE (eds) CRS Press, Boca Raton, London, New York, Washington, DC, pp 329–345

Sakai A, Kobayashi S, Oiyama I (1990) Cryopreservation of nucellar cells of navel orange (*Citrus sinensis* Osb. var. brasiliensis Tanaka) by vitrification. Plant Cell Rep 9:30–33

Salaj T, Blehová A, Salaj J (2007) Embryogenic suspension cultures of *Pinus nigra* Arn.: growth parameters and maturation ability. Acta Physiol Plant 29:225–231

Salaj T, Panis B, Klubicova K, Salaj J (2022) Cryopreservation of *Abies alba* embryogenic tissues by slow-freezing method. Not Bot Horti Agrobot Cluj-Napoca 50:12770

San José MC, Janeiro LV, Corredoira E (2015) Simple strategy for the in vitro conservation of *Alnus* glutinosa (L.) Gaertn germplasm. Trees 29:539–549

Sánchez Fernández C, Martínez-Santiago MT, Vidal González NP et al (2008) Preservation of *Quercus robur* germplasm by cryostorage of embryogenic cultures derived from mature trees and rapd analysis of genetic stability. Cryo Letters 29:493–504

Scocchi A, Vila S, Mroginski L, Engelmann F (2007) Cryopreservation of somatic embryos of paradise tree (*Melia azedarach* L.). Cryo Letters 28:281–290

Seidl R, Schelhaas M-J, Rammer W, Verkerk PJ (2014) Increasing forest disturbances in Europe and their impact on carbon storage. Nat Clim Chang 4:806–810

Sen-Rong H, Ming-Hua Y (2012) A simple and efficient protocol for cryopreservation of embryogenic calli of the medicinal plant *Anemarrhena asphodeloides* Bunge by vitrification. Plant Cell Tissue Organ Cult 109:287–296

Sharma S, Mukherjee P (2023) Improved cryopreservation protocol for tea (*Camellia sinensis* (L.) O. Kuntze) using preconditioning of shoot tip donor plants and V cryo-plate technique. In Vitro Cell Dev Biol Plant 59:285–297
Singh S, Agrawal A, Kumar R et al (2021a) Seed storage behavior of *Musa balbisiana* Colla, a wild progenitor of bananas and plantains – implications for ex situ germplasm conservation. Sci Hortic 280:109926–109926
Singh S, Thangjam R, Harish GD et al (2021b) Conservation protocols for *Ensete glaucum*, a crop wild relative of banana, using plant tissue culture and cryopreservation techniques on seeds and zygotic embryos. Plant Cell Tissue Organ Cult 144:195–209
Sisunandar RA, Turquay P et al (2010) Cryopreservation of coconut (*Cocos nucifera* L.) zygotic embryos does not induce morphological, cytological or molecular changes in recovered seedlings. Planta 232:435–447
Soejima K, Shimoda K, Takeuchi M et al (2012) Preliminary report of novel technique for cryopreservation—vacuum-assisted cryoprotectant infiltration. Cryobiology 65:21–26
Song J, Lee Y, Yi J et al (2020) Preconditioning for cryopreservation of in vitro grown bulblets of lily using droplet-vitrification. Korean J Plant Resour 33:689–695
Song D, Jaganathan GK, Li J et al (2024) The optimal moisture content for cryopreservation of Longjing tea seeds. Cryo Letters 45:380–387
Stanwood PC (1985) Cryopreservation of seed germplasm for genetic conservation. In: Cryopreservation of plant cells and organs. CRC Press, Boca Raton
Suranthran P, Gantait S, Sinniah UR, Subramaniam S, Alwee SSRS, Roowi SH (2012) Effect of loading and vitrification solutions on survival of cryopreserved oil palm polyembryoids. Plant Growth Regul 66:101–109
Suzuki M, Ishikawa M, Okuda H et al (2006) Physiological changes in gentian axillary buds during two-step preculturing with sucrose that conferred high levels of tolerance to desiccation and cryopreservation. Ann Bot 97:1073–1081
Tacán M, Tapia C, Pérez C (2017) Effects of accelerated ageing and cryopreservation on seeds and embryonic axes of *Phaseolus vulgaris* L and *Arachis hypogaea* L. germination and seedlings vigor. PeerJ Prepr 5:e3203v1
Taniguchi T, Konagaya K, Nanasato Y (2020) Somatic embryogenesis in artificially pollinated seed families of 2nd generation plus trees and cryopreservation of embryogenic tissue in *Cryptomeria japonica* D. Don (Sugi). Plant Biotechnol 37:239–245
Taratima W, Panglilad K, Trunjaruen A, Maneerattanarungroj P (2022) Seed and embryogenic callus cryopreservation of Thai rice (*Oryza sativa* L.) "Hom Mali Daeng". Int J Agric Technol 18:2663–2674
Teixeira da Silva JA, Kim H, Engelmann F (2015) Chrysanthemum low-temperature storage and cryopreservation: a review. Plant Cell Tissue Organ Cult 120:423–440
Toivonen PMA, Kartha KK (1989) Cryopreservation of cotyledons of nongerminated white spruce [*Picea glauca* (Moench) Voss] embryos and subsequent plant regeneration. J Plant Physiol 134:766–768
Tokgöz HB, Karakaş H, Kaya E et al (2023) Cryopreservation of *Lilium candidum* germplasm: analysis of pre- and post-freeze treatments. Cryo Letters 44:263–273
Us-Camas R, Rivera-Solís G, Duarte-Aké F, De-la-Peña C (2014) *In vitro* culture: an epigenetic challenge for plants. Plant Cell Tissue Organ Cult 118:187–201
Valdés YC, Shukla MR, González Vega ME, Saxena PK (2021) Improved conservation of coffee (Coffea arabica L.) germplasm via micropropagation and cryopreservation. Agronomy 11:1861
Valladares S, Toribio M, Celestino C, Vieitez AM (2004) Cryopreservation of embryogenic cultures from mature *Quercus suber* trees using vitrification. Cryo Letters 25:177–186
van der Walt K, Burritt DJ, Nadarajan J (2022) Impacts of rapid desiccation on oxidative status, ultrastructure and physiological functions of *Syzygium maire* (Myrtaceae) zygotic embryos in preparation for cryopreservation. Plants 11:1056
van der Walt K, Nadarajan J, Mathew L et al (2023) Advances in cryopreservation of Syzygium maire (swamp maire, maire tawake) zygotic embryos, a critically endangered tree species endemic to New Zealand. Front Conserv Sci 4:1269881

Varis S, Ahola S, Jaakola L, Aronen T (2017) Reliable and practical methods for cryopreservation of embryogenic cultures and cold storage of somatic embryos of Norway spruce. Cryobiology 76:8–17

Varis SA, Virta S, Montalbán IA, Aronen T (2022) Reducing pre- and post-treatments in cryopreservation protocol and testing storage at −80 °C for Norway spruce embryogenic cultures. Int J Mol Sci 23:15516

Vieitez A, San José M, Corredoira E (2011) Cryopreservation of zygotic embryonic axes and somatic embryos of European chestnut. Methods Mol Biol 710:201–213

Vilardo AFRM, Silva Lima GD, Cordeiro LS et al (2023) Application of d cryo-plate technique for the cryopreservation of in vitro-grown shoot tips of *Tarenaya spinosa* (Cleomaceae). Cryo Letters 44:333–342

Vollmer R, Villagaray R, Egusquiza V et al (2016) The potato cryobank at the International Potato Center (CIP): a model for long term conservation of clonal plant genetic resources collections of the future. Cryo Letters 37:318–329

Walters C, Pence VC (2021) The unique role of seed banking and cryobiotechnologies in plant conservation. Plants People Planet 3:83–91

Walters C, Touchell DH, Power P et al (2002) A cryopreservation protocol for embryos of the endangered species *Zizania texana*. Cryo Letters 23:291–298

Wang M-R, Bi W, Shukla MR et al (2021) Epigenetic and genetic integrity, metabolic stability, and field performance of cryopreserved plants. Plants 10:1889

Wasileńczyk U, Wawrzyniak MK, Martins JPR et al (2024) Cryopreservation of sessile oak (*Quercus petraea* (Matt.) Liebl.) plumules using aluminium cryo-plates: influence of cryoprotection and drying. Plant Methods 20:53

Wei Q, Shi P, Khan FS et al (2023) Cryopreservation and cryotolerance mechanism in zygotic embryo and embryogenic callus of oil palm. Forests 14:966

Winkelmann T, Mußmann V, Serek M (2004) Cryopreservation of embryogenic suspension cultures of *Cyclamen persicum* Mill. Plant Cell Rep 23:1–8

Wyse SV, Dickie JB (2017) Predicting the global incidence of seed desiccation sensitivity. J Ecol 105:1082–1093

Wyse SV, Dickie JB, Willis KJ (2018) Seed banking not an option for many threatened plants. Nat Plants 4:848–850

Yamamoto S, Rafique T, Saman Priyantha W et al (2011) Development of a cryopreservation procedure using aluminium cryo-plates. Cryo Letters 32:3

Yamamoto S, Wunna, Rafique T et al (2015) The aluminum cryo-plate increases efficiency of cryopreservation protocols for potato shoot tips. Am J Potato Res 92:250–257

Yong SH, Park KB, Seol YW et al (2023) Optimization conditions for cryopreservation of *Nelumbo nucifera* Gaertn., aquatic plant. Plant Biotechnol Rep 17:625–635

Zeliang PK, Pattanayak A (2024) Cryopreservation of embryogenic callus in *Oryza sativa* L.: assessment of impact of callus age on regeneration; morphological and genetic stability regenerants. Plant Cell Tissue Organ Cult 158:23

Zeliang PK, Pattanayak A, Iangrai B et al (2010) Fertile plant regeneration from cryopreserved calli of *Oryza rufipogon* Griff. and assessment of variation in the progeny of regenerated plants. Plant Cell Rep 29:1423–1433

Zhang Q, Xiang W, Yao J et al (2015) Direct proembryogenic masses initiation and plant regeneration from immature *Torreya grandis* embryos. Trees 29:1605–1612

Zhang Y, Deng S, Lin H et al (2024) The development of a procedure for the cryopreservation of the callus of *Anthurium andraeanum* by vitrification. Plants 13(21):3106

Zimnoch-Guzowska E, Flis B (2021) Over 50 years of potato parental line breeding programme at the plant breeding and acclimatization institute in Poland. Potato Res 64:743–760

Zimnoch-Guzowska E, Chmielarz P, Wawrzyniak MK et al (2022) Polish cryobanks: research and conservation of plant genetic resources. Acta Soc Bot Pol 91:9121

Zohner CM, Mo L, Renner SS et al (2020) Late-spring frost risk between 1959 and 2017 decreased in North America but increased in Europe and Asia. Proc Natl Acad Sci 117:12192–12200

Conserving Aesthetics: Cryopreservation of Ornamental Plant Genetic Resources

7

Ravi Gowthami, P. E. Rajasekharan, Subhash Chander, P. Magudeeswari, and Anju Mahendru-Singh

Abstract

The cryopreservation of ornamental crops or plants genetic diversity is essential for conserving biodiversity, cultural heritage, and economic value in horticulture. Cryopreservation, which involves conserving these genetic resources at extremely low temperatures, i.e., in liquid nitrogen (LN), provides a reliable solution for maintaining genetic integrity over long periods. Unlike traditional methods such as seed banks and field collections, which face risks from pest infestations, genetic drift, and environmental changes, cryopreservation offers indefinite storage with minimal maintenance. This method is particularly advantageous for ornamental species with recalcitrant seeds or those that rely on vegetative propagation, such as orchids, lilies, and rare flowering shrubs. Optimizing protocols for cryopreserving explants such as shoot tips, seeds, pollen, or somatic embryos is critical, along with pretreatment methods such as dehydration and vitrification, to increase success rates. Advances in cryoprotectants and recovery techniques further improve post-thaw viability. In an era of rapid urbanization and a growing demand for sustainable landscaping, cryopreservation safeguards the aesthetic and ecological value of ornamental plants. It also facilitates the restoration of endangered species, ensuring future breeding potential and cultural significance. This strategy supports biodiversity conservation and sustainable horticultural practices, contributing to ecological balance and innovative landscaping.

R. Gowthami (✉) · S. Chander · A. Mahendru-Singh
ICAR-National Bureau of Plant Genetic Resources (NBPGR), Pusa Campus, New Delhi, India
e-mail: Gowthami.R@icar.org.in

P. E. Rajasekharan
Formerly with ICAR-Indian Institute of Horticultural Research, Bengaluru, India

P. Magudeeswari
Vanavarayar Institute of Agriculture, Pollachi, Coimbatore, Tamil Nadu, India

S. Gantait, P. Chmielarz (eds.), *Conservation of Plant Genetic Resources*,
https://doi.org/10.1007/978-981-96-8331-4_7

Keywords

Biodiversity conservation · Cryopreservation · Dehydration · Genetic resources · Ornamental plants · Vitrification

7.1 Introduction

Ornamental plants contribute significantly to global biodiversity and play vital aesthetic, ecological, and cultural roles. Among the approximately 406,700 plant species on Earth, approximately 85,000–99,000 are categorized as ornamental (Gupta et al. 2019). These plants encompass a diverse range of forms, including shrubs, trees, creepers, climbers, ferns, palms, cacti, succulents, bamboos, and grasses, each of which is valued for its unique visual appeal and adaptability to various environments. Annuals, orchids, and bulbous plants add vibrant color to gardens and public spaces, whereas grasses and bamboos contribute texture and architectural elegance to landscapes.

Ornamental plants thrive across diverse ecosystems, from forests and wetlands to urban landscapes, where they enhance public parks, private gardens, and green spaces. In addition to their aesthetic appeal, they provide critical ecological benefits, such as supporting pollinators, improving air quality, and stabilizing soils. Many ornamental species hold cultural and economic significance, featuring prominent rituals, celebrations, and artistic traditions. Additionally, several ornamental plants possess medicinal properties that contribute to community well-being (Gupta et al. 2019).

The ornamental plant industry is experiencing rapid expansion, with annual growth rates ranging between 12% and 25%. The increasing demand for high-quality ornamental plants both for domestic use and export has fuelled nursery sector growth. However, this commercialization has led to concerns such as the loss of native species, reduced genetic diversity, and habitat degradation. Many ornamental plants are increasingly threatened by environmental stress, diseases, and climate change. The pressure from large-scale cultivation and trade has further strained natural populations, making conservation efforts imperative.

Conserving ornamental plants is vital for conserving genetic resources, protecting unique species, and maintaining their cultural and ecological contributions. Ornamental crop conservation relies on two main strategies: in situ and ex situ conservation. In situ conservation focuses on protecting ornamental plants within their natural or cultivated habitats, allowing them to thrive in their native ecosystems (protected areas or botanical gardens) and maintaining their genetic diversity and ecological roles. This method also promotes sustainable land management practices that prevent habitat destruction and maintain local biodiversity. On the other hand, ex situ conservation involves preserving plants outside their natural environment, often in controlled settings such as seed banks, field gene banks, nurseries, in vitro gene banks, and cryogene banks (Chowdhuri and Deka 2019).

Cryopreservation is a technique used to store living tissues/biological samples at ultralow temperatures, i.e., in LN either in liquid phase or vapor phase. By exposing biological samples to ultra-low temperature, cellular activity, including cell division and metabolic processes, is effectively halted. This cessation of activity allows the biological material to be preserved in a dormant state, preventing degradation and ensuring that the samples remain viable over long periods (Gowthami et al. 2024a, b, c; Nagel et al. 2024; Kang et al. 2025).

Cryoconservation is a widely recognized and remarkably effective approach for the long-term storage of plant germplasms in gene banks. By freezing plant cells, tissues, or even entire organs, cryopreservation allows plant genetic material to be preserved. Since cellular functions are paused at such low temperatures, the genetic material remains unchanged and can be revived when needed, making it a crucial tool for conserving economically important rare and/or endangered plant species (Sharma et al. 2020; Agrawal et al. 2022, 2024).

Cryopreservation is particularly useful for storing recalcitrant seeds, which are unable to withstand freezing or drying, as well as plant materials that are difficult to preserve via conventional techniques such as seed banking or field gene banks. With cryopreservation, plant genetic resources can be safely stored without the risk of degradation due to environmental factors such as temperature fluctuations or moisture loss. This ensures that valuable genetic diversity is maintained and available for future research, breeding programs, and ecological restoration efforts. As a result, cryopreservation has become a preferred method for safeguarding plant germplasm in gene banks, where it offers a reliable means of conserving genetic material for future use (Agrawal et al. 2022; Nagel et al. 2024).

7.2 Role and Importance of Cryopreservation

Cryopreservation is a powerful tool for conserving plant genetic material, offering numerous advantages over conventional conservation methods. This ensures prolonged viability of valuable genetic resources while retaining their genetic fidelity. Below are some of its key benefits (Fig. 7.1):

7.2.1 Conservation of Genetic Diversity

Cryopreservation enables the long-term storage of genetic material from wild, rare, and endangered plant species, playing a crucial role in biodiversity conservation. By maintaining genetic diversity, it supports breeding programs aimed at enhancing crop resilience, improving yields, and adapting plants to changing environmental conditions. This method helps preserve valuable heritable characteristics, such as resistance to different pests, diseases, and drought, ensuring their availability for future agricultural and horticultural advancements.

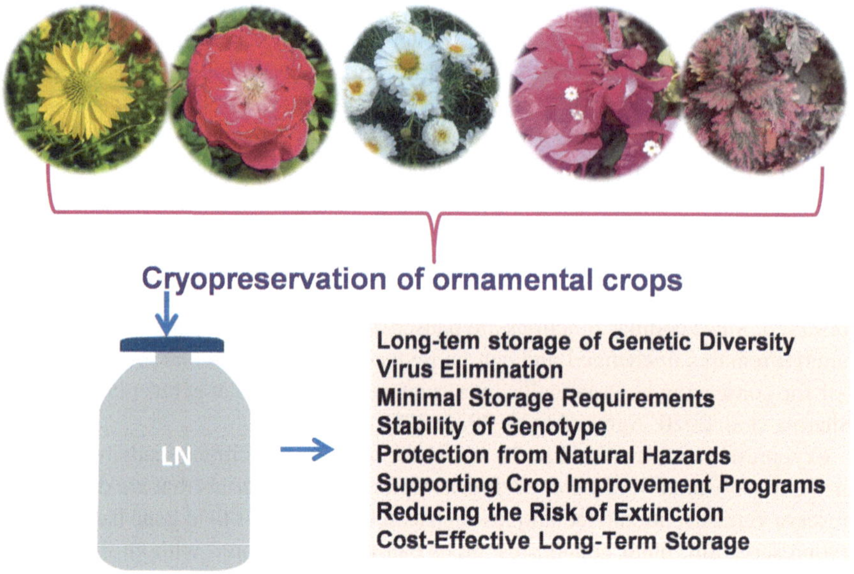

Fig. 7.1 Ornamental plant cryopreservation: need and importance. (Source: authors)

7.2.2 Virus Elimination

One of the unique benefits of cryopreservation is its ability to eliminate viruses from infected plants. Techniques such as shoot tip cryotherapy, which involves freezing shoot tips or meristematic tissues, allow for the virus elimination. This process ensures the propagation of virus-free plants, contributing to healthier plant stocks and improved crop quality.

7.2.3 Minimal Storage Requirements

Unlike traditional preservation methods that require large spaces and complex infrastructure, cryopreservation is highly efficient and space-saving. Plant material can be stored in standardized containers at ultralow temperatures, which require minimal space and resources.

7.2.4 Stability of Genotype

Cryopreservation halts all biochemical and metabolic processes by storing plant material at ultralow temperatures, ensuring that both genotypes (genetic makeup) remain stable over time. This stability ensures that the stored plant material can be revived and regrown without significant genetic alterations, maintaining its original traits. However, to maintain stability of the genotype, the cryopreservation protocol needs to be standardized and validated usind different methods.

7.2.5 Protection from Natural Hazards

Cryogenically stored plants are shielded from a wide range of environmental risks, including diseases, pests, and extreme climatic conditions. As a result, plants are protected from by natural hazards, and the stored germplasm remains secure and viable for long-term use. This makes cryopreservation a particularly reliable method for safeguarding plant species that may be vulnerable to changing environmental factors.

7.2.6 Elimination of the Need for Serial Subculturing and Prevention of Genetic Drift

In vitro conservation under in vitro gene banks often requires frequent subculturing of plant material, a process that can lead to genetic drift, reduced vigor, and loss of plant quality over time. Cryopreservation eliminates this need, ensuring that plant material is stored in a stable, undisturbed state, thus maintaining genetic integrity and plant vitality.

7.2.7 Long-Term Storage with Low Maintenance

Once cryopreserved, plant material can be stored indefinitely, requiring minimal maintenance. Unlike conventional methods, which may necessitate regular monitoring or reculturing, cryopreservation offers a long-term solution without the need for frequent intervention. This makes it a reliable and low-maintenance option for conserving plant genetic resources.

7.2.8 Supporting Crop Improvement Programs

Cryopreservation plays an essential role in supporting crop improvement programs by preserving valuable genetic traits for future breeding and preserving pollen. Cryopreserved pollen is available throughout the year and is devoid of seasonal influence on breeding activities.

7.2.9 Reducing the Risk of Extinction

For ornamental crops that face threats from habitat loss, overharvesting, or climate change, cryopreservation offers a safeguard against extinction. By preserving genetic material in cryobanks, there is a backup source of plant material that can be reintroduced into their natural habitats or used for restoration efforts, helping to prevent the loss of unique ornamental species.

7.2.10 Cost-Effective Long-Term Storage

The initial investment in cryopreservation infrastructure can be high, yet it offers significant long-term savings. Unlike traditional storage methods, which require ongoing maintenance, precaution, and space, cryopreserved material remains viable for decades or even centuries, drastically reducing the need for continual repropagation. Over time, cryopreservation has become a highly cost-effective strategy for conserving ornamental crops.

7.3 Ornamental Cryopreservation: Explants and Techniques

The cryopreservation of ornamental crops involves various explants, each with unique definitions, advantages, and disadvantages (Benson 2008). Seeds are the most common explants and are defined as mature ovules containing an embryo; they are advantageous because of their natural tolerance to desiccation and ease of storage, but some species produce recalcitrant seeds that cannot withstand low temperatures. Zygotic embryos, which are excised from seeds, are ideal for species with recalcitrant seeds, offering high regeneration potential, although their extraction can be labor intensive. Embryonic axes, consisting of the shoot and root meristems of the embryo, are useful because of their high regeneration capacity but are sensitive to cryoprotectant toxicity. Shoot tips, comprising the apical meristem, and axillary buds are advantageous for maintaining genetic integrity but require precise handling during cryopreservation. Somatic embryos, derived from somatic cells, offer high regeneration potential and genetic uniformity but are prone to somaclonal variation. Protocorms, early developmental stages of orchid seedlings, are valuable for orchid conservation but require specialized media for recovery. Embryonic tissues, such as calli or cell suspensions, are advantageous for mass propagation but may lose embryogenic potential during cryopreservation. Pollen can be cryopreserved and used for breeding purposes and can also be stored as haploid genetic resources in the case of threatened species. Desiccation-tolerant explants such as orthodox seeds and dormant buds can be cryopreserved without complex pretreatments, as their natural dehydration process aligns with cryopreservation requirements (Engelmann 2011). In contrast, explants with high cellular water contents, such as embryos, shoot tips, cell suspensions, and calluses, are highly susceptible to freezing damage. These require artificial dehydration to avoid intracellular ice formation and ensure cell/tissue survival during cryopreservation (Engelmann 2011). The following are the most commonly used cryopreservation techniques for ornamental plants (Table 7.1).

Table 7.1 Selected examples on cryopreservation of ornamental crops using different explants

Scientific name	Explant	Cryopreservation technique	References
Acer mono	Suspended cells	Vitrification	Park et al. (2005)
Aerides multiflora	Protocorms	Encapsulation–dehydration	Bunnag and Hongthongkham (2014)
Aesculus hippocastanum	Embryogenic callus	Vitrification	Lambardi et al. (2005)
Amorphophallus koratensis	Pollen	Direct freezing	Soonthornkalump et al. (2020)
Amaryllis belladonna	Zygotic embryos	Slow cooling	Sershen et al. (2011)
Anacamptis morio	Seeds	Encapsulation–dehydration	Wood et al. (2000)
Anthurium andraeanum	Callus	Vitrification	Zhang et al. (2024a)
Arundina graminifolia	Protocorms	D cryoplate	Cordova and Thammasiri (2016)
Asplenium cuneifolium	Gametophytes	Encapsulation–dehydration	Makowski et al. (2011)
Begonia x erythrophylla	Adventitious shoots	Encapsulation–dehydration	Burritt (2008)
Betula pendula	In vivo buds	Slow cooling	Ryyänen and Aronen (2007)
Bletilla striata	Immature seeds	Direct freezing, vitrification	Hirano et al. (2005a)
	Protocorms	Droplet vitrification	Jitsopakul et al. (2008a)
Buxus hyrcana	Shoot tips	Encapsulation–dehydration	Kaviani and Negahdar (2017)
Buxus sempervirens	Shoot tips	Encapsulation–dehydration	Negahdar et al. (2021)
Catharanthus roseus	Suspension cells	Encapsulation–dehydration	Bachiri et al. (1995)
Cattleya spp.	Seeds	Vitrification	Vettorazzi et al. (2019)
Cattleya walkeriana	Seeds	Vitrification	Galdiano et al. (2017)
Centaurium rigualii	Nodes	Encapsulation–dehydration	González-Benito et al. (1997)
Chrysanthemum morifolium	Apical and axillary shoot tips	Droplet vitrification	Lee et al. (2011)
	Shoot tip	Droplet vitrification	Wang et al. (2014)
	Shoot tip	Droplet vitrification	Bi et al. (2016)
	Shoot tip	D cryoplate	Tanaka et al. (2016)
Chrysanthemum sp.	Shoot tip	Encapsulation–dehydration	Kulus et al. (2019)

(continued)

Table 7.1 (continued)

Scientific name	Explant	Cryopreservation technique	References
Chrysanthemum cv. Chuju	Shoot tip	Vitrification	Su et al. (2022)
Chrysanthemum x grandiflorum	Shoot tip	Encapsulation–dehydration	Zalewska and Kulus (2013)
Cleisostoma arietinum	Protocorms	Encapsulation–dehydration	Maneerattanarungroj et al. (2007)
Cleome rosea	Shoot tip	V cryoplate	da Silva Cordeiro et al. (2015)
Clitoria ternatea	Somatic embryo	Encapsulation–dehydration	Nair and Reghunath (2008)
Coelogyne dayanum	Seeds	Vitrification	Hakim et al. (2015)
Crocus sativus	Shoot tips	Vitrification	Zadeh et al. (2009)
Crocus hyemalis	Embryogenic callus	Vitrification	Baghdadi et al. (2011)
Crocus moabiticus	Embryogenic callus	Vitrification	Baghdadi et al. (2011)
Cyclamen persicum	Embryogenic suspension	Vitrification	Mubmann et al. (2006)
Cymbidium eburneum	Protocorms	Vitrification, encapsulation–vitrification	Gogoi et al. (2012)
	Protocorms	Encapsulation–dehydration	Gogoi et al. (2013)
Cymbidium finlaysonianum	Protocorm-like bodies	Vitrification	Rittirat et al. (2019)
Cymbidium hookerianum	Protocorms	Encapsulation–dehydration	Gogoi et al. (2013)
Cyrtopodium hatschbachii	Immature seeds	Encapsulation–dehydration	Surenciski et al. (2012)
Dahlia cv	Shoot tips	Droplet vitrification	Gowthami et al. (2023c)
Dendrobium Bobby Messina	Protocorm-like bodies	Vitrification	Antony et al. (2010)
Dendrobium cariniferum	Protocorms	Encapsulation–dehydration, encapsulation–vitrification	Pornchuti and Thammasiri (2008)
Dendrobium cruentum	Seeds	Encapsulation–dehydration, encapsulation–vitrification, D cryoplate, V cryoplate	Prasongsom et al. (2020)
Dendrobium draconis	Protocorms	V cryoplate	Jitsopakul et al. (2020)

(continued)

Table 7.1 (continued)

Scientific name	Explant	Cryopreservation technique	References
Dendrobium nobile	Protocorm-like bodies	Encapsulation–dehydration, Encapsulation–vitrification	Mohanty et al. (2012)
	Protocorms	Vitrification	Vendrame and Faria (2011)
Dendrobium signatum	Pollinia	V cryoplate, D cryoplate	Jitsopakul et al. (2019)
Dendrobium sp.	Shoot tips	Encapsulation–dehydration	Lurswijidjarus and Thammasiri (2004)
Dianthus caryophyllus	Shoot tips	Encapsulation–vitrification	Halmagyi and Deliu (2007)
Doritis pulcherrima	Seeds	Vitrification	Thammasiri (2000)
Elateriospermum tapos	Zygotic embryos	Vitrification	Zaimah et al. (2007)
Fritillaria imperialis	Bulb scale	Encapsulation–dehydration and encapsulation–vitrification	Seydi et al. (2020)
Galanthus elwesii	Apical meristems	Droplet vitrification	Maślanka et al. (2013)
Galanthus elwesii	Somatic embryos	Encapsulation–dehydration	Pawłowska (2008)
Galanthus nivalis	Somatic embryos	Encapsulation–dehydration	Pawłowska (2008)
Gentiana cruciata	Proembryonic callus	Slow cooling	Mikuła et al. (2005)
	Proembryonic callus	Vitrification	
	Proembryonic callus	Droplet vitrification	Mikuła et al. (2011a)
	Proembryonic callus	Encapsulation–dehydration	Mikuła et al. (2008)
G. kurroo	Cell suspension	Vitrification	Mikuła et al. (2006)
	Cell suspension	Encapsulation–dehydration	Mikuła et al. (2011b)
	Shoot tips	Vitrification, droplet vitrification	Sharma et al. (2021)
G. macrophylla	Hairy roots	Encapsulation–dehydration	Xue et al. (2008)
G. scabra	Axillary buds	Encapsulation–dehydration	Suzuki et al. (2005)
	Shoot tips	Encapsulation–vitrification	Tanaka et al. (2004)
	Shoot tips	Vitrification	

(continued)

Table 7.1 (continued)

Scientific name	Explant	Cryopreservation technique	References
G. tibetica	Cell suspension cultures	Vitrification	Mikuła (2006)
	Proembryonic callus	Encapsulation–dehydration	Mikuła et al. (2008)
G. triflora	Shoot tips	Vitrification, encapsulation–vitrification	Tanaka et al. (2004)
G. pneumonanthe	Shoot tips	Vitrification, encapsulation–vitrification	Tanaka et al. (2004)
Grammatophyllum speciosum	Protocorms	Droplet vitrification, encapsulation–dehydration, and encapsulation–vitrification	Sopalun et al. (2010)
	Seeds	D cryoplate	Thanasuttanithi and Thammasiri (2022)
Helianthus tuberosus	Shoot tips	Vitrification	Volk and Richards (2006)
	Shoot tips	Droplet vitrification	Zhang et al. (2017)
Iris primula	Shoot tips	Vitrification	Jevremović et al. (2011)
Lamprocapnos spectabilis	Callus	Encapsulation–vitrification	Kulus and Miler (2021)
Lilium auratum	Meristems	Vitrification	Bouman et al. (2003)
Lilium formosanum	Meristems	Vitrification	Bouman et al. (2003)
Lilium henryi	Meristems	Vitrification	Bouman et al. (2003)
Lilium × *lancifolium*	Meristems	Droplet vitrification	Chen et al. (2011)
Lilium × *longiflorum*	Meristems	Droplet vitrification	Chen et al. (2011)
Lilium ledebourii	Seeds	Dehydration	Kaviani et al. (2009)
Lilium ledebourii	Seeds	Encapsulation–dehydration	Kaviani (2010)
Lilium ledebourii	Seeds	Vitrification, desiccation, encapsulation–dehydration	Mohajeri et al. (2014)
Lilium ledebourii	Small leaf squares (SLSs, 3 × 4 mm), each bearing at least one adventitious bud	Vitrification	Pan et al. (2018)
Lilium martagon	Meristems	Droplet vitrification	Urbaniec-Kiepura and Bach (2017)
Lilium spp.	Shoot tips	Droplet vitrification	Yin et al. (2014)
Lilium spp.	Shoot tips	Droplet vitrification	Yi et al. (2013)

(continued)

Table 7.1 (continued)

Scientific name	Explant	Cryopreservation technique	References
Lilium sp.	Meristems	Vitrification	Bouman et al. (2003)
Lilium × *siberia*	Meristems	Droplet vitrification	Chen et al. (2011)
Magnolia macrophylla	Shoot tips	Droplet vitrification	Folgado and Panis (2019)
Melia azedarach	Embryonic axes	Encapsulation–dehydration	Kaviani (2007, 2010)
	Somatic embryos	Desiccation, encapsulation–dehydration, pregrowth–dehydration	Scocchi et al. (2007)
Narcissus L 'Carlton'	Somatic embryos	Droplet vitrification	Maślanka et al. (2016)
Nelumbo nucifera	Seeds	Vitrification	Yong et al. (2023)
Oncidium bifolium	Seeds, protocorms	Encapsulation–dehydration	Flachsland et al. (2006)
Paeonia lactiflora	Zygotic embryos	Encapsulation–dehydration	Kim et al. (2004)
	Shoot tips	Desiccation	Seo et al. (2007)
Paphiopedilum exul	Seeds	Encapsulation–vitrification, encapsulation–dehydration	Imsomboon and Thammasiri (2020)
Petunia × *Calibrachoa*	Callus	Vitrification	Zhang et al. (2024b)
Phalaenopsis bellina	Protocorm-like bodies	Encapsulation–dehydration,	Khoddamzadeh et al. (2011)
Phlebodium aureum	Prothalia	Encapsulation–dehydration	Pawłowska (2008)
Pinus nigra	Embryogenic tissues	Slow freezing	Salaj et al. (2011)
Ponerorchis graminifolia	Immature seeds	Vitrification	Hirano et al. (2005b)
Rhododendron simsii	Shoot tips	Encapsulation–dehydration	Verleysen et al. (2005)
Rhynchostylis gigantea	Pollinia	V cryoplate	Jitsopakul et al. (2021)
Rosa agrestis	Shoot tips	Droplet vitrification	Pawłowska and Szewczyk-Taranek (2015)
Rosa canina	Shoot tips	Droplet vitrification	Pawłowska and Szewczyk-Taranek (2015), Pawłowska (2012)

(continued)

Table 7.1 (continued)

Scientific name	Explant	Cryopreservation technique	References
Rosa dumalis	Shoot tips	Droplet vitrification	Pawłowska and Szewczyk-Taranek (2015)
Rosa rubiginosa	Shoot tips	Droplet vitrification	Pawłowska and Szewczyk-Taranek (2015), Pawłowska (2012)
Rosa 'New Dawn'	Meristem	Encapsulation–dehydration	Pawłowska (2008)
	Shoot tips	Encapsulation–dehydration	Pawlowska and Bach (2011)
Rosa × hybrida	Shoot tips	Droplet vitrification, encapsulation–dehydration	Halmagyi et al. (2022)
Sterculia cordata	Zygotic embryos	Vitrification	Nadarajan et al. (2007)
Strelitzia reginae	Zygotic embryos	Desiccation	Figueiredo et al. (2021)
Taxus baccata	Apical buds	Encapsulation–dehydration	Moghaddam et al. (2024)
Tulipa tarda	Apices	Droplet vitrification	Maślanka and Szewczyk (2021)
Vanda coerulea	Seeds	Vitrification	Thammasiri and Soamkul (2007)
	Protocorms	Encapsulation–dehydration	Jitsopakul et al. (2008b)
Vanda tricolor	Immature seeds	Vitrification	Jitsopakul et al. (2012)
Vanilla planifolia	Shoot apices	Droplet vitrification	Gonzalez-Arnao et al. (2009, 2011)

7.3.1 Controlled-Rate Freezing or Slow Cooling

This process is also known as two-step freezing, which is a prominent method in cryopreservation, particularly for the preservation of the genetic diversity of plants. This technique involves gradually freezing/cooling samples (cells/tissues) at a rate of 0.5–2 °C per minute until they reach approximately −40 °C, after which the samples are immersed in LN (Nausch and Buyel 2021; Mosa et al. 2023). The slow freezing technique begins with a phase called subcooling, during which plant cells remain unfrozen even as the temperature decreases. In this phase, the high concentration of solutes in the cytoplasm prevents the cell's water from freezing, allowing the cytoplasm to remain in a liquid form. The cell wall plays a vital role by acting as a barrier, protecting the cell membrane from potential damage caused by ice crystal formation (Mazur 1984). With further decreases in temperature, the extracellular solutions surrounding the cells begin to freeze, increasing the osmotic pressure in the intercellular spaces. This relatively high osmotic pressure forces water out of the cells and into the extracellular environment, where it freezes, thus preventing the formation of ice crystals inside the cells. The goal of this process is to remove a greater part of the water from the plant cells before they are immersed in LN. By decreasing the intracellular water content, the ability to form harmful freezing crystals inside the cells is greatly reduced. The combination of subcooling, osmotic dehydration, and gradual cooling ensures effective cryopreservation of plant tissues, particularly those with high water contents, maintaining their structural integrity for successful long-term storage. The slow freezing method offers benefits such as efficient batch processing, consistency, and versatility across plant species, aiding genetic conservation. However, it requires precise control over cooling rates, expensive programmable freezers, and specialized expertise. Improper cooling can cause damage, and cryoprotectants must be carefully regulated to avoid toxicity, necessitating species- and tissue-specific protocols.

7.3.2 Vitrification

Vitrification is a cutting-edge cryopreservation method that emerged in the 1990s as an innovative alternative to traditional freezing techniques. The term "vitrification" refers to the dynamic conversion of liquid solute in the cell/tissue from a liquid state into a glassy/metastable state without the creation of ice nucleation (Benson 2008). In the vitrification process, biological material, such as plant tissues, is first treated with a mixture of highly concentrated cryoprotectants. These cryoprotectants include both penetrating agents, which enter the cells, and nonpenetrating agents, which remain outside the cells. Once the tissue is fully treated with cryoprotectants, it is then rapidly immersed in LN. This glassy state effectively suspends the biological material under stable conditions, preventing the creation of ice crystals that typically disintegrate the cell membranes and damage organelles.

One of the primary advantages of vitrification is its ability to inhibit the development of damaging ice nucleation. By increasing the viscosity of the cell interior, vitrification inhibits the crystallization of water present inside and outside the cell. This is particularly important for preserving cellular integrity, as ice crystals can physically rupture the cell membrane and disrupt the function of organelles. The formation of a glassy state further protects cell membranes by providing a gelatinous barrier that ensures that the cells remain intact during cryogenic storage. This protection at the cellular level is critical for the survival of plant tissues at cryogenic temperatures. Additionally, vitrification helps minimize water loss during the cryoconservation procedure, diminishing the risk of lethal dehydration. The cryoprotectants involved in the process form a stable, glassy matrix that not only prevents ice formation but also helps to retain cellular moisture, which is vital for the regeneration of the plant material after thawing. As a result, plant tissues generally have a higher survival rate and regeneration percentage than those preserved through traditional slow freezing methods.

Cryoprotective substances are classified into three major categories according to their ability to penetrate plant cells, and each type plays a unique role in the cryopreservation process. (a) Penetrating cryoprotectants, such as ethylene glycol (EG), propylene glycol (PG), glycerol (Gly), and dimethyl sulfoxide (DMSO), are small molecules capable of crossing both the cell wall and plasma membrane. These agents work by lowering the freezing point of the water inside the cells, preventing ice nucleation and stabilizing the cellular structures during freezing. (b) Cell wall-penetrating cryoprotectants, including oligosaccharides such as sucrose, sorbitol, and mannitol, as well as amino acids such as proline and low-molecular-weight polymers such as polyethylene glycol (PEG 1000), can penetrate the cell wall but not the plasma membrane. They primarily assist in dehydration by reducing osmotic pressure in the extracellular space before immersion in LN. (c) Nonpenetrating cryoprotectants, such as high-molecular-mass polymers such as soluble proteins, polysaccharides, and larger forms of polyethylene glycol (PEG 6000), cannot cross the plasma membrane or cell wall. Instead, they provide protection by stabilizing the extracellular environment and preventing the development of ice outside the cells. Each category of cryoprotectant serves a specific purpose, and their usage must be carefully calibrated for each plant species and tissue type to balance effectiveness and avoid toxicity, ensuring successful preservation (Zamecnik et al. 2021). Using a mixture of cryoprotectants and osmotic agents, plant vitrification solutions are prepared in various combinations and concentrations. These solutions protect plant tissues by reducing ice formation, minimizing cellular damage, stabilizing proteins and cellular components, and promoting controlled dehydration, ultimately improving the viability/survival cell/tissues of plants during freezing and thawing. Common vitrification solutions such as PVS2, PVS3, and PVS4, as well as customized variations, are formulated to suit different plant species, types of explants, and specific preservation needs. However, during exposure to these solutions, successful dehydration is necessary, as prolonged exposure to PVS may cause toxicity, and this toxicity can lead to cellular damage, ultimately reducing the effectiveness of the preservation process. To minimize the risks associated with exposure to these PVSs, it is essential to carefully determine both the temperature and duration of exposure. This ensures that the plant tissues are sufficiently dehydrated while preventing excessive toxicity, which can impair regrowth following cryopreservation (Kulus and Zalewska 2014; Bettoni et al. 2021; Kaviani and Kulus 2022; Gowthami et al. 2023a, b).

7.3.3 Desiccation

Desiccation refers to the controlled dehydration of plant tissues before they undergo cryopreservation. The process involves diminishing the water molecules of the plant cell/tissue to minimize the threat of ice nucleation amid freezing, which can otherwise cause significant damage to cellular structures. To achieve effective desiccation, tissues are generally exposed to sterilized air in a laminar airflow cabinet or workbench, wherein a continuous flow of sterile compressed air allows controlled and steady reduction in tissue moisture levels. Another technique involves the use of activated silica gel (Engelmann 2000). Silica gel absorbs moisture from the tissues, gradually reducing their water content without causing sudden dehydration stress. Precise control over the desiccation process is crucial because excessive dehydration can damage plant tissues, whereas insufficient drying may not prevent ice crystal formation during subsequent freezing. Once the desiccation process is complete, the germplasm is preserved by rapidly immersing the dehydrated samples in LN.

7.3.4 Encapsulation–Dehydration (ED)

This approach was derived from synthetic seed technology, which was developed by Fabre and Dereuddre (1990). This technique involves encapsulating plant explants in a protective gel, typically made from alginate-Ca beads, followed by a dehydration process that reduces the moisture content of the explants following both osmotic and physical dehydration. After encapsulation, these beads are then treated with a highly concentrated sucrose solution, typically ranging from 0.75 to 1.5 M, for several hours or even days. The sucrose solution helps facilitate osmotic dehydration by drawing out water from the explants, further preparing them for freezing. After osmotic dehydration, the beads undergo a physical dehydration process, where they are exposed to sterile airflow under a laminar flow workbench or placed in silica gel. The aim is to reduce the moisture level of the encapsulated beads to an optimal level of approximately 20–25%. This step is essential because it prepares the explants for freezing, ensuring that the moisture content is low enough to escape from ice nucleation, which could damage the plant cells/tissues. Once the beads have reached the appropriate moisture level, they are immersed in LN, effectively freezing the explants for long-term storage. For recovery, these cryopreserved encapsulated beads are unfrozen by placing them in a water bath at approximately 40 °C for 2–3 min, allowing them to rehydrate and recover their viability. While the encapsulation–dehydration (ED) process is labor intensive and time-consuming compared with other cryopreservation techniques, such as vitrification, it offers distinct advantages. A significant advantage of the ED method is that sucrose is used as the sole cryoprotectant, eliminating the need for more commonly used cryoprotectants such as glycerol or dimethyl sulfoxide (DMSO), which may be toxic to plant tissues at high concentrations. By relying on sucrose, a naturally occurring sugar, the ED method helps to avoid the chemical toxicity associated with other cryopreservation techniques, making it valuable mainly for plant/crop species that are sensitive to the harmful effects of other cryoprotectants. By encapsulating explants, the outer gel layer helps shield delicate tissues from mechanical damage during the dehydration process. Furthermore, the gel acts as a protective barrier during LN storage, minimizing the risk of cell rupture due to the expansion of the intracellular ice crystalline structure. Despite the labor and time needed, ED remains a valuable method for cryopreserving plant genetic material, particularly for species that might not tolerate other preservation methods.

7.3.5 Encapsulation–Vitrification (EV)

This is a refined cryopreservation technique that combines both vitrification and encapsulation–dehydration methods, effectively merging their benefits (Sakai and Engelmann 2007). In the first phase, the explant is encapsulated in a protective gel, often made from calcium alginate. The encapsulation step is pivotal in reducing the harmful effects of osmotic stress, which occurs when plant tissues lose or gain water too rapidly during the cryopreservation process. Encapsulation serves as a protective barrier against direct cryoprotectant toxicity to the explants, reducing the likelihood of damage to the cells. After encapsulation, the plant explants undergo vitrification. Vitrification involves treating the encapsulated material with cryoprotectants, which prevent ice crystal formation when the material is subjected to ultralow temperatures.

7.3.6 Droplet Vitrification (DV)

This is a refined version of the droplet-freezing method originally identified by Kartha et al. (1982), in which plant shoot tips are placed in a drop of a cryo-solution, typically DMSO, and then gradually freeze in a programmable apparatus to approximately −40 °C before being plunged into the LN. However, DV was designed to provide an efficient, rapid, and effective means of preserving small explants, viz., embryos or shoot tips (Panis et al. 2005, 2020; Sakai and Engelmann 2007). This technique involves placing small explants directly into a drop of a plant vitrification solution (PVS) on a strip of aluminum foil. Once the explants are positioned in the PVS droplet on the aluminum foil, and incubated for appropriate period, allowing the solution to permeate the cells and prepare them for rapid cooling. The aluminum foil, known for its excellent thermal conductivity, facilitates high cooling and warming rates. This ensures that the explants undergo a rapid transition from room temperature to the ultralow temperatures needed for preservation, minimizing the risk of ice crystal formation within the cells and ensuring optimal vitrification. The explants are then directly exposed to LN, completing the cryopreservation process.

The DV method offers several significant advantages, particularly in its efficiency and minimal risk of mechanical damage. One of the key benefits is that the explants are not placed into cryovials, as the foil strip is simply inserted into the cryovial with the explants already in position on the foil. This minimizes the risk of mechanical damage, which is a common issue with other methods that require more extensive handling of the plant material.

7.3.7 V and D Cryoplates

The cryoplate technique involves a specially designed aluminum plate with depressions (referred to as a cryoplate) to hold plant shoot tips. These plate-oriented cryopreservation methods utilize calcium alginate to securely attach shoot tips to aluminum cryoplates, which measure 7 × 37 × 0.5 mm. Each cryoplate is designed with 10 wells, which are tailored to hold the explants in place during the cryopreservation process. The V cryoplate and D cryoplate methods, developed by Yamamoto et al. (2011) and Niino et al. (2013), respectively, are advanced techniques for long-term conservation. The V cryoplate method is an amalgamation of two techniques—droplet vitrification and encapsulation vitrification. In the V cryoplate method, precultured explants are encapsulated in alginate gel on cryoplates, treated with loading solution (LS), and then exposed to PVS2 before cryopreservation. On the other hand, the D cryoplate method integrates V-cryoplate and encapsulation–dehydration methods. Both methods offer unique advantages, with the V cryoplate method emphasizing a more controlled vitrification process and the D cryoplate method focusing on dehydration, each addressing specific challenges associated with cryopreservation for various plant species. The D cryoplate method involves air-drying explants on cryoplates or silica gel desiccation under a laminar flow cabinet before freezing. These methods address many challenges commonly faced in traditional vitrification techniques, such as insufficient dehydration and material loss. A key benefit of these techniques is that the explants (shoot tips) are securely tightly sealed at the cryoplates, significantly reducing the risk of damage or loss throughout the cryopreservation process. Another major advantage is the greater viability and growth rate of cryopreserved explants. The rapid cooling and warming rates employed in these techniques further contribute to the successful preservation and regrowth of the treated materials.

7.4 Core Factors Determining Cryopreservation Efficacy in Ornamental Plants

The cryopreservation of ornamental crops using different explants is a complex, multistep process that requires careful execution of several critical stages to ensure successful preservation and recovery. Some of the important factors are discussed below:

7.4.1 Age of the Donor Plants

The age of shoot tip donor plants plays a critical role in the success of cryopreservation and subsequent post-thaw regrowth. Explants from cultures of optimal age exhibit greater dehydration tolerance, which is crucial for survival during cryopreservation. However, the ideal pregrowth period varies across plant species and explant types (Wang and Perl 2006; Condello et al. 2011; Feng et al. 2013). In general, the age of the donor plants used for cryopreservation ranges from 1 to 12 weeks from the last subculture, with the most common range being 2–8 weeks. For example, in *Dahlia*, a popular ornamental plant, studies have demonstrated that shoot tips excised from 6- to 8-week-old cultures present significantly increased regrowth rates (23.33%) compared with those taken from younger (4-week-old) or older (12-week-old) plants, whose regrowth rates are 13% and 6.67%, respectively (Gowthami et al. 2023c). Shoot tips from younger plants may lack sufficient cellular differentiation and metabolic stability, increasing their susceptibility to cryoinjury, whereas older tissues may experience reduced regenerative capacity due to cellular aging and accumulated oxidative stress. Shoot tips from intermediate-aged plants often possess an optimal balance of water content, cellular solutes, and hormonal profiles, which are crucial for surviving dehydration and freezing processes during cryopreservation.

7.4.2 Position of Explants

Bud position has been found to impact shoot regrowth in cryopreserved shoot tips (Dereuddre et al. 1988; Halmagyi et al. 2005). The viability of the axillary shoot tips during cryopreservation decreases as the distance from the apical shoot tip increases. The survival rates of shoot tips from the first four positions (the nearest to the apex) were comparable to those of apical shoot tips (90%). However, survival rates decreased from the fifth position (68%) and decreased sharply near the stem base (position 11), reaching below 10% (6–9%) in carnation (Dereuddre et al. 1988). Wang et al. (2014) reported that shoot tips excised from buds located at nodes 1–6 of 4-week-old stock shoots and nodes 3–7 of 6-week-old stock shoots presented greater shoot regrowth than those from other nodes in *Chrysanthemum*. Lee et al. (2011) tested both apical and axillary shoot tips. The findings indicated that apical shoot tips excised from 4-week-old *Chrysanthemum* in vitro cultures and axillary shoot tips from 5.5- to 7-week-old in vitro cultures achieved greater in vitro recovery. Notably, no noteworthy changes were observed in in vitro growth between the axillary and apical tips. Studies on *Dahlia* have shown that apical shoot tips showed higher survival after cryopreservation than axillary shoot tips (Fig. 7.2) (Gowthami et al. 2023c). This difference in sensitivity can be attributed to the physiological and developmental characteristics of the explants. Apical shoot tips, taken from vigorous mother in vitro cultures, contain actively growing proliferating meristematic tissues/cells, which are inherently more resilient to desiccation and LN exposure (Marković et al. 2014). These cells present greater metabolic activity, greater concentrations of cryoprotective compounds (such as sugars and antioxidants), and a more robust cellular structure, all of which contribute to their enhanced tolerance to the extreme conditions of cryopreservation. However, basal shoot tips are more differentiated and contain higher water contents, making them prone to ice crystal formation and cellular damage during freezing.

7.4.3 Pretreatment of Mother Cultures

Pretreatment of mother cultures is an essential step in enhancing the physiological tolerance of donor cultures to the challenging conditions associated with cryopreservation. These conditioning steps are applied before cryoprotection to prepare

Fig. 7.2 Isolation of explant for cryopreservation of *Dahlia* cv. (**a**) Donor plants; (**b**) in vitro shoots for isolation of shoot tips; (**c**) apical shoot tips (1.0 mm × 0.5 mm); (**d**) bulging of shoot tips 15 days after post-thaw recovery. (Source: authors)

the plant material for the stresses of the freezing process. The selection of pretreatment methods depends on the plant species and may include various techniques designed to increase the tolerance of the culture to cryoinjury. Common pretreatment strategies include:

(i) Cold hardening, where shoot cultures are exposed to temperatures below their active growing conditions. This process helps to induce freezing tolerance by acclimatizing plants to cold stress, thereby improving their ability to survive the freezing process.

(ii) The application of osmotic agents, such as sucrose, to reduce the water content of the cells (Reed 1996). This reduces the amount of freezable water, which can damage ice crystals during the freezing process.

Osmotic pretreatments reduce the amount of freezable water, safeguarding cell membranes and other cellular structures from ice-induced damage, thereby improving the chances of successful recovery post-thawing. These pretreatment methods are critical for improving the overall effectiveness of cryopreservation, particularly for those plant species that have higher water contents in tissues/cells and are more sensitive to lower temperatures.

The cold hardening of donor plants is a critical process that enhances their inherent tolerance to relatively low temperatures and dehydration by triggering genes associated with cold stress responses (Takagi 2000). This adaptation is particularly important for cryopreservation, as it prepares plant tissues to withstand the extreme conditions of freezing and thawing. For example, many *Gentiana* species are naturally cold hardy, have been shown to benefit significantly from extended cold-hardening periods. Studies have demonstrated that *Gentiana* axillary buds require 34 days of cold hardening (10 days at 15 °C followed by 24 days at 5 °C) for successful cryopreservation via the encapsulation–desiccation technique (Suzuki et al.

1998). Similarly, in *Gentiana scabra*, a 50-day cold-hardening period at 5 °C, with an 8-h photoperiod and a light intensity of 26 μmol/m^2/s, was found to be essential for improving post-thaw recovery and survival in 3-week-old cultures. Sharma et al. (2021) revealed that preconditioning 4-week-old mother cultures at 8 °C without light for 4 weeks significantly enhanced post-thaw recovery (40% regeneration). This improvement is attributed to the physiological and biochemical changes induced by cold hardening, such as the accumulation of cryoprotective compounds such as sugars, which help tissues to withstand the extreme stresses of freezing and thawing. These adaptations reduce cellular damage, particularly from ice crystal formation, and improve the ability of the tissue to recover after cryopreservation. Conversely, the study also revealed that increasing sucrose concentrations to 6% or 9% or modifying nitrate levels in the pregrowth medium did not further increase survival/regrowth in *G. kurroo* after thawing. The ability of mother in vitro cultures to harden at low temperatures (0–5 °C for 5–7 days) is vital for the successful cryopreservation of *Lilium* shoot tips according to several studies (Yi et al. 2013; Chen et al. 2011; Bouman et al. 2003). Pretreating shoots of *Begonia* × *erythrophylla* with a medium containing 0.75 M sucrose, 3.8 μM abscisic acid, and 2.15 mM proline led to more than 50% of the shoots surviving freezing conditions (Burritt 2008). For *Gentiana kurroo*, shoot multiplication medium supplemented with 3% sucrose for 2 days was highly effective at improving post-thaw survival. However, increasing the sucrose concentration to 6% or 9% in the pregrowth medium did not provide additional benefits for post-thaw survival and regrowth. In addition to sucrose, the nitrogen composition of in vitro medium also significantly impacts post-thaw recovery. Ammonium nitrate, a common nitrogen source, has been shown to adversely affect the regeneration of cryopreserved shoot tips in some species. For example, reducing or substituting ammonium nitrate with potassium nitrate has been reported to improve post-thaw recovery in plants such as *Holostemma annulare* (Decruse and Seeni 2002; Decruse et al. 2004).

7.4.4 Size of Explants

The size of shoot tips has been identified as a critical factor influencing the success of cryopreservation, particularly in achieving higher recovery and survival rates post-thaw (Gowthami et al. 2023b; Bettoni et al. 2019; Normah et al. 2019). Smaller shoot tips, typically ranging from 0.5 mm to 2 mm, are often preferred as explant in many species. This reduces the risk of intracellular ice crystal formation, a major cause of cellular damage during freezing. The decreased recovery of larger shoot tips after thawing may be attributed to unequal and inadequate desiccation, causing the death of the cells/tissues owing to the development of ice nucleation and less efficient heat exchange during cooling/thawing (Gowthami et al. 2023b; Jiang et al. 2019; Panis et al. 2005; Zhang et al. 2015a). However, small shoot tips are more likely to consist of actively dividing meristematic cells, which are more resilient to the stresses of cryopreservation and escape the incidence of temperature variation among internal and external cell layers, which is harmful during cryopreservation.

However, the optimal size of shoot tips must be determined for good recovery during cryopreservation (Engelmann 2014; Volk and Caspersen 2007). For example, the use of sufficiently small shoot tip sizes of 1–2 mm in DV protocols led to 40% in *Rosa* (Pawłowska 2012). In contrast, only 18% of shoot tips survived (no in vitro shoots developed) when the same protocols were applied to comparatively large shoot tips, i.e., 3–4 mm in size for *Rosa* shoot tips (Kwaśniewska et al. 2017). However, with the use of the same technique (DV, with few modifications) on other cultivars of rose, shoot tips of different sizes (1–2 mm or 3–4 mm) result in greater regeneration (72%) (Halmagyi and Pinker 2006; Halmagyi et al. 2022). However, the optimal explant size is not a one-size-fits-all value. It should be determined experimentally for each plant species and cryopreservation protocol.

7.4.5 Preculture

Preculture is a critical step in cryopreservation and is aimed at improving the physiological condition of in vitro shoots and increasing their viability rates after exposure to LN. This process is particularly essential for tissues with high water contents, such as cell suspensions, calli, and embryos, which require artificial dehydration to reduce the amount of intracellular water before freezing. A key component of preculture media is the addition of carbohydrates, typically sucrose, which serves as a nontoxic cryoprotectant. Sucrose promotes osmotic dehydration and promotes the accumulation of endogenous complexes such as abscisic acid (ABA), polypeptides, proline, sugars, and bound water. These compounds improve the tolerance of plants to dehydration and freezing stresses, thereby increasing their survival rates during cryopreservation (Suzuki et al. 2006). Shoot tips, one of the most commonly cryopreserved explants, are typically precultured in semisolid, liquid, or solid media comprising relatively high concentrations of sugars, usually ranging from 0.3 to 1.0 M. This sugar treatment induces osmotic dehydration and increases starch accumulation in the meristem, which helps protect cellular membranes under severe dehydration conditions (Bachiri et al. 2000; Towill 2002; Reed 2008; Kulus et al. 2018; Mathew et al. 2018). In addition to sucrose, other cryoprotective additives, such as glycerol or dimethyl sulfoxide (DMSO), may be included in preculture media, depending on the particular requirements of the plant species. For callus tissues, which have a greater water content than seeds or embryonic calluses, preculturing is especially crucial. Culturing calluses in high-sugar media enhances their resistance to osmotic stress and increases their survival by stabilizing their cellular membranes and proteins under low-temperature conditions. Sucrose, in particular, plays a vital role in protecting the lipid structure of cell membranes and maintaining protein stability during freezing, making it an indispensable component of preculture media (Bachiri et al. 2000). Meristems of *Euonymus koopmannii* precultured without sucrose in the control group presented a survival rate of only 20%. Increasing the sucrose concentration to 0.2 M doubled the survival rate to 43.3%. The highest survival rate of 56.7% was observed at a 0.3 M concentration, suggesting that this concentration provided an optimal osmotic environment. However, a slight decrease in the survival rate to 53.3% was noted when the sucrose concentration was further increased to 0.4 M (Gubaidullin et al. 2024). Preculturing petunia shoot tips with 0.2 M sucrose for 0–3 days significantly increased the average regrowth, ranging from 68% to 81%. In contrast, a longer preculture duration of 4 or 5 days resulted in lower regrowth. Among these, the 2-day preculture duration achieved the highest average regrowth levels (Zhang et al. 2015b). Lee et al. (2011) reported that the greatest shoot regrowth in cryopreserved *Chrysanthemum* shoot tips occurred after stepwise preculture with 0.3 M sucrose for 31 h, 0.5 M sucrose for 17 h, and 0.7 M sucrose for 7 h. In contrast, direct preculture with 0.5 M sucrose for 1 day resulted in the greatest shoot regrowth in *Chrysanthemum* (Halmagyi et al. 2004; Wang et al. 2014). Embryogenic cell suspensions of *Anthurium andraeanum* were precultured on half-strength MS media enriched with 0.5 M sorbitol for 2 days (Wang et al. 2010).

7.4.6 Water/Moisture Content

The water/moisture content is one of the most critical factors determining the success of cryopreservation. Plant cells are highly sensitive to freezing temperatures, and improper moisture levels can lead to ice crystal formation, cellular damage, and ultimately the failure of cryopreservation. Precise control of the water content is essential to ensure the viability, survival, and regeneration of explants after cryopreservation (Stanwood 1985; Benson et al. 2013). During the cryopreservation process, excessive water can result in lethal intracellular ice formation, whereas insufficient dehydration can cause osmotic stress or incomplete vitrification. Therefore, optimizing water status through techniques such as controlled drying, osmotic dehydration, or the use of cryoprotective agents is vital to minimize cellular damage and enhance post-thaw recovery. For example, in shoot tip cryopreservation, reducing the water content to a specific range (e.g., 20–30%) has been shown to improve survival rates by promoting vitrification and preventing ice crystallization. Similarly, in seed cryopreservation, achieving an optimal moisture content (e.g., 10–15%) is crucial to maintain seed viability during storage in LN. Zygotic birds-of-paradise were cryoconserved in LN following dehydration for various durations (0, 30, and 60 min) with silica gel, resulting in moisture contents of 16.79%, 11.38%, and 11.49%, respectively. Among the various desiccation periods, embryos dehydrated for 30 min presented the greatest percentage of seedling development, at 66.13%. No variation in survival was observed during acclimatization, and no anatomical deviations were detected in the photosynthetic tissues. Additionally, flow cytometry analysis confirmed the cytogenetic fidelity of the in vitro plantlets (Figueiredo et al. 2021). Rodrigues et al. (2014) investigated the ideal moisture level for the cryopreservation of seeds from ten endangered bromeliad species of the genera *Alcantarea*, *Nidularium*, *Pitcairnia*, *Vriesea*, and *Wittrockia*. The study revealed that seed viability was compromised at water contents below 7%, with the ideal range for cryopreservation being 6–7% (dry mass basis, DMB).

The optimal moisture content for pollen cryopreservation varies depending on the developmental stage of the pollen, which can be categorized into two-celled or three-celled pollen on the basis of the number of cells present during dispersal (Apel and Hirt 2004). Cryopreserved pollens with relatively high water contents can lead to ice crystal formation, causing membrane rupture and cell collapse, which ultimately results in the loss of germination capacity. On the other hand, excessively low moisture content can impair the post-thaw germination ability of pollen (Barnabás and Rajki 1981).

7.4.7 Loading Solution

Osmoprotective solutions, such as loading solutions (LSs), are important solutions in plant cryopreservation because they help the cell/tissues avoid osmotic stress and prepare explants for subsequent exposure to highly concentrated cryoprotective agents (CPAs). The composition of LS includes 2.0 M glycerol and 0.4 M sucrose, which has been successfully used in the cryopreservation of various ornamental crops. For example, in *Euonymus koopmannii*, the control group had a survival rate of 6.7%, whereas after 15 min of osmopriming, survival increased fivefold to 33.3%. A 20-min treatment yielded the highest survival rate of 60%, whereas extending the treatment to 30 min led to a decrease of 46.7% (Gubaidullin et al. 2024). Similarly, osmoprotection for 20 or 30 min led to greater average regrowth (72–76%) of petunia shoot tips, outperforming other tested durations (Zhang et al. 2015b). The effectiveness of LS is attributed to the synergistic action of sucrose, which provides osmotic support, and glycerol, which penetrates cells to stabilize cellular structures. The duration and composition of LS treatment are often optimized on the basis of the specific requirements of the plant species and explant type. For example, extending the loading solution exposure time from 20 to 80 min in *Dahlia* significantly improved post-thaw regrowth, increasing it from 23.33% to 30% (Gowthami et al. 2023c). The embryogenic calli of *Anemarrhena asphodeloides* treated with a loading solution for 30 min presented greater recovery (67.8%) and produced healthy in vitro shoots post-cryopreservation (Sen-Rong and Ming-Hua 2012).

7.4.8 Cryoprotection and Cryoprotectants

Cryoprotection is the key to minimizing physical and chemical injuries during cryopreservation, and it largely depends on the treatments applied before the final immersion of plant tissues in LN. By stabilizing cell structures and reducing ice formation inside cells, cryoprotectants play crucial roles in preventing damage. In many cryopreservation protocols, an amalgamation of both penetrating and nonpenetrating cryoprotectants is used to enhance the protective effects during freezing and thawing (Panis et al. 2005; Gowthami et al. 2024a, b, c). This approach leverages the benefits of both types of cryoprotectants to optimize the preservation and recovery of plant material. A precise balance is necessary to avoid toxicity and damage from excessive cryoprotectant concentrations while ensuring that sufficient protection is provided against freezing-induced damage.

The duration of exposure during cryopreservation plays a critical role in the regrowth of cryopreserved explants, and this effect varies depending on the specific cryopreservation technique employed. In the case of *Lamprocapnos spectabilis*, shorter PVS3 treatment durations yielded better recovery for explants via droplet vitrification techniques (30–60 min) and vitrification (60 min) (Kulus 2020). These techniques benefit from shorter exposure times, likely because they rely on rapid dehydration and vitrification processes, which are optimized by faster treatment durations. On the other hand, encapsulation vitrification, which involves embedding the explants in a bead matrix, requires a longer dehydration duration (approximately 150 min) to achieve optimal results. The bead matrix plays a significant role in this process by providing a physical structure that supports the explant and facilitates slower, more controlled dehydration, which is essential for the successful cryopreservation of larger or more complex tissues. Halmagyi et al. (2004) reported that exposure to 100% PVS2 for 5 min or 60% PVS2 for 15 min at room temperature resulted in the greatest shoot regrowth in *Chrysanthemum*. In contrast, Lee et al. (2011) reported that dehydration with PVS2 at room temperature led to the lowest regrowth, whereas PVS3 gave the best results for axillary and apical shoot tips; the optimal exposure times were 60 and 90 min, respectively, with no significant decrease in regrowth after 120 min at 0 °C. Wang et al. (2014) reported the highest regrowth with 30 min of dehydration in PVS2 at 0 °C.

7.4.9 Cryopreservation Methods

The selection of the cryopreservation method depends on various factors, including the crop being preserved, available equipment, technical expertise, and budget. Therefore, there is no universally applicable or "standard" cryopreservation protocol. Instead, the method is typically chosen on the basis of a review of the literature to identify successful approaches used for similar species (Reed et al. 2001, 2004). The selection of an optimal cryopreservation technique is species dependent (Kulus 2020). Presently, common cryopreservation techniques for shoot tips in vegetatively propagated crops include vitrification, droplet vitrification, encapsulation–vitrification, encapsulation–dehydration, two-step cooling methods (or their modifications), and the use of cryoplates (V and D cryoplates). These methods are well documented in various research articles (Sakai and Engelmann 2007; Reed 2008; Engelmann 2004; Nagel et al. 2024; Aryakia 2025). Among the three cryopreservation techniques examined by Kulus (2020) in *Lamprocapnos spectabilis*, encapsulation-vitrification offer several clear advantages over vitrification and droplet vitrification. The survival rate after cryopreservation using EV was highest (73.1%), followed by that after droplet vitrification (52.4%) and vitrification (25.8%). However, for *Chrysanthemum morifolium* 'Escort', vitrification was better than slow cooling, EV, and droplet vitrification methods (Halmagyi et al. 2004). *L. ledebourii* seeds were cryopreserved via different cryopreservation techniques with various survival/

Fig. 7.3 Cryopreservation of *Gentiana kurroo*. (**a**) Post-thaw regrowth of shoot tips using droplet vitrification technique. (**b**) Post-thaw regeneration of shoot tips. C control, V vitrification, DV droplet vitrification

germination percentages, e.g., 50% (Kaviani et al. 2008; Kaviani 2010) and 69% (Mohajeri et al. 2014) via ED, 10% via EV (Urbaniec-Kiepura and Bach 2014), 75% pregrowth–desiccation (Kaviani et al. 2009), 97% via vitrification, and 98% via desiccation (Mohajeri et al. 2014). Sharma et al. (2021) reported that compared with vitrification, droplet vitrification performed well in terms of post-thaw regrowth in *Gentiana kurroo* (Fig. 7.3).

7.5 Pollen Cryopreservation

Pollen cryopreservation is an advanced biotechnological method for the long-term storage of pollen, maintaining its viability and fertilization potential, which is essential for ornamental crop breeding, hybridization, and conservation. A key benefit is overcoming temporal and geographic barriers, allowing pollen from early-flowering varieties to be used with late-flowering varieties. This ensures year-round pollen availability, is vital for plants with brief flowering periods, and supports continuous breeding. Cryopreservation also conserves genetic diversity, particularly for rare and endangered species such as orchids, by storing pollen at −196 °C. By enabling crossbreeding between species with mismatched flowering times, it facilitates polyploid breeding and distant hybridization. Additionally, cryopreserved pollen enhances controlled pollination efficiency, as it is unaffected by environmental factors, improving fertilization success. This reduces the need for fresh pollen collection, cutting labor and costs in commercial flower production. Moreover, it aids in the study of pollen biology and germination while increasing seed production and hybrid vigor in crops (Dinato et al. 2020; Gowthami et al. 2024d; Nagar et al. 2023; Rajasekharan and Rohini 2023; Shankar et al. 2023, 2024). The primary goal of pollen cryopreservation is to maintain pollen viability and fertility at levels comparable to those of fresh pollen, ensuring its usability for future breeding, research, and conservation efforts in ornamental crops.

Pollen grains can be broadly categorized into two types on the basis of their cellular structure and desiccation tolerance: bicellular and tricellular. Bicellular pollen, often referred to as orthodox pollen, is highly resistant to desiccation and can tolerate moisture levels as low as 11.1%, making it well suited for cryopreservation. Pollen cryopreservation has been reported in several ornamental crops, such as gladiolus (Rajasekharan et al. 1994), rose cultivars (Rajasekharan and Ganeshan 1994; Marchant et al. 1993), and *Amorphophallus koratensis* (Soonthornkalump et al. 2020). Xu et al. (2014) cryopreserved pollen from 60 species/cultivars (85% of those tested), and the pollen remained viable after 1 year of cryopreservation. Among these, 36 (51% of the total studied) maintained high germination rates comparable to those of fresh pollen, demonstrating their suitability for long-term cryopreservation. Ren et al. (2019) assessed the viability of pollen from 102 ornamental species/cultivars after 8–10 years of cryopreservation. The viability ranged from 1% to 58%, with 11.76% increasing viability, 16.67% stabilizing, and 71.57% decreasing. Rajasekharan and Rohini (2023) compiled pollen cryopreservation protocols for gladiolus, *Gloriosa superba*, *Tagetes*spp., orchid species, *Polianthes tuberosa*, *Aloe vera*, and rose.

Orchid pollen is uniquely structured, ranging from single grains (monads) to groups of four (tetrads) or multiple tetrads. These pollen units are aggregated into specialized structures known as pollinia, which consist of a cohesive mass of pollen grains attached to a stalk and a viscidia—a sticky pad that facilitates attachment to pollinators. Pollinia cryopreservation has been reported in *Cohniella cepula* (Dolce et al. 2016). In addition to the desiccation technique, pollinia are also cryopreserved via different cryotechniques, such as vitrification in *Dendrobium* (Vendrame et al. 2008), V cryoplate in *Rhynchostylis gigantea* (Jitsopakul et al. 2021), *Luisia macrantha pollinia* via air-drying, vitrification, and silica gel methods (Ajeeshkumar and Decruse 2013), and V and D cryoplation in *Dendrobium signatum* (Jitsopakul et al. 2019).

The ICAR-Indian Institute of Horticultural Research, Bengaluru, Karnataka, conserves pollen from a diverse array of species, including fruits, vegetables, ornamental plants, and rare, endangered, and threatened (RET) medicinal plants. Recognized for its groundbreaking efforts, the IIHR pollen cryobank was honored in the *Limca Book of Records* in 2001. Protocols have been developed for pollen cryopreservation of different ornamental species, such as *Agave sisalana, Asclepias curassavica, Barleria obtusa, Bauhinia purpurea, Brownea grandiceps, Cassia fistula, Clitoria ternatea, Costus* spp., *Crinum* spp., *Gladiolus* spp., *Gloriosa superba, Hemerocallis fulva, Hydrocera triflora, Jatropha* spp., *Kigelia africana, Millingtonia hortensis, Nymphaea* spp., *Nymphoides* spp., *Orchid* spp., *Polianthes tuberose, Rosa* spp., *Spathiphyllum wallisii, Spathoglottis plicata, Strelitzia reginae, Tagetes* spp., and *Torenia* spp.

7.6 Thawing, Rewarming, and Recovery

Thawing and recovery are critical phases in plant cryopreservation, as improper handling can lead to ice recrystallization, devitrification, and cellular damage. During the thawing and rewarming phases, ice recrystallization can occur, where smaller ice crystals merge into larger aggregates. This process can be lethal to cells, so it is essential to rewarm cryopreserved explants rapidly (Sakai and Engelmann 2007).

Seed thawing is typically achieved by immersing cryovials in a water bath or placing them at room temperature. Shoot tip warming is typically achieved by immersing cryovials in a water bath at 35–42 °C (typically approximately 38 °C) for 2–3 min, especially in methods such as vitrification (Gowthami et al. 2023a, b) and encapsulation–dehydration (Feng et al. 2013). This rapid thawing ensures a quick transition from the vitrified (glass-like) state to a liquid state, minimizing the risk of ice crystal formation that can harm cellular structures. Alternatively, explants can be placed directly into an unloading solution at room temperature, which helps gradually reduce osmotic stress and supports better post-storage recovery.

After thawing, the removal of cryoprotectants is crucial to minimize toxicity and osmotic stress. This is typically achieved by washing explants in liquid Murashige and Skoog (MS) media, often supplemented with 1.2 M sucrose, for approximately 20 min. This step helps to gradually dilute and remove cryoprotectants, reducing the risk of cellular damage. Specifically, for droplet vitrification, encapsulation–vitrification, and V cryoplate methods, explants are often rinsed in an unloading solution containing 0.3–1.2 M sucrose at room temperature (22–25 °C) for a brief period before being plated onto recovery media to help with the recovery process (Sakai and Engelmann 2007; Gowthami et al. 2023a, b).

Once the cryoprotectants are removed, the explants are transferred to a recovery medium (RM), which plays a vital role in promoting regeneration and regrowth. For long-term germplasm storage, an effective regeneration system is essential, as the recovery process must support germination, embryo development, or the formation of plantlets. The RM must be carefully designed to support the rehydration of explants, which typically begins within a few hours of placing the tissues onto the medium. To increase nutrient uptake and accelerate recovery, explants are often inoculated onto semisolid or liquid media/full-strength MS or half-strength MS media. Additionally, the RM is supplemented with growth regulators, such as cytokinins and auxins, to stimulate cell division and regeneration. However, callus formation should be avoided to prevent somaclonal variation, which compromises genetic stability. The removal or substitution of ammonium in the regrowth medium has been shown to significantly increase the regeneration of cryopreserved shoot tips in *Chrysanthemum morifolium* var. 'Borami' (Yi et al. 2018) and *Aster altaicus* var. *uchiyamae* (Choi et al. 2019). In contrast, Jitsopakul et al. (2008a) reported greater survival and improved development of cryopreserved *Bletilla striata* protocorms when a regrowth medium containing ammonium was used than when ammonium-free medium was used.

Furthermore, incorporating antioxidants such as polyvinylpyrrolidone, ascorbic acid, tocopherol, catalase, pyruvate dehydrogenase, malate dehydrogenase, and melatonin into the recovery medium can mitigate oxidative stress, which is often induced during dehydration/rehydration and freezing/thawing processes. These antioxidants help counteract the accumulation of harmful reactive oxygen species (Cripps 2006). Light exposure is another factor that can influence recovery. Gradual exposure to light after rewarming can help improve the recovery rate by alleviating stress and preventing photooxidation, which facilitates tissue repair (Yin and Hong 2009).

Chen et al. (2011) used an RM consisting of MS supplemented with 0.5 mg/L 6-benzylaminopurine, 0.1 mg/L α-naphthaleneacetic acid, 0.3 mg/L gibberellic acid, 30 g/L sucrose, and 7 g/L agar. Cultures were initially maintained in the dark for 2 weeks at 20 °C, followed by transfer to a 16-h photoperiod with a light intensity of 35 $\mu mol\ m^{-2}\ s^{-1}$ at 20 °C. Yin et al. (2014) used MS media supplemented with 1 mg/L NAA and 0.2 mg/L TDZ (pH 5.8). The cultures were kept in the dark at 23 ± 2 °C for 3 days and then transferred to light conditions to evaluate survival and shoot regrowth.

7.7 Evaluation of Cryopreservation Efficiency

The effectiveness of a cryopreservation method is evaluated via specific parameters tailored to the type of tissue or organ being preserved. These parameters include the survival rate, regrowth rate, and other species-specific metrics. Below is a detailed breakdown of the evaluation methods for different plant materials. For cryopreserved seeds, the primary evaluation metrics are the germination rate and seedling formation. Survival is assessed by the percentage of seeds that germinate and produce healthy seedlings. Regrowth is evaluated on the basis of the development of normal seedlings with intact shoots and roots. The most common approach is the standard germination test, where seeds are placed on germination paper (roll-towel paper) or Petri plates lined with germination paper under optimal conditions. For seeds with low germination rates or dormancy issues, the tetrazolium (TTC) test is widely used. Hosomi et al. (2012) reported that both the TTC and germination results confirmed that seed viability remained intact after storage in *Cattleya* sp. Zygotic embryos, embryonic axes, shoot tips, protocorms, and axillary buds post-thaw viability is generally assessed through in vitro regeneration. TTC, FDA staining, and in vitro germination/in vitro regeneration are used to assess the viability of protocorms, pollen, somatic embryos, and embryogenic tissue. FDA staining was used to evaluate the post-cryopreservation survival of *Caladenia latifolia* protocorms (Watanawikkit et al. 2012). Bustam et al. (2016) reported that the use of asymbiotic germination media (½ MS + 5% coconut water) for primary protocorms, followed by protocorm proliferation media (½ MS + 5 μM NAA + 2 μM BA) for secondary protocorms after 1 week on standard recovery media, significantly increased.

7.8 Conclusion and Future Prospects

Cryopreservation has emerged as an essential strategy for the long-term conservation of ornamental crops, safeguarding their genetic diversity and ensuring the availability of valuable germplasm for future breeding programs and research initiatives. The success of cryopreservation hinges on optimizing key factors such as explant selection, preconditioning treatments, cryoprotectant application, precise freezing and thawing protocols, and efficient post-thaw recovery strategies. Recent innovations, including vitrification, droplet vitrification, ED, EV, and V and D cryo-plate techniques, have significantly improved survival and regeneration rates across a diverse range of ornamental species. However, challenges remain, particularly in standardizing protocols for recalcitrant species and minimizing cellular damage during freezing and thawing.

In the future, advancing cryopreservation techniques will require interdisciplinary approaches to overcome these challenges. The integration of omics technologies such as genomics, transcriptomics, and proteomics can provide deeper insights into the molecular mechanisms governing cryotolerance and recovery, enabling the development of more precise and effective protocols. Nanotechnology holds promise for enhancing cryoprotectant delivery, whereas artificial intelligence-driven optimization of protocol parameters can further improve efficiency and reproducibility. Strengthening collaboration among researchers, conservationists, and horticulturists will be key to extending cryopreservation techniques to a broader range of ornamental species, particularly those that are endangered or economically significant.

By addressing existing limitations and leveraging emerging technologies, cryopreservation will continue to play a pivotal role in preserving the beauty, diversity, and genetic wealth of ornamental crops, ensuring their sustainability for future generations.

Competing Interests The authors have no relevant financial or nonfinancial interests to disclose.

References

Agrawal A, Gowthami R, Chander S, Srivastava V (2022) Sustainability of *in vitro* genebanks and cryogenebanks. Indian J Plant Genet Resour 35:180–184

Agrawal A, Gowthami R, Chander S, Srivastava V, Shankar M, Deepak DA (2024) Biotechnological tools for conservation of plant genetic diversity. In: Al-Khayri JM, Jain SM, Penna S (eds) Sustainable utilization and conservation of plant genetic diversity. Springer, Singapore, pp 711–752

Ajeeshkumar S, Decruse SW (2013) Fertilizing ability of cryopreserved pollinia of Luisia macrantha, an endemic orchid of Western Ghats. Cryo Letters 34(1):20–29

Antony JJ, Keng CL, Rathinam X, Marimuthu S, Subramaniam S (2010) Preliminary study on cryo preservation of Dendrobium Bobby Messina protocorm- like bodies by vitrification. Afr J Biotechnol 9(42):7064–7070

Apel K, Hirt H (2004) Reactive oxygen species: metabolism, oxidative stress, and signal transduction. Annu Rev Plant Biol 55:373–399

Aryakia E (2025) Cryobanking of rare and endangered plant species: key propagules and genetic stability assessments. J Nat Conserv 23:126844

Bachiri Y, Bajon C, Sauvanet A, Gazeau C, Morisset C (2000) Effect of osmotic stress on tolerance of air-drying and cryopreservation of *Arabidopsis thaliana* suspension cells. Protoplasma 214:227–243

Bachiri Y, Gazeau C, Hansz J, Morisset C, Dereuddre J (1995) Successful cryo preservation of suspension cells by encapsulation-dehydration. Plant Cell Tissue Organ Cult 43:241–248

Baghdadi SH, Makhadmeh I, Syouf M, Arabiat A, Shibli RA, Shatnawai MA (2011) Cryopreservation by vitrification of embryogenic callus of wild crocus (*Crocus hyemalis* and *Crocus moabiticus*). Acta Hortic 908:239–247

Barnabás B, Rajki E (1981) Fertility of deep-frozen maize (*Zea mays* L.) pollen. Ann Bot 48:861–864

Benson EE (2008) Cryopreservation of phytodiversity: a critical appraisal of theory & practice. Crit Rev Plant Sci 27:141–219

Benson EE, Betsou F, Fuller BJ, Harding K, Kofanova O (2013) Translating cryobiology principles into transdisciplinary storage guidelines for biorepositories and biobanks: a concept paper. Cryo Letters 34:277–312

Bettoni JC, Bonnart R, Shepherd AN, Kretzschmar AA, Volk GM (2019) Modifications to a *Vitis* shoot tip cryopreservation procedure: effect of shoot tip size and use of cryoplates. Cryo Letters 40:103–112

Bettoni JC, Bonnart R, Volk GM (2021) Challenges in implementing plant shoot tip cryopreservation technologies. Plant Cell Tissue Organ Cult 144:21–34

Bi W-L, Pan C, Liu J, Wang Q-C (2016) Greenhouse performance, genetic stability and biochemical compounds in *Chrysanthemum morifolium* 'Hangju' plants regenerated from cryopreserved shoot tips. Acta Physiol Plant 38:1–10

Bouman H, Tiekstra A, Petutschnig E, Homan M, Schreurs R (2003) Cryopreservation of *Lilium* species and cultivars. Acta Hortic 612:147–154

Bunnag S, Hongthongkham J (2014) Cryopreservation by encapsulation-dehydration of *Aerides multiflora* roxb. protocorms (protocorms). Jordan J Agri Sci 173(3834):1–7

Burritt DJ (2008) Efficient cryopreservation of adventitious shoots of Begonia x erythrophylla using encapsulation-dehydration requires pre-treatment with both ABA and proline. Plant Cell Tissue Org Cult 95:209–215

Bustam BM, Dixon K, Bunn E (2016) A cryopreservation protocol for *ex situ* conservation of terrestrial orchids using asymbiotic primary and secondary (adventitious) protocorms. In Vitro Cell Dev Biol Plant 52:185–195

Chen X-L, Li J-H, Xin X, Zhang Z-E, Xin P-P, Lu X-X (2011) Cryopreservation of in vitro-grown apical meristems of *Lilium* by droplet-vitrification. South Afr J Bot 77:397–403

Choi CH, Popova E, Lee H et al (2019) Cryopreservation of endangered wild species, *Aster altaicus var. uchiyamae Kitam*, using droplet-vitrification procedure. Cryo Letters 40(2):113–122

Chowdhuri TK, Deka K (2019) Biodiversity and conservation of ornamental crops. In: Rajasekharan PE, Rao VR (eds) Conservation and utilization of horticultural genetic resources. Springer Singapore, Singapore, pp 139–216

Condello E, Caboni E, Andrè E, Piette B, Druart P, Swennen R, Panis B (2011) Cryopreservation of apple *in vitro* axillary buds using droplet- vitrification. Cryo Letters 32:175–185

Cordova LB, Thammasiri K (2016) Cryopreservation on a cryo-plate of *Arundina graminifolia* protocorms, dehydrated with silica gel and drying beads. Cryo Letters 37(2):68–76

Cripps RF (2006) Effect of anti-oxidants and iron on the recovery of cryopreserved gametophytic tissue from fern *Lepisorus longifolius*. Cryobiology 53:367–446

da Silva Cordeiro L, Simões-Gurgel C, Albarello N et al (2015) Cryopreservation of *in vitro*-grown shoot tips of *Cleome rosea* Vahl (Cleomaceae) using the V cryo-plate technique. In Vitro Cell Dev Biol Plant 51:688–695

Decruse SW, Seeni S (2002) Ammonium nitrate in the culture medium influences regeneration potential of cryopreserved shoot tips of *Holostemma annulare*. Cryo Letters 23:55–60

Decruse SW, Seeni S, Nair GM (2004) Preparative procedures and culture media effect on the success of cryostorage of *Holostemma annulare* shoot tips. Plant Cell Tissue Organ Cult 76:179–182

Dereuddre J, Fabre J, Bassaglia C (1988) Resistance to freezing in LN of carnation (*Dianthus caryophyllus* L. var *Eolo*) apical and axillary shoot tips excised from different aged *in vitro* plantlets. Plant Cell Rep 7:170–173

Dinato NB, Imaculada Santos IR, Zanotto Vigna BB, de Paula AF, Fávero AP (2020) Pollen cryopreservation for plant breeding and genetic resources conservation. Cryo Letters 41(3):115–127

Dolce NR, Medina RD, Mroginski LA, Rey HY (2016) Storage of *Cohniella cepula* (Orchidaceae) Pollinia: fertilizing ability and subsequent fruit and seed formation. HortSci 51(10):1265–1269

Engelmann F (2000) Importance of cryopreservation for the conservation of plant genetic resources. In: Engelmann F, Takagi H (eds) Cryopreservation of tropical plant germplasm: current research progress and application. IPGRI, Tsukuba, pp 8–20

Engelmann F (2004) Plant cryopreservation: progress and prospects. In Vitro Cell Dev Biol Plant 40:427–433

Engelmann F (2011) Use of biotechnologies for the conservation of plant biodiversity. In Vitro Cell Dev Biol Plant 47:5–16

Engelmann F (2014) Cryopreservation of clonal crops: a review of key parameters. Acta Hort 1039:31–39

Fabre J, Dereuddre J (1990) Encapsulation-dehydration: a new approach to cryopreservation of *Solanum* shoot-tips. CryoLetters 11:413–426

Feng CH, Cui ZH, Li BQ, Chen L, Ma YL, Zhao YH, Wang QC (2013) Duration of sucrose preculture is critical for shoot regrowth of *in vitro*-grown apple shoot-tips cryopreserved by encapsulation-dehydration. Plant Cell Tissue Organ Cult 112:369–378

Figueiredo JRM, Paiva PDO, Silva DPC, Renato Paiva R, Mesquita R, Souza RR, Reis MV (2021) Anatomical changes and cytogenetic stability in bird of paradise plants after zygotic embryo cryopreservation by desiccation method. In Vitro Cell Dev Bio Plant 57:272–280

Flachsland E, Terada G, Scocchi A, Rey H, Mroginski L, Engelmann F (2006) Cryopreservation of seeds and in vitro-cultured protocorms of *Oncidium bifolium* Sims. (Orchidaceae) by encapsulation-dehydration. Cryo Letters 27:235–242

Folgado R, Panis B (2019) Cryopreservation of Ashe magnolia shoot-tips by droplet-vitrification. Acta Hortic 1234:233–240

Galdiano RF, Vendrame WA, Moretto C, Faria RT, de Macedo Lemos EG (2017) Seed cryopreservation, in vitro propagation and ex vitro growth of *Cattleya walkeriana* Gardner, a vulnerable ornamental orchid. Aust J Crop Sci 11(04):485–490

Gogoi K, Kumaria S, Tandon P (2012) A comparative study of vitrification and encapsulation-vitrification for cryopreservation of protocorms of *Cymbidium eburneum* L., a threatened and vulnerable orchid of India. Cryo Letters 33(6):443–452

Gogoi K, Kumaria S, Tandon P (2013) Cryopreservation of *Cymbidium eburneum* Lindl. and *C. hookerianum* Rchb.f., two threatened and vulnerable orchids via encapsulation–dehydration. In Vitro Cell Dev Biol Plant 49:248–254

Gonzalez-Arnao MT, Durán-Sánchez B, Jiménez-Francisco B, Lázaro-Vallejo CE, Valdés-Rodríguez SE, Guerrero A (2011) Cryopreservation and proteomic analysis of vanilla (*V. planifolia* A.) apices treated with osmoprotectants. Acta Hortic 908:67–72

Gonzalez-Arnao MT, Lazaro-Vallejo CE, Engelmann F, Gamez-Pastrana R, Martinez-Ocampo YM, Pastelin-Solano MC, Diaz-Ramos C (2009) Multiplication and cryopreservation of vanilla (*Vanilla planifolia* 'Andrews'). In Vitro Cell Dev Biol Plant 45:574–582

González-Benito ME, Pérez C, Viviani AB (1997) Cryopreservation of nodal explants of an endangered plant species (*Centaurium rigualii* Esteve) using the encapsulation-dehydration method. Biodivers Conserv 6:583–590

Gowthami R, Chander S, Gupta S, Srivastava V, Puneeth GM, Deepak DA, Shankar M, Singh K (2024c) Cryogenic conservation of medicinal plants importance and techniques. In: Husen A, Pant M (eds) Tissue culture techniques and medicinal plants. CRC Press, Taylor & Francis, Boca Raton, pp 248–274

Gowthami R, Chander S, Pandey R, Shankar M, Agrawal A (2023a) Development of cryoconservation protocol for In-vitro shoot bases of *Allium ampeloprasum* L. for long-term conservation. Indian J Plant Genet Resour 36(3):396–401

Gowthami R, Chander S, Pandey R, Shankar M, Agrawal A (2023c) Development of efficient and sustainable droplet-vitrification cryoconservation protocol for shoot tips for long-term conservation of *Dahlia* germplasm. Sci Hortic 321:112329

Gowthami R, Rajasekharan PE, Chander S, Shankar M, Srivastava V, Agrawal A (2024d) Cryopreservation of two-celled pollen: a model system for studying the cellular mechanisms of cryoinjury and recovery. 3 Biotech 14(12):1–9

Gowthami R, Rajashekaran PE, Martinez-Montero ME (2024a) Cryoconservation of antiallergic medicinal plants: techniques and applications. In: Gantait S, Majumder J, Sharangi AB (eds) Biotechnology of medicinal plants with antiallergy properties. Springer, Singapore, pp 611–634

Gowthami R, Rajashekaran PE, Martinez-Montero ME, Chander S, Singh AM, Puneeth GM, Shankar M, Deepak DA (2024b) *In vitro* techniques for safeguarding medicinal plants with antiallergic properties. In: Gantait S, Majumder J, Sharangi AB (eds) Biotechnology of medicinal plants with antiallergy properties. Springer, Singapore, pp 635–672

Gowthami R, Sharma N, Chandra R, Kurian JS, Malhotra EV, Agrawal A (2023b) Development of an improved and simple shoot tip cryoconservation protocol for cryobanking of *Bacopa monnieri* (L.) Wettst. germplasm. In Vitro Cell Dev Biol Plant 59:744–756

Gubaidullin N, Kali B, Tussipkan D, Manabayeva SA (2024) A promising strategy for conservation of Central Asian endemic plant *Euonymus koopmannii*. Biodivers J Biol Divers 25:3114–3120

Gupta V, Gore PG, Kak A (2019) Collection, conservation and utilisation of indigenous ornamental crops. Indian Hortic 64(4):76–81

Hakim MH, Elwon CAM, Norzahan MN, Ripin R, Aziz ZA (2015) Cryopreservation of *Coelogyne dayanum* seeds by vitrification. Acta Hortic 1078:53–60

Halmagyi A, Deliu C (2007) Cryopreservation of carnation (*Dianthus caryophyllus* L.) shoot tips by encapsulation-vitrification. Sci Hortic 113:300–306

Halmagyi A, Deliu C, Coste A (2005) Plant regrowth from potato shoot tips cryopreserved by a combined vitrification-droplet method. Cryo Letters 26:313–322

Halmagyi A, Fischer-Klüver G, Mix-Wagner G, Schumacher M (2004) Cryopreservation of *Chrysanthemum morifolium* (*Dendranthema grandiflora* Ramat.) using different approaches. Plant Cell Rep 22:371–375

Halmagyi A, Pinker I (2006) Plant regeneration from *Rosa* shoot tips cryopreserved by a combined droplet vitrification method. Plant Cell Tissue Org Cult 84:145–153

Halmagyi A, Vălimăreanu S, Şovărel G, Coste A (2022) Cryo-technologies for ex situ conservation of *Rosa* germplasm. Plan Theory 11(8):1095

Hirano T, Godo T, Mii M, Ishikawa K (2005a) Cryopreservation of immature seeds of *Bletilla striata* by vitrification. Plant Cell Rep 23:534–539

Hirano T, Ishikawa K, Mii M (2005b) Cryopreservation of immature seeds of *Ponerorchis graminifolia* var. suzukiana by vitrification. Cryo Letters 26:139–146

Hosomi ST, Custódio CC, Seaton PT, Marks TR, Machado-Neto NB (2012) Improved assessment of viability and germination of Cattleya (Orchidaceae) seeds following storage. In Vitro Cell Dev Plant 48:127–136

Imsomboon T, Thammasiri K (2020) Cryopreservation of *Paphiopedilum exul* (Ridl.) Rolfe seeds using encapsulation vitrification and encapsulation-dehydration methods. Acta Hortic 1298:195–204

Jevremović S, Subotić A, Benelli C, De Carlo A, Lambardi M (2011) Cryopreservation of *Iris pumila* shoot tips by vitrification. Acta Hortic 908:355–359

Jiang X-R, Ren R-F, Di W, Jia M-X, Li Z-D, Liu Y, Gao R-F (2019) Hydrogen peroxide and nitric oxide are involved in programmed cell death induced by cryopreservation in *Dendrobium* protocorm like bodies. Plant Cell Tissue Organ Cult 137:553–563

Jitsopakul N, Homchan P, Thammasiri K (2020) Cryopreservation of *Dendrobium draconis* Rchb. f. protocorms using V cryo plate method. Acta Hortic 1298:213–218

Jitsopakul N, Homchan P, Thammasiri K (2021) Cryopreservation orchid pollinia using the V cryo-plate. Cryo Letters 42(1):25–32

Jitsopakul N, Sangyojarn P, Homchan P, Thammasiri K (2019) Efficiency of aluminum cryo-plates for cryopreservation of *Dendrobium signatum* Rchb. f. pollinia. Acta Hortic 1234:279–286

Jitsopakul N, Thammasiri K, Ishikawa K (2008b) Cryopreservation of *Vanda coerulea* protocorms by encapsulation-dehydration method. Cryo Letters 29:253–260

Jitsopakul N, Thammasiri K, Ishikawa K (2008a) Cryopreservation of *Bletilla striata* mature seeds, 3-day germinating seeds and protocorms by droplet-vitrification. Cryo Letters 29(6):517–526

Jitsopakul N, Thammasiri K, Yukawa C, Ishikawa K (2012) Effect of cryopreservation on seed germination and protocorm development of *Vanda tricolor*. Sci Asia 38:244–249

Kang P, Kim SJ, Park HJ, Han SJ, Kim IC, Lee H, Yim JH (2025) Trends and challenges in plant cryopreservation research: a meta-analysis of cryoprotective agent development and research focus. Plan Theory 14(3):447

Kartha KK, Leung NL, Mroginski LA (1982) *In vitro* growth responses and plant regeneration from cryopreserved meristems of cassava (*Manihot esculenta* Crantz). Z Pflanzenphysiol 107:133–140

Kaviani B (2007) Effects of salicylic acid and encapsulation on enhancing the resistance of embryonic axes of persian lilac (*Melia azedarach* L.) against cryo preservation. Int J Agri Biol 9(4):625–627

Kaviani B (2010) Cryopreservation by encapsulation–dehydration for long-term storage of some important germplasm: seed of lily [*Lilium ledebourii* (Baker) Bioss.], embryonic axe of persian lilac (*Melia azedarach* L.), and tea (*Camellia sinensis* L.). Plant Omics 3:177–182

Kaviani B, Abadi DH, Torkashvand AM, Hoor SS (2009) Cryopreservation of seeds of lily (*Lilium ledebourii* Baker Bioss.): Use of sucrose and dehydration. Afr J Biotechnol 8:3809–3810

Kaviani B, Dehkaei MNP, Safarimotlagh MR, Darabi AH, Rafzadeh A (2008) Cryopreservation of lily [*Lilium ledebourii* (Baker) Bioss.] germplasm by encapsulation–dehydration. J Biotechnol 136:S273

Kaviani B, Kulus D (2022) Cryopreservation of endangered ornamental plants and fruit crops from tropical and subtropical regions. Biology 11(6):847

Kaviani B, Negahdar N (2017) Propagation, micropropagation and cryopreservation of *Buxus sempervirens* Pojark., an endangered ornamental shrub. S Afr J Bot 111:326–335

Khoddamzadeh AA, Sinniah UR, Lynch P, Kadir MA, Kadzimin SB, Mahmood M (2011) Cryopreservation of protocorm-like bodies (PLBs) of *Phalaenopsis bellina* (Rchb. f.) Christenson by encapsulation-dehydration. Plant Cell Tissue Org Cult 107:471–481

Kim HM, Shin JH, Sohn JK (2004) Cryopreservation of zygotic embryos of herbaceous peony (*Paeonia lactiflora* Pall.) by encapsulation-dehydration. Korean J Crop Sci 49(4):354–357

Kulus D (2020) Shoot tip cryopreservation of *Lamprocapnos spectabilis* (L.) Fukuhara using different approaches and evaluation of stability on the molecular, biochemical, and plant architecture levels. Int J Mol Sci 21:3901

Kulus D, Miler N (2021) Application of plant extracts in micropropagation and cryopreservation of bleeding heart: an ornamental-medicinal plant species. Agriculture 11:542

Kulus D, Rewers M, Serocka M et al (2019) Cryopreservation by encapsulation-dehydration affects the vegetative growth of chrysanthemum but does not disturb its chimeric structure. Plant Cell Tissue Organ Cult 138:153–166

Kulus D, Serocka M, Mikuła A (2018) Effect of various preculture and osmotic dehydration conditions on cryopreservation efficiency and morphogenetic response of chrysanthemum shoot tips. Acta Sci Pol Hortorum Cultus 17:139–147

Kulus D, Zalewska M (2014) Cryopreservation as a tool used in long-term storage of ornamental species—a review. Sci Hortic 168:88–107

Kwaśniewska E, Dziedzic E, Pawłowska B (2017) Integration of cryopreservation and tissue culture for germplasm conservation and propagation of *Rosa pomifera* 'Karpatia'. Not Bot Horti Agrobot 45:208–214

Lambardi M, De Carlo A, Capuana M (2005) Cryopreservation of embryogenic callus of *Aesculus hippocastanum* L. by vitrification/one-step freezing. Cryo Letters 26:185–192

Lee YG, Popova E, Cui HY, Kim HH, Park SU, Bae CH, Lee SC, Engelmann F (2011) Improved cryopreservation of Chrysanthemum (*Chrysanthemum morifolium*) using droplet-vitrification. Cryo Letters 32:487–497

Lurswijidjarus W, Thammasiri K (2004) Cryopreservation of shoot tips of Dendrobium Walter Oumae by encapsulation/ dehydration. Sci Asia 30:293–299

Makowski D, Mikuła A, Rybczynski JJ (2011) Cryopreservation of *Asplenium cuneifolium* gametophyte and regeneration of plant material in post-rewarming culture. In: Cryopreservation of Crop Species in Europe. Eur Coop Sci Technol, Angers, pp 151–153

Maneerattanarungroj P, Bunnag S, Monthatong M (2007) In vitro conservation of *Cleisostoma arietinum* (Rchb.f.) Garay, rare Thai orchid species by an encapsulation-dehydration method. Asian J Plant Sci 6:1235–1240

Marchant R, Power JB, Davey MR, Chartier-Hollis JM, Lynch PT (1993) Cryopreservation of pollen from two rose cultivars. Euphytica 66(3):235–241

Marković Z, Chatelet P, Preiner D, Sylvestre I, Kontić JK, Engelmann F (2014) Effect of shooting medium and source of material on grapevine (*Vitis vinifera* L.) shoot tip recovery after cryopreservation. Cryo Letters 35(1):40–47

Maślanka M, Panis B, Bach A (2013) Cryopreservation of *Galanthus elwesii* Hook. apical meristems by droplet vitrification. Cryo Letters 34:1–9

Maślanka M, Panis B, Malik M (2016) Cryopreservation of *Narcissus* L 'Carlton' somatic embryos by droplet vitrification. Propag Ornam Plants 16:28–35

Maślanka M, Szewczyk A (2021) Droplet-vitrification cryopreservation of *Tulipa tarda* Stapf. apical meristems. Plant Cell Tissue Organ Cult 144:91–95

Mathew L, McLachlan A, Jibran R, Burritt DJ, Pathirana R (2018) Cold, antioxidant and osmotic pre-treatments maintain the structural integrity of meristematic cells and improve plant regeneration in cryopreserved kiwifruit shoot tips. Protoplasma 255:1065–1077

Mazur P (1984) Freezing of living cells: mechanisms and implications. Am J Physiol Cell Physiol 247:C125–C142

Mikuła A (2006) Comparison of three techniques for cryopreservation and re-establishment of long-term *Gentiana tibetica* suspension culture. Cryo Letters 27:268–282

Mikuła A, Fiuk A, Rybczynski JJ (2005) Induction, maintenance and preservation of embryogenic competence of *Gentiana cruciata* L. cultures. Acta Biol Crac Bot 47:227–236

Mikuła A, Niedzielski M, Rybczyński JJ (2006) The use of TTC reduction assay for assessment of *Gentiana* spp. cell suspension viability after cryopreservation. Acta Physiol Plant 28:315–324

Mikuła A, Olas E, Sliwinska RJJ (2008) Cryopreservation by encapsulation of Gentian spp. cell suspensions maintains regrowth, embryogenic competence and DNA content. Cryo Letters 29:409–418

Mikuła A, Tomiczak K, Rybczyński JJ (2011a) Cryopreservation enhances embryogenic capacity of *Gentiana cruciata* (L.) suspension culture and maintains (epi)genetic uniformity of regenerants. Plant Cell Rep 30:565–574

Mikuła A, Tomiczak K, Wójcik A, Rybczyński JJ (2011b) Encapsulation-dehydration method of cryopreservation elevates embryogenesity of *Gentiana kurroo* cell suspension and carrying on genetic stability of its regenerants after cryopreservation. Acta Hortic 908:143–154

Moghaddam AS, Kaviani B, Torkashvand AM, Abdossi V, Eslami AR (2024) In vitro propagation, cold preservation and cryopreservation of *Taxus baccata* L., an endangered medicinal and ornamental shrub. Acta Sci Pol Hort Cult 23(1):13–28

Mohajeri MR, Zare AG, Shahab MAN, Jari SK (2014) Seed germination of *Lilium ledebourii* (Baker) Boiss. after cryopreservation. J Rangel Sci 4:279–285

Mohanty P, Das MC, Kumaria S, Tandon P (2012) High-efficiency cryopreservation of the medicinal orchid *Dendrobium nobile* Lindl. Plant Cell Tissue Organ Cult 109(2):297–305

Mosa KA, Ahmed AE, Hazem Y, Kanawati IS, Abdullah A, Hernandez-Sori L, Ali MA, Vendrame W (2023) Insights into cryopreservation, recovery and genetic stability of medicinal plant tissues. Fitoterapia:105555

Mubmann V, Serek M, Winkelmann T (2006) Cryopreservation of embryogenic suspension cultures of *Cyclamen persicum* Mill. Acta Hortic 725:391–396

Nadarajan J, Staines HJ, Benson EE, Mansor M, Krishnapillay B, Harding K (2007) Optimization of cryopreservation for *Sterculia cordata* zygotic embryos using vitrification techniques. J Trop For Sci 19(2):79–85
Nagar A, Gowthami R, Sureja AK, Munshi AD, Verma M, Singh AK, Mallick N, Singh J, Chander S, Shankar M, Pathania P, Rajkumar S (2023) Simple cryopreservation protocol for *Luffa* pollen: enhancing breeding efficiency. Front Plant Sci 14:1268726
Nagel M, Pence V, Ballesteros D, Lambardi M, Popova E, Panis B (2024) Plant cryopreservation: principles, applications, and challenges of banking plant diversity at ultralow temperatures. Annu Rev Plant Biol 75:797–824
Nair DS, Reghunath BR (2008) Effective use of encapsulation-dehydration technique in cryopreserving somatic embryos of butterfly pea (*Clitoria ternatea* L.). J Herb Spices Med Plants 13:83–95
Nausch H, Buyel JF (2021) Cryopreservation of plant cell cultures – diverse practices and protocols. New Biotechnol 25:86–95
Negahdar N, Hashemabadi D, Kaviani B (2021) In vitro conservation and cryopreservation of *Buxus sempervirens* L., a critically endangered ornamental shrub. Rus J Plant Physiol 68:661–668
Niino T, Yamamoto S, Fukui K et al (2013) Dehydration improves cryopreservation of mat rush (*Juncus decipiens* Nakai) basal stem buds on cryo-plates. Cryo Letters 34:549–560
Normah MN, Sulonga N, Reed B (2019) Cryopreservation of shoot tips of recalcitrant and tropical species. Adv Strat Cryobiol 87:1–14
Pan C, Liu J, Bi WL, Chen H, Engelmann F, Wang QC (2018) Cryopreservation of small leaf squares-bearing adventitious buds of *Lilium* Oriental hybrid 'siberia' by vitrification. Plant Cell Tissue Organ Cult 133:159–164
Panis B, Nagel M, Van den Houwe I (2020) Challenges and prospects for the conservation of crop genetic resources in field genebanks, in *in vitro* collections and/or in LN. Plan Theory 9(12):1634
Panis B, Piette B, Swennen R (2005) Droplet vitrification of apical meristems: a cryopreservation protocol applicable to all Musaceae. Plant Sci 168:45–55
Park Y-G, Kwon G-S, Tay D (2005) Cryopreservation for gene conservation of *Acer mono* Max. Propagat Ornam Plants 5(2):78–83
Pawłowska B (2008) Employment of encapsulation-dehydration method for LN cryopreservation of ornamental plant explants propagated in vitro. Folia Hortic 20(1):61–71
Pawłowska B (2012) Cryopreservation of *Rosa canina* and *R. rubiginosa* apical buds by the droplet vitrification method. Acta Hortic 937:905–909
Pawlowska B, Bach A (2011) Cryopreservation by encapsulation-dehydration of in vitro grown shoot buds of Rosa 'new Dawn'. Acta Hortic 908:303–307
Pawłowska B, Szewczyk-Taranek B (2015) Efficient cryopreservation by droplet vitrification of pentaploid roses and the phenotype of regenerated plants. Acta Soc Bot Pol 84:439–442
Pornchuti W, Thammasiri K (2008) Cryopreservation of protocorms of *Dendrobium cariniferum* Rchb. f. Acta Hortic 788:63–68
Prasongsom S, Thammasiri K, Chuenboonngarm N, Panvisavas N, Narangajavana J, Thitamadee S (2020) Conservation of *Dendrobium cruentum* Rchb. f. Acta Hortic 1298:187–194
Rajasekharan P, Rao T, Janakiram T, Ganeshan S (1994) Freeze preservation of gladiolus pollen. Euphytica 80(1–2):105–109
Rajasekharan P, Rohini M (2023) Pollen cryopreservation protocols, Springer protocols handbooks. Humana, New York
Rajasekharan PE, Ganeshan SS (1994) Freeze preservation of rose pollen in LN: feasibility, viability and fertility status after long-term storage. J Hortic Sci 69(3):565–569
Reed BM (1996) Pretreatment strategies for the cryopreservation of plant tissues. In: Normah MN, Narimah MK, Clyde MM (eds) *In vitro* conservation of plant genetic resources, Bangi Selangor, pp 73–87
Reed BM (2008) Plant cryopreservation: a practical guide. Springer, New York

Reed BM, Dumet DJ, Denoma JM, Benson EE (2001) Validation of cryopreservation protocols for plant germplasm conservation: a pilot study using *Ribes* L. Biodivers Conserv 10:939–949

Reed BM, Kovalchuk I, Kushnarenko S, Meier-Dinkel A, Schoenweiss K, Pluta S, Straczynska K, Benson EE (2004) Evaluation of critical points in technology transfer of cryopreservation protocols to international plant conservation laboratories. Cryo Letters 25:341–352

Ren R, Li Z, Li B, Xu J, Jiang X, Liu Y, Zhang K (2019) Changes of pollen viability of ornamental plants after long-term preservation in a cryopreservation pollen bank. Cryobiology 89:14–12

Rittirat S, Klaocheed S, Suppapan J, Chaithada P, Kalawong S, Thammasiri K (2019) Cryopreservation of an endangered pharmaceutically important orchid, *Cymbidium finlaysonianum* Lindl. using vitrification technique. Acta Hortic 1234:125–132

Rodrigues ARP, Forzza RC, Andrade ACS (2014) Physiological characteristics underpinning successful cryopreservation of endemic and endangered species of Bromeliaceae from the Brazilian Atlantic Forest. Bot J Linn Soc 176(4):567–557

Ryyänen L, Aronen T (2007) Phenotypic expression of leaf variegation in two *Betula pendula* Roth genotypes following micro propagation, cryopreservation and grafting. Propag Ornam Plants 7(1):23–28

Sakai A, Engelmann F (2007) Vitrification, encapsulation-vitrification and droplet-vitrification: a review. Cryo Letters 28:151–117

Salaj T, Matusikova I, Fraterova L, Pirselova B, Salaj J (2011) Regrowth of embryogenic tissues of *Pinus nigra* following cryopreservation. Plant Cell Tissue Org Cult 106:55–61

Scocchi A, Vila S, Mroginski L, Engelmann F (2007) Cryopreservation of somatic embryos of paradise tree (*Melia azedarach* L.). Cryo Letters 28:281–290

Sen-Rong H, Ming-Hua Y (2012) A simple and efficient protocol for cryopreservation of embryogenic calli of the medicinal plant *Anemarrhena asphodeloides* Bunge by vitrification. Plant Cell Tissue Org Cult 109:287–296

Seo MJ, Shin JH, Sohn JK (2007) Cryopreservation of dormant herbaceous peony (*Paeonia lactiflora* Pall.) shoot-tips by desiccation. Cryo Letters 28(3):207–213

Sershen N, Berjak P, Pammenter NW, Wesley-Smith J (2011) The effects of various parameters during processing for cryopreservation on the ultrastructure and viability of recalcitrant zygotic embryos of *Amaryllis belladonna*. Protoplasma 249(1):155–169

Seydi S, Sedaghathoor S, Kaviani B (2020) In vitro storage techniques of *Fritillaria imperialis* Lubra Maxima, a wild rare and critically endangered ornamental species. CMU J Nat Sci 19:612–628

Shankar M, Gowthami R, Tripathi K, Chander S, Deepak DA, Agrawal A (2024) Cryopreservation protocol for cowpea pollen storage. Cryo Letters 45(6):340–348

Shankar M, Gowthami R, Tripathi K, Deepak DA, Barpete S, Agrawal A (2023) In vitro germination and cryopreservation technique for long-term pollen conservation of underutilized legume: Grasspea (*Lathyrus sativus*). Indian J Agric Sci 93(2):205–209

Sharma N, Gowthami R, Devi SV, Malhotra EV, Pandey R, Agrawal A (2021) Cryopreservation of shoot tips of *Gentiana kurroo* Royle–a critically endangered medicinal plant of India. Plant Cell Tissue Organ Cult 144:67–72

Sharma N, Pandey R, Gowthami R (2020) *In vitro* conservation and cryopreservation of threatened medicinal plants of India. In: Rajasekharan PE, Wani SH (eds) Conservation and utilization of threatened medicinal plants. Springer, India, pp 181–228

Soonthornkalump S, Meesawat U, Sutthinon P (2020) Pollen viability and the preliminary study of pollen preservation method of *Amorphophallus koratensis* Gagnep. (Araceae), the potential exotic ornamental plant. Acta Hortic 1298:183–186

Sopalun K, Kanchit K, Ishikawa K (2010) Vitrification-based cryopreservation of *Grammatophyllum speciosum* protocorm. Cryo Letters 31:347–357

Stanwood PC (1985) Cryopreservation of seed germplasm for genetic conservation. In: Kartha KK (ed) Cryopreservation of plant cells and organs. CRC Press, Boca Raton, pp 199–226

Su P, Wang D, Kan W, Yao Y, Ding S, Chen X, Chen X, Hou J, Wu L (2022) Melatonin different-increasing the cryopreservation recovery rate of shoot tips of *Chrysanthemum morifolium* cv. *Chuju* by regulating the level of oxidative stress. Plant Cell Tissue Organ Cult 149:785–797

Surenciski MR, Flachsland ED, Terada G, Mroginski LA, Rey HY (2012) Cryopreservation *of Cyrtopodium hatschbachii* Pabst (Orchidaceae) immature seeds by encapsulation-dehydration. Biocell 36:31–36

Suzuki M, Akihama T, Ishikawa M (2005) Cryopreservation of encapsulation gentian axillary buds following 2 step-preculture with sucrose and desiccation. Plant Cell Tissue Organ Cult 83:115–121

Suzuki M, Ishikawa M, Akihama T (1998) A novel preculture method for the induction of desiccation tolerance in gentian axillary buds for cryopreservation. Plant Sci 135:69–76

Suzuki M, Ishikawa M, Okuda H, Noda K, Kishimoto T, Nakamura T, Ogiwara I, Shimura I, Akihama T (2006) Physiological changes in gentian axillary buds during two-step preculturing with sucrose that conferred high levels of tolerance to desiccation and cryopreservation. Ann Bot 97:1073–1081

Takagi H (2000) Recent developments in cryopreservation of shoot apices of tropical species. In: Engelmann F, Takagi H (eds) Cryopreservation of tropical plant germplasm. Current research progress and application. Japan International Research Center for Agricultural Sciences/ International Plant Genetic Resources Institute, Tsukuba/Rome, pp 178–193

Tanaka D, Akimoto-Kato A, Kusaba M, Taniguchi K, Niino T, Maki S, Matsumoto T (2016) Cryopreservation of chrysanthemum shoot tips by D cryo-plate method. JJS Agri Tech M 23:1–7

Tanaka D, Niino T, Isuzugawa K, Hikage T, Uemura M (2004) Cryopreservation of shoot apices of *in-vitro* grown Gentian plants: comparison of vitrification and encapsulation-vitrification protocols. CryoLetters 25:167–176

Thammasiri K (2000) Cryopreservation of seeds of a Thai orchid (*Doritis pulcherrima* Lindl.) by vitrification. Cryo Letters 21:237–244

Thammasiri K, Soamkul L (2007) Cryopreservation of *Vanda coerulea* Griff. Ex Lindl. seeds by vitrification. Sci Asia 33:223–227

Thanasuttanithi D, Thammasiri K (2022) Cryopreservation of *Grammatophyllum speciosum* Blume seeds by D cryo-plate method. Acta Hortic 1339:111–116

Towill LE (2002) Cryopreservation of plant germplasm: introduction and some observations. In: Towill LE, Bajaj YPS (eds) Biotechnology in agriculture and forestry 50: cryopreservation of plant germplasm II. Springer, New York, pp 3–21

Urbaniec-Kiepura M, Bach A (2014) Effects of pre-storage on Lilium martagon L. seed longevity following cryopreservation. Cryo Letters 35:462–472

Urbaniec-Kiepura M, Bach A (2017) Cryopreservation of *Lilium martagon* L. meristems by droplet-vitrification and evaluation of their physiological stability. Cryo Letters 38:78–89

Vendrame WA, Carvalho VS, Dias JM, Maguire I (2008) Pollination of *Dendrobium* hybrids using cryopreserved pollen. HortScience 43(1):264–267

Vendrame WA, Faria RT (2011) Phloroglucinol enhances recovery and survival of cryopreserved *Dendrobium nobile* protocorms. Sci Hortic 128(2):131–135

Verleysen H, Bockstaele EV, Debergh P (2005) An encapsulation-dehydration protocol for cryopreservation of the azalea cultivar 'Nordlicht' (*Rhododendron simsii* Planch.). Sci Hortic 106:402–414

Vettorazzi RG, Silva Carvalho V, Carvalho Teixeira M, Campostrini E, Da Cunha M, de Monteiro ME, Facio Viccini L (2019) Cryopreservation of immature and mature seeds of Brazilian orchids of the genus *Cattleya*. Sci Hortic 256:108603

Volk G, Richards K (2006) Preservation methods for Jerusalem artichoke cultivars. HortScience 41:80–83

Volk GM, Caspersen AM (2007) Plasmolysis and recovery of different cell types in cryoprotected shoot tips of *Mentha x piperita*. Protoplasma 231(3–4):215–226

Wang GL, Xu CY, Wang GD (2010) Study on vitrification cryopreservation condition of *Anthurium andraeanum* embryonic suspension cells. J Plant Res Environ 19(1):26–31

Wang Q, Perl A (2006) Cryopreservation in floricultural plants. Floric Ornam plant Biotechnol:523–539

Wang R-R, Gao X-X, Chen L, Huo L-Q, Li M-F, Wang Q-C (2014) Shoot recovery and genetic integrity of *Chrysanthemum morifolium* shoot tips following cryopreservation by droplet-vitrification. Sci Hortic 176(11):330–339

Watanawikkit P, Tantiwiwat S, Bunn E, Dixon KW, Chayanarit K (2012) Cryopreservation of *in vitro*-propagated protocorms of Caladenia for terrestrial orchid conservation. Bot J Linn Soc 170:277–282

Wood CB, Pritchard HW, Miller AP (2000) Simultaneous preservation of orchid seed and its fungal symbiont using encapsulation dehydration is dependent on moisture content and storage temperature. Cryo Letters 21:125–136

Xu J, Li BL, Li Q, Shi Y, Peng JG, Jia M, Liu Y (2014) Wide-scale pollen banking of ornamental plants through cryopreservation. Cryo Letters 35(4):312

Xue SH, Luo XJ, Wu ZH, Zhang HL, Wang XY (2008) Cold storage and cryopreservation of hairy root cultures of medicinal plant *Eruca sativa* Mill., *Astragalus membranaceus* and *Gentiana macrophylla* Pall. Plant Cell Tissue Organ Cult 92:251–260

Yamamoto S, Rafique T, Priyantha WS et al (2011) Development of a cryopreservation procedure using aluminium cryo-plates. Cryo Letters 32:256–265

Yi JY, Balaraju K, Baek HJ, Yoon MS, Kim HH, Lee YY (2018) A successful regeneration from shoot tips of *Chrysanthemum morifolium* (Ramat.) following cryopreservation by droplet-vitrification. Korean J Plant Res 31:675–683

Yi JY, Lee GA, Chung JW, Lee SY, Lim KB (2013) Efficient cryopreservation of *Lilium* spp. shoot tips using droplet-vitrification. Plant Breed Biotechnol 1:131–136

Yin M, Hong S (2009) Cryopreservation of *Dendrobium candidum* Wall. ex Lindl. protocorm-like bodies by encapsulation-vitrification. Plant Cell Tissue Organ Cult 98:179–185

Yin ZF, Bi WL, Long C, Zhao B, Volk GM, Wang QC (2014) An efficient, widely applicable cryopreservation of Lilium shoot tips by droplet-vitrification. Acta Physiol Plant 36:1683–1692

Yong SH, Park KB, Seol YW, Kim DH, Jeong MJ, Choi MS (2023) Optimization conditions for cryopreservation of *Nelumbo nucifera* Gaertn. aquatic plant. Plant Biotechnol Rep 17:625–635

Zadeh MS, Khosrowshahli M, Taeb M (2009) Cryopreservation of the axial meristem of *Crocus sativus* L. Cryobiology 59:370–418

Zaimah NA, Benson EE, Marzalina M (2007) Viability of *Elateriospermum tapos* (Perah) embryo after storage in LN. J Trop For Sci 19(1):1–5

Zalewska M, Kulus D (2013) Cryopreservation of in vitro-grown shoot tips of chrysanthemum by encapsulation-dehydration. Folia Hort 25:133–140

Zamecnik J, Faltus M, Bilavcik A (2021) Vitrification solutions for plant cryopreservation: modification and properties. Plants 10:2623

Zhang D, Ren L, Chen GQ, Zhang J, Reed BM, Shen XH (2015a) ROS induced oxidative stress and apoptosis-like event directly affect the cell viability of cryopreserved embryogenic callus in *Agapanthus praecox*. Plant Cell Rep 34:1499–1513

Zhang JM, Han L, Lu XX, Volk GM, Xin X, Yin GK, He JJ, Wang L, Chen XL (2017) Cryopreservation of Jerusalem artichoke cultivars using an improved droplet-vitrification method. Plant Cell Tissue Organ Cult 128:577–587

Zhang JM, Huang B, Zhang XN, Volk GM, Zhou YC, Chen XL (2015b) Identification of a highly successful cryopreservation method (droplet-vitrification) for petunia. In Vitro Cell Dev Biol Plant 51(4):445–451

Zhang Y, Deng S, Lin H, Chu Y, Huang J, Li S, Lin F, Zhang S, Jiang W, Ren L, Chen H (2024a) The development of a procedure for the cryopreservation of the callus of *Anthurium andraeanum* by vitrification. Plants 13(21):3106

Zhang Y, Guo X, Song J, Chen G, Shen X (2024b) Enhancing cryopreservation survival in *Petunia* × *Calibrachoa* 'Light Yellow' callus: insights into material characteristics and genetic integrity. Cryobiology 114:104846

8 Progress, Challenges, and Prospects of In Vitro Conservation of Tree Species Using Nonembryonic Tissue and Organs

Dariusz Kulus

Abstract

Protecting tree species is essential for maintaining biodiversity, ensuring ecosystem stability, and supporting future reforestation and adaptation to climate change. Trees play a crucial role in global ecological balance, carbon sequestration, and provide habitat for countless organisms. The in vitro conservation of trees using nonembryonic tissues and organs has emerged as a vital strategy for protecting genetic diversity, yet it faces significant challenges. These methods are grounded in controlled, sterile environments that minimize space requirements and allow for the preservation of germplasm regardless of season and climate regimes. Slow-growth storage (SGS) techniques have proven effective for medium-term preservation because they reduce metabolic activity and minimize the need for subculturing, therefore lowering contamination risks and operating costs. Synthetic seeds, i.e., the encapsulation of plant propagules within protective matrices, offer an innovative approach for storing and transporting germplasm, particularly for large-scale conservation programs. Cryopreservation, involving ultralow-temperature storage, further complements these methods by enabling the long-term preservation of diverse tree germplasms in the form of somatic tissue samples (e.g., buds, meristems, and nodes). Significant advancements have been made in this field, including the development of modern techniques such as droplet vitrification, encapsulation–dehydration, and cryoplate-based combinations. Despite the potential of tissue culture techniques to preserve genetic diversity, issues such as low explant survival rates, endophyte outbreaks, and ethylene production remain critical obstacles. Innovative solutions, including the use of silver thiosulfate to inhibit ethylene production or

D. Kulus (✉)
Laboratory of Horticulture and Landscape Architecture, Department of Biotechnology, Faculty of Agriculture and Biotechnology, Bydgoszcz University of Science and Technology, Bydgoszcz, Poland

S. Gantait, P. Chmielarz (eds.), *Conservation of Plant Genetic Resources*,
https://doi.org/10.1007/978-981-96-8331-4_8

screening for endophyte-control methods, are being explored to improve the final outcomes. However, cryopreservation remains species-specific and complex, with challenges related to recalcitrant tissues, post-thaw recovery, and protocol optimization. The need for user-friendly, cost-effective protocols further complicates implementation. Advances in genetic engineering, such as CRISPR-Cas9, and knowledge from postgenomic research hold promise for enhancing dehydration tolerance and improving the viability of cryopreserved tissues. Additionally, the integration of high-throughput screening methods facilitates the identification of optimal cryoprotocols and storage conditions. Prospects include establishing global cryobank networks for effective germplasm exchange and engaging indigenous communities to incorporate traditional knowledge into conservation strategies. This chapter summarizes the present knowledge on the in vitro-based conservation methods applied to trees.

Keywords

Biodiversity · Cryobiotechnology · In vitro slow-growth storage · Synthetic seeds · Recalcitrant species · Vitrification

8.1 Introduction: Why Do Trees Matter

According to recent studies, there are approximately 73,300 tree species on Earth, with about 9200 of them remaining undiscovered (14% more than previously thought). These undiscovered tree species are likely rare and hidden in South America's biodiversity hotspots, including the Amazon and the tropical Andes (Cazzolla Gatti et al. 2022). Trees, more than any other life form, need special conservation efforts because of the diverse ecosystem services they provide, the high risk of extinction, and their significant socioeconomic and cultural importance for humankind (Pritchard et al. 2014). Approximately 43% of the assessed tree species are threatened to perish, according to a study that used a computational approach to estimate extinction risk for over 21,000 tree species (Silva et al. 2022). For example, *Populus nigra*, a pioneer tree species in riparian ecosystems, is threatened with extinction because of the loss of its natural habitat (Storme et al. 2004). The wet forests in the Amazon Basin are predicted to largely disappear, leading to significant biodiversity loss. In West Africa and Central Africa, tropical forests are greatly impacted by climate change, with 50–80% of species facing drastically reduced ranges, potentially leading to extinction (Zegeye 2017). The preservation of tree genetic resources is crucial for sustainable forestry, horticulture, and biodiversity conservation worldwide. Genetic diversity is essential for the long-term viability of plant populations, enabling them to adapt to changing environmental conditions and resist pests and diseases (Potter et al. 2017). This is particularly important with climate change, which poses significant threats to forest ecosystems by altering temperature and rainfall patterns, increasing the frequency of extreme weather events, and facilitating the spread of invasive species (Salgotra and Chauhan 2023). The conservation of tree genetic resources supports robust ecosystems and helps prevent landscape degradation, ensuring that forests continue to provide essential ecosystem services such as carbon sequestration, timber production, and habitat provision for wildlife (FAO 2018).

The importance of genetic conservation extends beyond national borders, with global initiatives such as the FAO Global Plan of Action for Forest Genetic Resources and the Pan-European strategy for genetic conservation of forest trees underscoring the need for coordinated efforts (FAO 2025; de Vries et al. 2015). For example, the African Forest Landscape Restoration Initiative (AFR100) aims to restore 100 million hectares of degraded landscapes by 2030, which will require the conservation and sustainable use of tree genetic resources (AFR100 2025). Similarly, in Asia, countries such as China and India are implementing large-scale afforestation programs that rely on the availability of genetically diverse tree species (Tambe et al. 2022; Zheng et al. 2023). In Nordic countries, efforts are coordinated through NordGen Forest to conserve forest genetic resources, emphasizing the importance of national and international cooperation (NordGen 2025). The USDA Forest Service also highlights the importance of genetic conservation for tree species, particularly in North America. This involves strategies such as seed banking, in situ and ex situ conservation to ensure the genetic integrity of the tree populations (Leach 2006).

Genetic conservation is crucial not only in forests but also in fruit trees, which play a significant role in horticulture and agriculture. Fruit trees contribute to food security and support local economies and cultural heritage (Madhu et al. 2024). Quantifying the exact amount of food produced from fruit trees and the number of people it supports is complex due to the vast diversity of fruit trees and their global distribution. However, fruit trees are known to contribute significantly to food security and nutrition, especially in rural areas (Kaviani and Kulus 2022). Fruit trees such as mangoes, papayas, and oranges are rich in vitamins and help meet basic daily micronutrient requirements, contributing to dietary diversity and nutritional well-being (Harris et al. 2022). The growth of trees, especially fruit trees, was associated with improvements in both total household consumption and nutritional outcomes, especially in low- and middle-income countries (Miller et al. 2020). Moreover, traditional orchards are recognized as biodiversity hotspots, offering a range of habitats that support a wide range of wildlife. The genetic diversity of fruit trees, especially those of local forms, is vital for maintaining their productivity and adaptability to environmental changes. Moreover, it is essential for breeding purposes and the creation of new cultivars, especially since consumer preferences are changing rapidly (Awodoyin et al. 2015). Unfortunately, the contribution of trees on farms to the livelihoods of many farmers often remains unacknowledged in national reporting, reflecting a lack of official interest in these "Cinderella species" (Miller et al. 2020).

The conservation of tree genetic resources also supports other essential ecosystem services, including carbon sequestration, timber and medicine production, and trees grown for nonindustrial purposes, e.g., fuelwood, fodder, shade, fences, etc. (Rittenhouse and Rissman 2005). Protecting these resources benefits people in many ways. For example, *Tectona grandis*, or teak, plays a significant role in the human economy because of its highly valued timber. It is used in furniture production, construction, and shipbuilding, generating substantial revenue and employment opportunities in regions where it is cultivated (Chávez-Salgado et al. 2022). However, the demand for *T. grandis* has led to severe overexploitation of natural teak forests, particularly in tropical regions (Pachas et al. 2019). The loss of genetic diversity in tree species can lead to reduced forest productivity and increased vulnerability to pests and diseases, which can have significant economic impacts (Jactel et al. 2017). This chapter focuses on various strategies related to the conservation of tree species, with particular attention focused on cryopreservation and nonembryogenic explants derived in vitro.

8.2 Main Conservation Strategies for Trees

Circa situm, in situ, and ex situ conservation methods are complementary strategies used to protect woody plants, each offering distinct advantages and challenges (Fig. 8.1). Circa situm conservation involves preserving trees in farmland, where natural forests once existed but have been lost or modified significantly through agricultural expansion (Dawson et al. 2013). It plays an important conservation role in agricultural ecosystems, when in situ and ex situ methods cannot be employed easily, e.g., forest remnants in cleared agricultural landscapes, sacred groves, and coffee plantations (Maheswarappa et al. 2022).

In situ conservation involves preserving species in their natural habitats, which is crucial for maintaining ecological interactions and natural processes (Oldfield 2009). This approach ensures that the entire ecosystem, including thousands of species and their interactions, is protected, along with the ecosystem services they provide (Yang and Yeh 1992). For example, Hamann et al. (2005) successfully presented a comprehensive approach for assessing the in situ conservation of forest trees by using botanical and ecological data to identify gaps in protected areas in British Columbia, with methods to prioritize species-specific data collection and conservation efforts. However, in situ conservation may be insufficient if the remaining

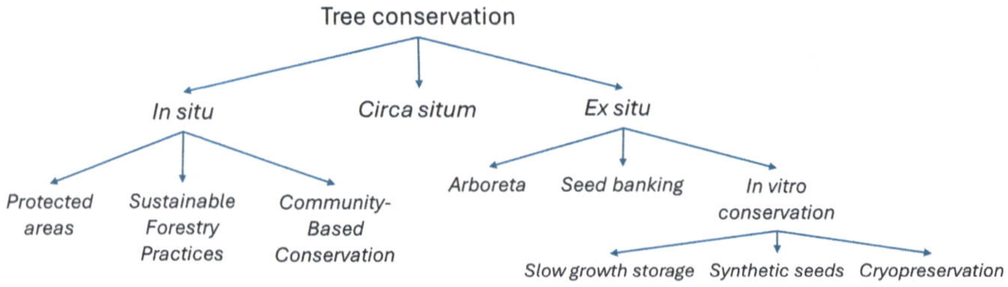

Fig. 8.1 Main conservation strategies applied for trees

populations are too small or continue to decline despite conservation efforts. In such cases, ex situ conservation becomes a vital alternative.

Ex situ conservation involves removing species or their germplasms from their natural environment and maintaining them under controlled conditions (Arnet et al. 2015). This approach is essential for protecting species threatened by habitat destruction, climate change, and other environmental challenges (Thomas et al. 2022). Moreover, ex situ strategies can be used to support gradually diminishing wild populations or reintroduce species into areas, where they have become extinct. Key facilities for ex situ conservation of trees include botanical gardens, seed banks, and in vitro-based banks.

8.3 Living Collections and Their Role in Creating Tissue Banks

Living collections in botanical gardens and arboreta play a significant role in ex situ conservation (Philpott et al. 2022a). Over the past 15 years, more than 470 arboreta have been accredited in 35 countries (Turner-Skoff et al. 2021). For example, Eastwoodhill Arboretum in New Zealand is preserving endangered (EN) and critically endangered (CR) trees, and hosts over 25,000 exotic and native woody species across 135 hectares. A total of 15 CR and 52 EN gymnosperm taxa can be found in the arboretum, representing approximately 38% and 40% of the world's CR and EN gymnosperms, respectively, and 1% of CR and EN angiosperms (Arnet et al. 2015). ArbNet, on the other hand, a global network of arboreta founded in 2011, has played a crucial role in developing, expanding, and connecting arboreta worldwide while raising awareness about the significance of trees and their conservation status (Turner-Skoff et al. 2021). The network aims to increase the number and types of accredited arboreta. It will also enhance support for regional network leaders in biodiversity hotspots and strengthen partnerships with other organizations focused on trees, such as Botanic Gardens Conservation International (BGCI) and the American Public Gardens Association (APGA), to foster collaboration and the sharing of knowledge, experiences, and ideas. Moreover, it supports urban forestry by facilitating community tree planting, organizing trainings, and resources (Turner-Skoff et al. 2021). Living collections require careful management to maintain genetic diversity and prevent genetic drift, often using pedigree-based management tools to track and maintain diversity (Foster et al. 2022). Nevertheless, such collections provide a rich source of explants—plant tissues used to initiate in vitro cultures. This includes young shoots, roots, or other plant parts that can be cultured to regenerate whole plants or specific tissues (Pawłowska and Ptak 2023). The availability of such high-quality primary explants (i.e., those obtained in vivo) significantly enhances the success rates of tissue culture techniques. Moreover, desiccated leaves or other tissues can also be stored at room temperature in silica gel for short-term conservation or further extraction of DNA for plant breeding, biotechnology, and biodiversity evaluation (Hodkinson et al. 2007). When explants are contributed to tissue banks, metadata such as taxon name, collection location, and environmental conditions are essential. Living collections typically maintain detailed records that facilitate this process, ensuring that the collected tissues meet research-grade standards (Kew 2019).

The drawbacks of living collections include several challenges, such as the maintenance of organisms in artificial habitats, the risk of inbreeding depression, the accumulation of deleterious alleles, and demographic stochasticity in small populations (Kasso and Balakrishnan 2013). Financial constraints are significant, as maintaining these collections requires funding for maintenance, staff, and infrastructure. Space and resource limitations also pose issues, as living collections need specific conditions to exist, which can be difficult to maintain. Moreover, management problems arise from balancing conservation goals with visitor access and educational programs. Finally, environmental risks such as climate change, pests, and diseases can threaten the health and survival of plants (Dosman 2006). For these reasons, the collection and storage of smaller propagules are attracting increasing attention.

8.4 Seed Banks: Why Isn't It Enough?

Seed banks are particularly effective at conserving large amounts of genetic material in a small space at a relatively low cost (Hanson and Ellis 2020). They store seeds in refrigerated environments with low humidity, which helps maintain seed viability over long periods (Lennon et al. 2021). An example program at the Commonwealth Forestry Institute, UK, involves the collection of seeds of nonindustrial woody species native to Central America and neighboring regions of South America. The program has successfully collected seeds from 25 species, 21 of which belong to the Leguminosae family (Hughes and Styles 1984).

However, many species cannot be conserved via traditional seed banking because of their desiccation sensitivity or other dormancy issues (Alvarez-Buylla and Martínez-Ramos 1990). Tweddle et al. (2003) reported that all tropical and subtropical habitat taxa contain some species with desiccation-sensitive seeds, reaching 47% of woody species in evergreen rainforests (mostly at the mature growth phase) and, to a lesser extent, in temperate and Mediterranean ecosystems. Recalcitrant seeds are defined as sensitive to drying below a threshold of 0.2–0.3 g $H_2O \cdot g^{-1}$ DW or a relative humidity <90%. If stored at subzero temperatures at such high moisture contents, lethal intracellular ice crystals are formed, killing the embryo (Walters 2015). Among the recalcitrant tree species, one can mention maple, elm, and oak, as well as many tropical species, e.g., rubber trees (Srivastava 2002). Examples of horticultural recalcitrant species include avocado, cacao, coconut, coffee, durian, jackfruit, lychee, mango, and rambutan (Righetti et al. 2023). The seeds of these species cannot be stored for long periods without losing viability. Compared with those considered fully tolerant (<0.07 g $H_2O \cdot g^{-1}$ dry weight), other nonorthodox species can withstand only partial desiccation (up to 0.10–0.15 g $H_2O \cdot g^{-1}$ dry weight). Seeds with such intermediate storage behaviors in terms of drying and freezing sensitivity or very short longevity (e.g., *Coffea*) also present challenges during their conventional storage in seed banks and are considered "exceptional species" (Pence et al. 2020). In addition, species that are seedless, e.g., *Musa* spp. (edible banana), or fruit crops propagated vegetatively as their seeds are not true to type, e.g., *Malus domestica* (apple) and *Citrus* spp., are not storable through seeds (O'Brien et al. 2021; Pence et al. 2022). The development of alternative storage methods based on in vitro cultures and cryopreservation is needed.

8.5 In Vitro Storage

In vitro culture plays a crucial role in the preservation of tree genetic resources by providing a controlled environment for the maintenance and propagation of plant material (Ochatt et al. 2021). This controlled environment allows for the precise regulation of ambient conditions; such as temperature, light, and nutrient availability, which are essential for optimizing growth and development. This technique is particularly beneficial for species that are difficult to conserve through traditional methods, i.e. those that do not produce or have short-lived seeds (Fernández et al. 2023). By utilizing tissue culture, researchers can establish disease-free collections, ensuring the genetic integrity of clonal microshoots (Arora et al. 2022). Furthermore, in vitro culture allows for the efficient multiplication of rare and endangered species, facilitating their conservation and potential reintroduction into their natural habitats (Fenning et al. 2022; Parab et al. 2021). For example, Abd Elaziem et al. (2022) used in vitro propagation for conservation of the rare date palm (*Phoenix dactylifera*) using immature inflorescence, whereas Martinez et al. (2023) successfully reproduced, hardened, and acclimatized *Artemisia tridentata*, a slow-growing woody species emblematic to western North America. Chmielarz et al. (2023) demonstrated that in vitro culture can be successfully utilized for the propagation of even 800-year-old pedunculate oak trees. This highlights the versatility of in vitro techniques across a wide range of species and explant types. Moreover, in addition to their rapid multiplication potential, in vitro cultures can also be used to slow the growth pace of plants for the purpose of medium-term storage.

8.5.1 In Vitro Slow-Growth Storage

In vitro slow-growth storage (SGS), or minimal growth, is an innovative and effective technique for the conservation of tree species, allowing for the maintenance of genetic resources with minimal metabolic activity. This method involves manipulating environmental and medium conditions, such as temperature, light intensity, nutrient composition, and the addition of plant growth regulators (PGRs), osmotic agents, or inhibitors, to significantly reduce growth rates while preserving viability (Tarraf et al. 2024). Combining these physical and chemical factors creates a progressive approach that can be utilized to establish general methods for the medium-term preservation of trees (Cha-um and Kirdmanee 2007).

Slow-growth storage is based on a reduction of metabolic processes in plant tissues, which extends the time intervals between necessary subcultures. By maintaining cultures at a low temperature (typically 2–5 °C for temperate and 10–15 °C for tropical species) often combined with limited light intensity (10–50 $\mu mol{\cdot}m^{-2}{\cdot}s^{-1}$) or total darkness, accompanied by adjustments in sucrose, mannitol, and other medium components levels, researchers can minimize labor costs and reduce the risk of contamination associated with frequent transfers. Samples, including *Castanea sativa*, *Ficus carica*, *Prunus* spp., *Pyrus* spp., and *Malus* spp., highlight the effectiveness of this method in maintaining genetic diversity in economically important fruit trees (Benelli et al. 2022). For example, Arbeloa et al. (2017) reported the successful conservation of 138 fruit tree clones from 18 species via slow-growth techniques. These researchers achieved a survival rate of 98.6% after 12 months under low-temperature conditions. Banana plantlets survived after 140 days of in vitro SGS and showed little browning (Rodrigues et al. 2022). The oil palm cultures were successfully stored for 4 months under a controlled atmosphere with an oxygen level of 1%. Following this storage period, they demonstrated rapid proliferation when transferred back to standard conditions (Engelmann 1990). According to Reed and DeNoma (2016), *Pyrus* germplasm for a wide range of species and cultivars can remain at 4 °C for 2–4 years. Similarly, *Castanea sativa* shoot cultures stored at 8 °C presented high survival (82%) and normal growth after 48 months, whereas those stored at 4 °C presented significantly lower survival rates (56% after 12 months) and no recovery after 24 months. The recovery of healthy shoots required BA at 0.44 μM, while low light conditions during storage positively affected shoot survival over extended periods (Capuana and Di Lonardo 2013). The mean storage duration of sweet cherry and plum shoots was 9–12 months, whereas that of cherry shoots ranged from 3 to 12 months if they were cultured on PGR-free Murashige and Skoog media at 4°C with 7–15 $\mu mol/m^2/s^1$ light and a 10-h photoperiod. The cold storage duration could be extended over 21 months after 0.22 μM BA and 0.49 μM IBA were added to the culture medium (Turdiyev et al. 2020). Lowering the content of macroelements by half, especially nitrogen, is also often applied to slow the growth of explants (Benelli et al. 2022; Sota and Kongjika 2014a). Interestingly, in most studied tree species, MS media (Murashinge and Skoog 1962) are effective for SGS (Benelli et al. 2022). WPM (woody plant medium) was used only for *Castanea sativa* (Capuana and Di Lonardo 2013) and *Citrus jambhiri* (Carvalho et al. 2022; McCown and Lloyd 1981). In *Prunus mahaleb*, MS basal media are superior to WPM and LP media (Quoirin and Lepoivre 1977). The highest survival and regeneration percentages (93.36% and 83.72%, respectively) were found in cultures stored in ½ MS media without sucrose for 3 months (Sota and Kongjika 2014b). Unlike angiosperms, no information on the application of slow-growth techniques to conifer species exists. The effectiveness of SGS is genotype dependent, requiring species-specific protocols owing to their variations in response, whereas innovations in container technology, such as gas-permeable plastic bags, have increased storage possibilities, as observed in stone-fruit species (Benelli et al. 2022). The effects of various factors on SGS storage efficiency are given in Table 8.1.

Table 8.1 Effect of various factors on the slow-growth storage (SGS) efficiency

Factor	Description	Effect on SGS	Example	References
Temperature	Low temperatures (2–5 °C for temperate, 10–15 °C for tropical) reduce metabolic activities	Prolongs storage time, reduces contamination risk	*Castanea sativa* at 8 °C, *Arbutus unedo* at 18 °C	Capuana and Di Lonardo (2013), Mascarello et al. (2017)
Light conditions	Darkness or reduced light intensity (10–40 μmol $m^{-2}.s^{-1}$) affects growth rates	Darkness is cost-effective; low light affects recovery rates	*Crataegus monogyna* in darkness, *Prunus avium* under low light	Arbeloa et al. (2017), Sedlak et al. (2019)
Medium composition	Includes plant growth regulators, sucrose, and osmotic agents	Optimizes growth control and survival rates	MS with BA and sucrose for *Prunus avium*, WPM with BA and sucrose for *Castanea sativa*, MS with reduced nitrogen content and mannitol for *Pyrus* spp.	Capuana and Di Lonardo (2013), Sedlak et al. (2019), Turdiyev et al. (2020)
Storage containers	Affects humidity and contamination risk	Proper sealing reduces moisture loss and contamination	Sealed tubes and jars were more effective than plastic bags with *Malus* and *Musa* spp.	Kovalchuk et al. (2009), Irish et al. (2013)
Shoot size and quality	Shoots of 2–4 cm length, healthy and free of disorders, are preferred	Ensures high survival and recovery rates with reduced risk of somaclonal variation	*Prunus* spp. and *Malus domestica* show high survival with healthy shoots	Benelli et al. (2022), Gianní et al. (2015), Monticelli et al. (2020)

Abbreviations: *BA* benzyl adenine, *MS* Murashige and Skoog, *WPM* woody plant medium

Despite its benefits, standardizing SGS protocols for diverse genotypes remains a challenge, necessitating continued research and collaboration among global institutions to improve the method and expand its application. Slow-growth storage also carries risks of genetic instability. In most cases, the genetic homogeneity of the in vitro-derived plant material is maintained, as reported for *Ficus carica* (Parab et al. 2021), *Dimocarpus longan* (Ala El-naby Ahmed 2022), and *Plukenetia volubilis* (Restrepo-Osorio et al. 2020). However, prolonged exposure to in vitro conditions, particularly the use of cytokinins, can increase the possibility of somaclonal variation—spontaneous genetic mutations that occur during tissue culture (Leva 2012). For example, *Pistacia lentiscus* presented genetic variation after 6 months of cold storage at 4 °C, with genetic similarity values ranging from 0.66 to 0.84, indicating potential genetic instability due to transposon activation (Koç et al. 2014). A high level of polymorphisms was also detected in in vitro-raised *Phoebe goalparensis*, an endemic valuable timber tree of Northeast India (Mudoi et al. 2023). These changes highlight the need for careful monitoring and management of in vitro cultures to minimize genetic alterations and ensure the integrity of conserved plant material.

8.5.2 Synthetic Seeds

The production of synthetic seeds offers a promising approach to tree biotechnology. Synthetic seeds (or synseeds, artificial seeds, manufactured seeds) are created by encapsulating any meristematic tissue (such as shoot apical meristem, axillary buds, or callus) in a nutrient-rich alginate matrix that supports regrowth when needed (Kulus 2019a). This technology has numerous applications in forest and fruit tree species, including the rapid propagation of elite genotypes, conservation of endangered species, production of disease-free plant material, and germplasm storage. Additionally, it can significantly reduce labor, costs, and space requirements (Tailor and Joshi 2024). To date, synthetic seed technology has been successfully applied for the propagation of various tree species, including conifers such as pine, spruce, and fir, as well as broad-leaved species such as eucalypts, acacias, oak, and willow, for the purpose of reforestation, reintroduction, and degraded land rehabilitation (Kundu et al. 2019).

Synthetic seeds can be preserved under short-, medium-, and long-term storage conditions. Short- and medium-term preservation involves maintaining slow but steady growth under in vitro culture conditions for a limited duration. This can be achieved through various strategies, either individually or in combination, including the use of minimal growth media, storage at low temperatures or reduced light intensity, lowering oxygen levels, and incorporating growth-retarding compounds such as abscisic acid (ABA) and osmotic agents such as mannitol or sorbitol (Tailor and Joshi 2024). The most commonly recommended storage temperature for synthetic seeds of tree species is 4 °C. For example, in the study by Ikhlaq et al. (2010), in vitro-derived shoots of the olive Moraiolo were encapsulated in alginate to create synthetic seeds, which were then stored at either 4 °C or 21 °C for varying intervals. The results showed that cold storage (4 °C) significantly increased the regrowth capacity compared with that at room temperature, with optimal results observed at a 45-day storage interval, indicating that this approach is promising for olive germplasm exchange. Similarly, encapsulated nodal segments of *Ficus carica* stored at 4 °C for 1–6 weeks presented variable sprouting frequencies, and the resulting plantlets acclimatized successfully in the field without any detectable variations after 6 months (Sharma et al. 2015). On the other hand, synseeds of *Azadirachta* presented higher germination rates and regeneration potential at 24 °C than did those stored at 4 °C, with the best survival rate of genetically stable plants achieved in a soil-organic manure mixture during the hardening phase (Kader et al. 2022). The encapsulation of bud explants from in vitro cultures of *Amygdalus scoparia* in alginate enabled efficient short-term storage at 25 °C and germplasm exchange,

Table 8.2 Factors affecting synthetic seed production and storage in trees

Factor	Description	Effects on synthetic seed production
Explant selection	The type of plant tissue used for encapsulation. Common explants include shoot tips, axillary buds, nodal explants, and callus	Controls the effectiveness of synthetic seed production. Meristematic explants promote genetic stability, shoot tips, and axillary buds have better viability. Different explants may require different methods for encapsulation and germination
Encapsulating agent	The material used to cover the explant provides a suitable environment for growth. Alginate is a common encapsulating agent. Others include carboxymethyl cellulose, carrageenan, ethyl cellulose, Gerlite, guar gum, nitrocellulose, Polyox, polyacrylamide, and sodium pectate	Plays a crucial role in the viability, germination, shape, physiology, and biochemistry of synthetic seeds. Optimal concentrations for firm, clear beads are typically 3% sodium alginate and 100 mM calcium chloride for polymerization. The encapsulating agent's concentration affects gel strength, porosity, and permeability
Matrix composition	The matrix surrounds the explant and influences the physical, chemical, and biological features of the synthetic seed. Key factors include concentrations of sodium alginate and calcium chloride, exposure time to calcium chloride, and the addition of growth regulators. Additives like activated charcoal, polyethylene glycol, sucrose, amino acids, vitamins, and other basal culture media salts can be included	Affects gel strength, viscosity, porosity, water content, pH, osmotic potential, and nutrient supply. Influences seed formation, germination, further plant growth, morphology, physiology, and biochemistry. The concentration of the encapsulating agent can be changed to make the gel stronger or more fluid. The presence of additives can improve the qualities of the matrix and the development of explants
Temperature	The temperature during storage and germination	Temperature significantly impacts the storage of synthetic seeds, with lower temperatures enhancing longevity by reducing metabolic activity and degradation, while higher temperatures accelerate deterioration and loss of viability
Light	Presence or absence of light. Light intensity and quality (spectral composition)	Cold-stored explants are kept in conditions of low light intensity or in total darkness to reduce the growth rates. Alternatively, for manipulation, control, and handling, synthetic seeds can be kept in light spectra outside the photosynthetically active radiation (PAR) (400–700 nm). Using yellow (570–590 nm) and green (500–570 nm) lights minimizes photosynthetic stimulation, reduces oxidative stress, and helps maintain dormancy, making it beneficial for long-term storage and preservation of synthetic seeds

(continued)

Table 8.2 (continued)

Factor	Description	Effects on synthetic seed production
Other factors	Germination conditions. Presence of an additional fungicidal/antimicrobial layer	Culture medium composition needs to be optimized to stimulate the growth of explants. Microbial contamination can affect the quality and performance of synthetic seeds if they are to be sown directly ex vitro

resulting in a high conversion rate of 88.96% into plantlets. This approach was particularly effective when synthetic seeds were transferred to peat and perlite substrates, providing a nonrefrigerated alternative for conserving this endangered species (Alirezaei et al. 2020). For *Juglans regia*, synthetic seeds containing shoot tips develop plantlets after medium-term storage, regardless of the temperature regime (8 °C or 25 °C). While incubation at 8 °C enabled a 4-month conservation period, slight shoot necrosis was observed. Conversely, incubation at 25 °C offered a shorter conservation duration (3.5 months) but resulted in better shoot quality (Sota et al. 2023). A summary of the factors affecting synthetic seed production and storage can be found in Table 8.2.

For the long-term conservation of synthetic seeds, cryopreservation is a reliable method that has been widely adopted in many tree species to extend their storage periods.

8.5.3 Cryopreservation

Cryopreservation is a powerful tool for ex situ conservation, allowing for the long-term storage of plant tissues in Dewar flasks at ultralow temperatures (Fig. 8.2). This approach is reliable to protect the genetic diversity of tree species, ensuring that genetic diversity is maintained and can be used for future reforestation efforts (Gantait et al. 2016). This method reduces the space and labor needed for maintaining living collections and extends the lifespan of stored materials (Roque-Borda et al. 2021). Additionally, cryopreservation supports the management of in vitro cultures, reducing the risk of contamination and genetic drift, which is crucial for the breeding and development of improved tree cultivars (Ballesteros et al. 2024).

8.6 Principles of Tree Cryopreservation

Cryopreservation is an advanced technique for the long-term preservation of tree genetic resources, allowing for the storage of living cells, tissues, or organs at ultralow temperatures, typically in liquid nitrogen (LN; −196 °C) or its vapor phase (Barra-Jiménez et al. 2015). Due to the high fluctuation in the LN vapor temperature (ranging between −135 °C and −190 °C), the latter approach is rarely used. The

Fig. 8.2 Dewar flasks, invented in 1892 by Scottish scientist Sir James Dewar, revolutionized cryogenics by using a double-walled vacuum design to minimize heat transfer, leading to the development of cryopreservation research and enabling long-term conservation of plant genetic resources

fundamental objective of cryopreservation is to prevent ice crystal formation within cells (both during freezing and thawing), which can lead to lethal cellular damage and loss of viability (Gantait et al. 2016). Consequnetly, cryopreservation is based on physical events such as vitrification, during which water inside the cells is transformed into a glass-like state without the formation of harmful ice crystals. This process is facilitated by the use of cryoprotective agents (CPAs or cryoprotectants), which are crucial for protecting cells during the freezing and thawing processes. Their primary function is to inhibit both intracellular and extracellular ice formation (Fahy and Wowk 2015). CPAs achieve this through various mechanisms that stabilize cellular structures and plasma membranes. For example, permeating cryoprotectants such as dimethyl sulfoxide (DMSO), glycerol, and ethylene glycol can cross cell membranes and modulate the rate of cellular dehydration during freezing. These compounds disrupt the formation of ice crystals by competing with water molecules for hydrogen bonding, effectively lowering the freezing point of intracellular fluids and promoting a vitrified state instead of crystalline ice formation (Al-Jubori and Mohammed 2024). Conversely, nonpermeating cryoprotectants, e.g., sucrose and trehalose, do not enter cells but influence their osmotic environment, affecting water movement during the dehydration and rehydration phases. They increase the viscosity of the extracellular solution, which helps prevent ice crystal growth outside the cell while also adsorbing to cell membranes to provide additional protection against ice formation (Bhandari and Nayyar 2014). Furthermore, these nonpermeating agents help maintain membrane integrity by lowering lipid phase transition temperatures, thus preserving membrane fluidity during freezing stress (Marcantonini et al. 2022). The choice of cryoprotectant type and concentration plays an essential role in cell survival and function (Li et al. 2025). The combination of permeating and nonpermeating CPAs in cryopreservation protocols has been shown to increase the survival rates of trees (e.g., gray poplar) by reducing toxicity and osmotic shock during cooling and warming processes (Pokorná et al. 2020). For example, the combination of 0.2 M sorbitol with 5% DMSO was shown to be an effective cryoprotectant for the tissues of *Pinus roxburghii* (Malabadi et al. 2006).

The efficiency of cryopreservation is influenced by various factors related to the type, size, and age of the explants used. A broad range of explant types can be used for cryostorage in most tree species, including embryogenic (zygotic embryos and their fragments, somatic embryos, friable embryogenic tissues, and cell suspensions) and nonembryogenic propagules (pollen, nodular segments, in vitro shoot tips, and in vivo dormant buds) (Nguyen et al. 2023). These different explant types exhibit varying survival rates and regrowth potentials after LN-storage. Nonembryogenic organs are commonly employed because of their high viability and regeneration capacity and because they are genetically more stable than other explants, such as calli and embryogenic tissues (Li et al. 2018). Among these, shoot tips and dormant buds are the most popular. Shoot tips are small explants, obtained in vitro, measuring 0.5–2.0 mm in length and consisting of an apical dome and 3–6 leaf primordia. In contrast, buds are larger explants (2.0–10.0 mm) that include both shoot tips and older tissues. Dormant buds refer specifically to winter buds collected from temperate trees (Jung et al. 2013). Although dormant buds are easier to obtain directly from trees, in vitro-derived shoot tips can be produced in large quantities in a short time and at any point of the year, making them a versatile option for cryopreservation. Moreover, shoot tips are considered the best candidates for cryopreservation because they are composed of organized tissues consisting of meristematic cells with dense cytoplasm and small vacuoles that have the highest chance of survival at ultralow temperatures (Normah et al. 2019). For example, in a study on apple, in vitro-derived shoot tips presented much greater regeneration potential (83%) than did dormant buds (16%) (Wu et al. 2001). However, in some *Pyrus communis* (pear) cultivars, dormant buds are slightly more effective (Bilavcik et al. 2021). Other nonembryogenic in vitro-derived explants (such as axillary buds and nodal segments) can also be used for cryopreservation, although they are less popular. Vidyagina et al. (2021) reported that the use of axillary buds as explants after the removal of apical buds increased recovery, on average, by 8% in *Populus tremula*. San José et al. (2015) reported that there was no effect of explant type (shoot apex or nodal segment) on cryopreservation efficiency in *Alnus glutinosa*. Adventitious root tips can also be effectively cryoconserved, as reported in several plant species, although not with trees (Popova et al. 2021; Zhang et al. 2023).

The size of the explants plays a crucial role; smaller explants often demonstrate better survival rates because they can be more effectively penetrated by cryoprotectants and subjected to rapid cooling techniques, although few studies on this aspect in trees exist (Baek et al. 2003). For example, in date palm, only meristems smaller than 3 mm in diameter were found to survive cryogenic treatments (Fki et al. 2013). Additionally, the age of explants affects their physiological state, with younger tissues generally showing greater tolerance to cryogenic conditions than mature tissues. In silver birch (*Betula pendula*), the recovery of shoot tips derived from less than 20-month-old cultures was significantly better than that from old cultures (at least 55 months old) (Ryynänen and Häggman 2001). This can be explained by their greater metabolic activity and lower levels of cellular damage during the freezing process.

Notably, the performance of in vitro explants is significantly influenced by genotype. Numerous studies have shown that the outcomes of tissue culture can vary widely among different cultivars (Nin et al. 2019). For example, the same cryopreservation protocol applied to eight plum genotypes resulted in a regrowth rate ranging from 0% to 33.3% (Vujović et al. 2023). Sometimes, however, the genotype effect is less evident, as in the case of *Syringa vulgaris* 'Aucubaefolia' and 'Polina Osipenko'. Explants of both cultivars withstand the various steps of the protocol, resulting in high survival and regrowth rates after exposure to liquid nitrogen, with percentages ranging from 67% to 100% for survival and 63% to 88% for regrowth (Koroleva et al. 2023).

The number of explants needed for the effective maintenance of cryopreserved germplasm depends on its intended use. Typically, cryostored accessions serve as long-term backup collections that support active collections such as tissue cultures or arboreta. If a moderate survival rate of approximately 40% is accepted, approximately 20 vials (each with 10 propagules) per genotype would be necessary to sustain a cryobank (five vials are used for an initial viability test, five are stored at a remote location for backup, five are kept as part of the primary base collection, and one vial is retrieved annually to fulfill community requests) (Tsai and Hubsher 2004).

Once suitable tissues are selected, they undergo pretreatment with CPAs to increase their tolerance to freezing stress. This may involve dehydration or osmotic treatment to reduce the water content to optimal levels (usually between 10% and 40%), which is essential for effective vitrification (Bivalcik et al. 2021; Ribeiro de Souza et al. 2023). Typically, explants are precultured on sucrose-enriched media, often supplemented with ABA and/or proline. Including the preculture step can increase the regrowth success of in vitro-derived explants, as reported in olive (Bradaï et al. 2023). A 7-day preculture at 4 °C with 0.5 M sucrose significantly improved regrowth for all studied genotypes of *Prunus domestica* (plum) (Vujović et al. 2023). For *Eriobotrya*, an important fruit tree in the Rosaceae, supplementation of the preculture medium with 5% DMSO resulted in significantly greater preservation than the control (Liu et al. 2023). The concentration of osmotically active components and preculture timing need to be optimized for each species. The physical hardening of donor plants can also be effective. Cold acclimation of donor in vitro cultures for 8–12 weeks at 0 °C significantly enhanced shoot recovery from cryopreserved shoot tips of pear (Scottez et al. 1992). The explants are subsequently loaded with diluted CPAs to prepare them for more intense dehydration. The most commonly applied loading solution (LS) consists of 2 M glycerol and 0.4 M sucrose. Finally, the explants can be properly dehydrated with a plant vitrification solution (PVS; a mixture of CPAs). Various types of PVSs significantly influence the cryopreservation efficiency of tree species. For example, a study on the cryopreservation of in vitro-derived shoot tips of pears using various PVS solutions revealed that PVS2 (30% glycerol, 15% dimethyl sulfoxide, 15% ethylene glycol, and 0.15 M sucrose) generally resulted in low regrowth rates, with most cultivars showing rates of less than 20%, although 'Moravka' performed better, with 62.5% regrowth after 40 min of treatment. The 90% PVS2 variant improved the results for 'Moravka' at the same duration and yielded moderate regrowth rates of 37.5% for 'Belošljiva' and 33.3% for 'Požegača'. In contrast, PVS3 (50% glycerol and 50% sucrose) was more effective, with 'Sitnica' and 'Crnošljiva' achieving 45.8% and 42.9% regrowth rates, respectively, after 80 minutes of treatment, whereas 'Moravka' reached a regrowth rate of 48.1% after 60 minutes (Vujović et al. 2023). Vujović et al. (2021) developed a modified PVS2 solution (22.5% sucrose, 37.5% glycerol, 15% ethylene glycol, and 15% DMSO) that resulted in considerably higher survival and regrowth rates after cryopreservation of apple shoot tips than did the original PVS2. Paul et al. (2000) reported, with apple axillary shoot tips, that PVS4 containing 180% sucrose and 120% ethylene glycol presented the highest shoot regrowth rates, which varied from 64% to 77%, depending on the cultivar. The duration of PVS treatment is a genotype- and explant-related issue. Typically, PVSs that do not contain DMSO require longer exposure. DMSO, however, may be toxic to some cells (Zhang et al. 2016).

Cooling rates during the freezing process must be controlled to avoid intracellular ice formation, especially in older cryopreservation techniques (Yi et al. 2014). Alternatively, rapid direct immersion in LN can be applied via more modern protocols. However, controlling the temperature of CPA solutions is still important, as cold solutions are usually more stable and effective (Benelli 2021).

Thawing and unloading (removal of CPAs) procedures also require careful management. Vitrification is metastable; therefore, rapid thawing is generally preferred to prevent ice recrystallization, which can compromise cell integrity. The remaining CPAs are usually removed during the rewarming stage with 1,2 M sucrose unloading solution. The composition of the recovery medium is also essential (Popova et al. 2023). Oxidative stress poses a significant challenge in tree cryopreservation, primarily because of the accumulation of reactive oxygen species that result from tissue wounding, excessive dehydration, and the freezing process (Krishna et al. 2008). This leads to browning and cell damage as phenolic compounds are oxidized when the cytoplasmic and vacuolar contents are mixed. To prevent oxidative stress, the incorporation of antioxidants such as ascorbic acid, activated charcoal, melatonin, or tocopherol into unloading solutions or recovery media is essential. These antioxidants effectively scavenge harmful radicals, including singlet oxygen and hydrogen peroxide, thereby increasing tissue viability and minimizing browning after cryopreservation. This approach supports better recovery of cryopreserved tissues, ensuring higher survival and regeneration rates (O'Brien et al. 2021).

8.7 Progress in Tree Cryopreservation Techniques

The history of plant cryopreservation has advanced significantly since its origination in the twentieth century, driven by the need to preserve the genetic diversity of woody species. A pivotal figure in this field was Prof. Akira Sakai from Japan, who demonstrated the survival of mulberry twigs after exposure to liquid nitrogen (Sakai 1956). He later described the successful cryopreservation of dormant vegetative buds from cold-hardy woody plants by removing freezable water through extracellular freezing at −30 °C (Sakai 1960). These studies laid the foundation for further developments. Between the 1970s and 1990s, early efforts focused primarily on herbaceous plants; however, researchers soon recognized the unique challenges posed by woody species (Normah et al. 2019).

8.7.1 Two-Stage Cooling

A two-stage cooling technique was first developed. In this approach, explants after incubation in a CPA solution (usually DMSO) undergo an initial slow cooling phase (approximately 1 $°C \cdot min^{-1}$) to a critical temperature (often −40 °C), followed by rapid immersion in liquid nitrogen (Jung et al. 2013; Yi et al. 2014). This technique was successfully applied with cold-hardy species, i.e., shoot tips of apples (Wu et al. 2001).

8.7.2 Vitrification

The cryopreservation of tropical trees has been a critical problem for many years, as these species are naturally susceptible to low temperatures. The discovery of PVS2 has played a key role in advancing plant cryopreservation techniques, particularly for tropical woody species. Excellent results have been reported, with survival rates of approximately 80% for cryopreserved nucellar cells of navel oranges (*Citrus sinensis* Osb. var. brasiliensis), which can subsequently regenerate into plantlets (Sakai et al. 1990). This innovative vitrification method involves the use of a highly concentrated cryoprotective solution, which allows for sufficient dehydration of cells before they are plunged directly into liquid nitrogen. Therefore, slow cooling was no longer needed. Since the 2000s, cryopreservation methods have evolved into more advanced techniques, enabling the successful long-term storage of diverse woody plants. These modern techniques, which are based on rapid cooling, include droplet vitrification, encapsulation–vitrification, encapsulation–dehydration, and modifications such as D-cryoplate and V-cryoplate.

8.7.3 Droplet Vitrification

This method, developed by Panis et al. (2005) for Musaceae species, significantly improved cryopreservation efficiency by using small droplets of vitrification solution placed on aluminum foil strips. Aluminum has high thermal conductivity, allowing ultrarapid cooling and warming, which are essential for successful cryopreservation. Owing to better heat exchange, the formation of damaging ice crystals is prevented, increasing the chance of survival of plant tissues postcryopreservation

(O'Brien et al. 2021). Droplet-vitrification cryotherapy has also proven to be an effective method for virus eradication in quince (*Cydonia oblonga*) and apple shoot tips. In quince ('QA' rootstock and 'Neyshabour' cultivar), cryotherapy successfully eliminated all apple chlorotic leaf spot virus (ACLSV) and 67% apple stem grooving virus (ASGV) (Farhadi-Tooli et al. 2022). For apple rootstock (*Malus prunifolia* 'Marubakaido'), cryotherapy with optimal PVS2 exposure (20–40 min) resulted in 100% eradication of apple stem pitting virus (ASPV) and 70% removal of ASGV (Souza et al. 2020).

8.7.4 Encapsulation–Vitrification and Encapsulation–Dehydration

Encapsulation–vitrification and encapsulation–dehydration are effective cryopreservation techniques for woody crop species such as apple, pear, mulberry (Niino and Sakai 1992; Paul et al. 2000), and Chinese bitter orange (*Poncirus trifoliata* × *Citrus sinensis*) (Wang et al. 2002a, b). In both methods, dissected shoot tips are coated with sodium alginate beads (4–5 mm in diameter) and solidified in $CaCl_2$ solution. Encapsulation–vitrification involves further dehydration of the beads in vitrification solutions before plunging them into liquid nitrogen. Encapsulation–dehydration, on the other hand, replaces chemical dehydration with air-drying in a laminar flow hood or silica gel before LN storage. Air desiccation is longer, but can be performed in a more sustainable pace (O'Brien et al. 2021).

8.7.5 D-Cryoplate and V-Cryoplate

The V-cryoplate and D-cryoplate methods utilize specially designed aluminum cryoplates (37 mm length, 7 mm width, and 0.5 mm thickness with 10 wells). In both methods, samples are placed in wells on a cryoplate after being embedded in an alginate solution. The cryoplate was initially coated with a 2–3% sodium alginate solution in a calcium-free basal medium supplemented with sucrose (0.2–0.4 M). The samples are then placed within the wells of the plate, and a further layer of the alginate solution and calcium chloride is applied to encapsulate them completely. The V-cryoplate method employs PVS for dehydration, whereas the D-cryoplate method uses an air current from a laminar flow cabinet or silica gel to reduce the water content. Following dehydration, the cryoplates are immersed in liquid nitrogen (O'Brien et al. 2021). Vujović et al. (2023) used V-cryoplate and D-cryoplate methods with plum shoot tips. After cryopreservation, five genotypes achieved regrowth rates of over 40% in at least one of the applied protocols, whereas two genotypes presented regrowth rates of approximately 10%. These techniques are also effective in cryopreservation and virus eradication from apple (Wang et al. 2024). In the recently developed cryo-mesh technique, explants are placed on stainless-steel mesh strips before further treatments (Funnekotter et al. 2016). The major steps of the most common cryopreservation procedures are shown in Fig. 8.3.

To date, cryopreservation has been successfully applied to a diverse range of in vitro-derived tree explants. In addition to conifers such as *Pinus* and *Abies* (Hazubska-Przybył et al. 2024), successful protocols have been developed for deciduous species such as *Quercus* (Lambardi and Shaarawi 2017; Pence and Chaiken 2021). Furthermore, tropical species such as *Carya illinoinensis* (pecan) and fruit trees have also been used for cryopreservation because of their economic importance and vulnerability (Morrissey et al. 1990; Benelli et al. 2013; Wang et al. 2018). Nevertheless, despite their benefits, cryopreservation protocols are lacking for many other species, and further research is needed to develop effective techniques for a broader range of tree propagules (Philpott et al. 2022a, b). For example, in conifer species, embryogenic tissue is usually cryostored rather than shoot tips or axillary buds (Ballesteros et al. 2024). Unfortunately, no single cryopreservation method is universally applicable to all species and explants. Moreover, various cryopreservation techniques may have different efficiencies within the same species or even cultivar. For example, two-step freezing resulted in the greatest in vitro growth recovery (83%) of cryopreserved shoot tips in the apple 'Golden Delicious', followed by vitrification (60%) and encapsulation–dehydration (33%) (Wu et al. 2001). For citrange (*Citrus insitorum*; a hybrid of the sweet orange and the trifoliate orange), both encapsulation–dehydration and encapsulation–vitrification yielded good survival rates of cryopreserved shoot tips, yet the latter was more effective by approximately 20% (Wang et al. 2002a, b). This highlights the need for meticulous studies when optimizing a cryoprotocol for new germplasms. The advantages and disadvantages of the most common cryopreservation techniques are presented in Table 8.3.

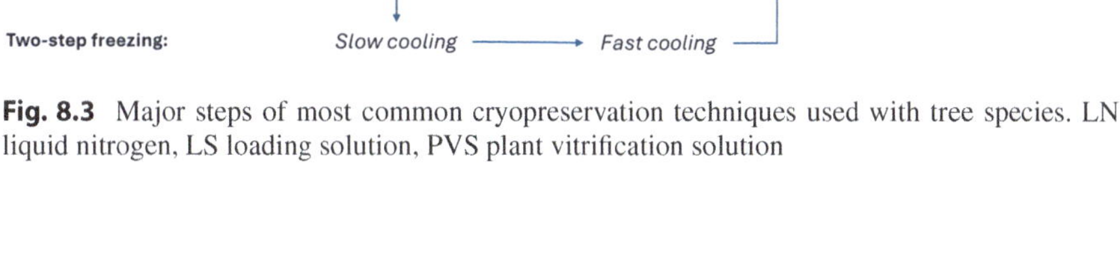

Fig. 8.3 Major steps of most common cryopreservation techniques used with tree species. LN liquid nitrogen, LS loading solution, PVS plant vitrification solution

Table 8.3 Comparison of various cryopreservation techniques based on their advantages and disadvantages

Technique	Advantages	Disadvantages
Controlled rate cooling (slow cooling, two-step freezing)	Allows for the precise control of the process (temperature distribution is more regular and stabilized; more control over ice crystal formation) Effective with cold-hardy species	Requires specific cooling rates and intermediate cooling temperatures Comparatively longer duration than vitrification Requires specialized and expensive equipment Not effective with tropical species
Vitrification	Avoids intracellular ice crystallization by transitioning the cytosol into an amorphous, (biological) glass state, protecting plant tissues from damage Fast cooling, with rates of −100 °C per minute, preserving the integrity of the biological material Routinely used for successful cryopreservation of temperate and tropical crop species Easy to apply, low cost, and less time-consuming (working time less than 1 hour)	Some plants can be sensitive to the highly concentrated PVS2 solution used, even for a short duration, requiring careful regulation of exposure time and selection of proper PVS Small explants can be lost or damaged during the transfer from one solution to another Temperature distribution may be less regular compared to slow freezing
Encapsulation–dehydration Encapsulation–vitrification	Encapsulation enables easier manipulation of the explant and provides additional physical protection limiting the loss of cryopreserved explants The encapsulation procedure gives a faster recovery of the culture suspension than vitrification and ensures culture homogeneity and organogenic potential Does not require the use of toxic cryoprotectants (ED)	An additional encapsulation step is needed; therefore, the entire procedure is longer and more complex Longer dehydration is required due to the physical coating barrier ED can be especially time-consuming if an additional osmotic dehydration (in sucrose solution) is needed
Droplet vitrification	High velocities of cooling and rewarming are achieved in this method due to the use of aluminum strips with high thermal conductivity	The number of explants that can be placed on a strip is limited The strips are more difficult to handle than encapsulated explants
D-cryoplate V-cryoplate	Handling of explants is easy and quick because only the cryoplates are manipulated Cryopreservation of a higher number of samples simultaneously is achieved	Cryoplate techniques necessitate the use of special plates The adhesion of samples to cryoplates is affected by the size of the wells and explants Adhesion of explants is affected by the surface tension of the cryoplate

The overall problem with plant cryopreservation is increased complexity in larger volumes, so only small volumes are commonly cryostored

8.7.6 DNA Storage

DNA availability is essential in botanical and evolutionary research, impacting plant breeding, biotechnology, and biodiversity studies. For tree conservation, genomic DNA samples can be stored in specialized biorepositories under conditions that ensure long-term stability and prevent degradation. Purified DNA is often stored at −20 °C to −80 °C in ultralow-temperature freezers or in liquid nitrogen at −196 °C for indefinite preservation (GGBN 2025). While DNA is easily extracted and stored long term in DNA banks, curation methods remain underdeveloped compared with those of herbaria, seed banks, and botanic gardens and require efficient and well-regulated sample exchange procedures (Hodkinson et al. 2007).

8.8 Challenges and Prospects of Tree Conservation

In vitro-based storage methods present both opportunities and challenges in the conservation of tree species. While they allow for long-term preservation of germplasm under controlled conditions, issues such as tissue stress, contamination, somaclonal variation, significant expenditure, and the need for highly specialized staff and equipment must be addressed to improve survival rates and overall effectiveness. Using innovative methods and encouraging collaboration among various stakeholders can make the protection of tree germaplsm diversity more effective.

8.8.1 In Vitro SGS

Despite their great potential, the in vitro conservation of trees faces several challenges. The initiation of tissue culture (excision from the mother plant, disinfection, and transfer to aseptic conditions) can pose a significant challenge in establishing an in vitro bank (Philpott et al. 2022a). Explants collected from wild samples do not always adapt well to in vitro culture, and their survival rate can be quite low. Cutting plants from their mother plant and exposing them to sterilizing agents such as sodium hypochlorite can induce stress, often resulting in necrosis. This issue is particularly critical for rare or threatened species, as only a few shoots are available for collection each year (Philpott et al. 2022b). Another problem is the abundant production of ethylene, a stress hormone, in response to in vitro conditions or the exudation of harmful polyphenols into the culture medium. Moreover, even if initial shoot survival rates are high, they may decline considerably over time, as in the case of oak (Brennan et al. 2017). To address this, Winkeljohn et al. (2022) collected shoot cuttings from five oak species (*Q. alba*, *Q. bicolor*, *Q. macrocarpa*, *Q. muehlenbergii*, and *Q. palustris*) and cultured them on media with or without 50 µM silver thiosulfate (STS), which inhibits ethylene. After 1 month, shoot survival was significantly greater in the STS-treated group than in the control. Additionally, primary explants are vulnerable to contamination by bacteria and fungi. These contaminants can hinder the successful initiation of culture and may persist as latent endophytes (microorganisms living within plant tissues which are not eradicated during surface sterilization), potentially causing harm later (Gómez-Lama Cabanás et al. 2021). Volk et al. (2022) provided a detailed examination of methods for controlling and eliminating contamination, with a particular focus on endophytes in *Citrus* plants. Their research included two case studies, using the MinION Oxford Nanopore system and Kirby-Bauer disc diffusion assays, that illustrated the process of identifying endophytes and determining which antibiotics are most effective against them, which can help mitigate contamination issues.

8.8.2 Implementation of Genetic Engineering in Cryostorage

Cryopreservation techniques are complex and species/cultivar-specific. Many trees require specialized protocols due to their unique physiological traits, such as recalcitrant tissues that cannot be dehydrated or frozen conventionally. To address these challenges, a promising solution involves the application of genetic engineering techniques, such as CRISPR-Cas9, to increase dehydration tolerance. This approach could enable the development of tissues that can be (cryo)preserved and transported in a dry state, thereby improving the viability of these species under conservation programs (Farrant and Hilhorst 2022). Two ABA-controlled switches induce vegetative dehydration tolerance and longevity. By manipulating ABA signaling pathways, it may be possible to increase the dehydration tolerance of tree tissues before the application of CPAs. ABA influences cell division modes, switching from asymmetric to symmetric cell division under stress conditions. This switch can help in the production of stress-resistant cells, which are more suitable for cryopreservation (Beier et al. 2023). Vegetative dehydration tolerance may also be epigenetically induced. The LAFL transcription factor network, crucial for maintaining seed desiccation tolerance, is regulated by ABA and epigenetic mechanisms such as histone modifications and Polycomb Repressive Complexes. Studies suggest that similar epigenetic priming can induce desiccation tolerance in vegetative tissues through reversible activation of a LAFL-like network (Farrant and Hilhorst 2022). Unfortunately, genetic engineering of desiccation tolerance in tree tissues that are presently desiccation sensitive has not yet been performed. Another issue is related to the divergence between the survival and regrowth potential of cells. Sometimes, even when plant tissues survive the cryopreservation process, the subsequent regrowth rate is significantly lower, indicating sublethal damage, as reported in apple shoot tips (Vujović et al. 2021).

8.8.3 Tree Cryopreservation in the Postgenomic Era

In the postgenomic era, tree cryopreservation has progressed significantly, leveraging advancements in genomics and biotechnology to increase the preservation of genetic diversity (Staszak and Pawłowski 2012). The integration of genomic, transcriptomic, proteomic, and metabolomic data allows for a more precise selection of cryopreservation protocols adjusted to individual species, improving survival rates and viability after thawing (Tsai and Hubscher 2004). Additionally, the application of high-throughput screening techniques (DNA sequencing platforms and high-throughput phenotyping) could facilitate the selection of optimal cryoprotectants and storage conditions, minimizing cellular damage during the freezing process. By enabling researchers to analyze large datasets quickly, these techniques could allow the identification of specific molecular and physiological traits associated with cryoprotectant efficacy and cellular resistance during freezing (Feng et al. 2024; Naqvi et al. 2022). As researchers better understand the molecular mechanisms underlying stress responses in trees, they can develop innovative strategies to enhance cryopreservation outcomes (Tsai and Hubscher 2004). This synergy between genomic insights and cryopreservation techniques not only strengthens conservation efforts but also supports the sustainable management of forest resources in the face of climate change and habitat loss.

8.8.4 Broader Conservation Framework

Creating global networks of cryobanks for tree species is fundamental for preserving plant genetic diversity and addressing the challenges of local threats, such as climate change, pests, and habitat destruction. These networks could facilitate the exchange of plant material between institutions, ensuring that genetic resources are not restricted to a single location and remain accessible for research, restoration, and breeding programs. By distributing germplasm across multiple cryobanks, the risks associated with natural disasters or geopolitical instability affecting any region can be mitigated (Oldfield and Newton 2012). For example, collaborative efforts such as the Atlantic Forest Restoration Pact in Brazil have demonstrated how coordinated actions across institutions can enhance restoration success (Viani et al. 2020). Nonetheless, international germplasm exchange must comply with legal frameworks such as the Convention on Biological Diversity (CBD) and the Nagoya Protocol, which regulate access to genetic resources and benefit-sharing agreements (CBD 2025).

A broader conservation framework also includes engaging indigenous communities to ensure that conservation efforts respect traditional knowledge systems while gaining social acceptance of these initiatives. Understanding national threat "hotspots" and conservation priorities enables more effective initiatives and the strategic allocation of resources (Carrero et al. 2022). Moreover, collaboration with indigenous communities can provide valuable insights into local plant species and conservation methods, potentially reducing the need for costly research. Additionally, combining cryopreservation with traditional methods provides a robust framework for preserving tree species.

8.8.5 Cost Reduction

Establishing and maintaining in vitro or cryobank networks is resource intensive, requiring investments in infrastructure, skilled personnel, and continuous research to optimize protocols. Exploring lower-cost alternatives for expensive components of culture media can help reduce overall expenses. For example, using substitutes for gelling agents or carbon sources, such as isabgol and market sugar, can maintain effectiveness while lowering expenditures, as reported in edible banana (Agrwal et al. 2010). Additionally, the use of low-energy appliances can greatly reduce costs, as most of them are typically related to high energy consumption (Miler et al. 2019). Sharing best practices among institutions can facilitate learning and improve conservation methods without high research costs. Finally, implementing automated systems can reduce labor costs and improve efficiency by minimizing human error in maintaining optimal conditions for germplasm storage (Kulus 2019b).

8.8.6 Extraterrestrial Storage of Diversity

Transporting and storing plant biodiversity across space represents an innovative and promising strategy for global biodiversity conservation. While liquid nitrogen storage is commonly employed on Earth for preserving tree germplasms, it presents challenges related to atmospheric leakage in crewed spacecraft. Therefore, exploring LN-free cryogenic freezing methods or SGG protocols could serve as a viable alternative (Visscher et al. 2023). By combining terrestrial and extraterrestrial approaches, biodiversity conservation can move toward a more effective framework that addresses the challenges of habitat loss, climate change, and human pressures on ecosystems.

8.9 Conclusions

Both in situ and ex situ methods are essential components of a comprehensive conservation strategy for woody plants. They address different aspects of species protection and complement each other to ensure the long-term survival of plant germplasm. The ex situ conservation of tree genetic resources plays a crucial role in maintaining biodiversity, particularly in the face of environmental threats, as climate change does not affect the genotypes conserved in vitro for many generations. Techniques such as slow-growth storage, synthetic seeds, and cryopreservation are key methods used in in vitro conservation. SGS involves controlling environmental conditions such as temperature, light, and medium composition to slow plant growth, reducing the need for frequent subcultures and minimizing the risk of contamination. It is particularly useful in the short- and medium-term storage of germplasm. SGS is used for vegetatively propagated species and can be applied in commercial micropropagation to produce high-quality plantlets at reduced costs. On the other hand, cryopreservation involves storing plant material at ultralow temperatures, providing an effective long-term preservation approach. However, specialized and precise protocols for successful recovery are needed. The integration of high-throughput screening techniques accelerates the identification of optimal cryopreservation conditions, enabling more efficient protocol development. Advances in genetic engineering, particularly the use of CRISPR-Cas9, alongside findings from the post-genomic research, can promote dehydration tolerance and improve tissue hardiness during cryopreservation. All these techniques are essential for safeguarding tree diversity, but their application can vary on the basis of species-specific requirements and the availability of resources. Continued research is necessary to increase the efficiency and applicability of SGS and cryopreservation across a broader range of tree species. Collaboration among research institutions, engaging indigenous communities to incorporate traditional ecological knowledge, and the creation of global cryobank networks are vital for sharing best practices and improving these conservation startegies.

Competing Interests The authors have no relevant financial or nonfinancial interests to disclose.

References

Abd Elaziem TM, Ahmed MEsAEn, Abou El-Dis GR (2022) *In vitro* propagation for conservation of the rare date palm (*Phoenix dactylifera* L.) 'Amri' using immature inflorescence. In vitro Cell Dev Biol-Plant 58:1048–1056

AFR100 2 African forest landscape restoration initiative. https://afr100.org/

Agrawal A, Sanayaima R, Tandon R, Tyagi K (2010) Cost-effective *in vitro* conservation of banana using alternatives of gelling agent (isabgol) and carbon source (market sugar). Acta Physiol Plant 32:703–711

Ala El-naby, Ahmed ME (2022) *In vitro* propagation for conservation and genetic fidelity of the near threatened *Dimocarpus longan* plant. J Genet Eng Biotechnol 20(1):130

Alirezaei Z, Afazel M, Bidabadi SS (2020) *In vitro* conservation tactics for a sweet genotype of *Amygdalus scoparia*, an endangered medicinally important tree species by propagation using synthetic seeds. J Biosci Med 8:208–219

Al-Jubori MT, Mohammed RR (2024) The importance of cryoprotective agents (CPAs) in the cryopreservation process for the preservation of plant resources. IOP Conf Ser Earth Environ Sci 1371:042052

Alvarez-Buylla ER, Martínez-Ramos M (1990) Seed bank versus seed rain in the regeneration of a tropical pioneer tree. Oecologia 84:314–325

Arbeloa A, Marín JA, Andreu P, García E, Lorente P (2017) *In vitro* conservation of fruit trees by slow growth storage. Acta Hortic 1155:101–106

Arnet M, Santos B, Brockerhoff EG, Pelser BP, Ecroyd C, Clemens J (2015) Importance of arboreta for *ex situ* conservation of threatened trees. Biodivers Conserv 24:3601–3620

Arora K, Rai MK, Sharma AK (2022) Tissue culture mediated biotechnological interventions in medicinal trees: recent progress. Plant Cell Tissue Organ Cult 150:267–287

Awodoyin R, Olubode O, Ogbu J, Balogun R, Nwawuisi J, Orji K (2015) Indigenous fruit trees of tropical Africa: status, opportunity for development and biodiversity management. Agric Sci 6:31–41

Baek HJ, Kim HH, Cho EG, Chae YA, Engelmann F (2003) Importance of explant size and origin and of preconditioning treatments for cryopreservation of garlic shoot apices by vitrification. CryoLetters 24(6):381–388

Ballesteros D, Martínez MT, Sánchez-Romero C, Montalbán IA, Sales E, Moncaleán P, Arrillaga I, Corredoira E (2024) Current status of the cryopreservation of embryogenic material of woody species. Front Plant Sci 14:1337152
Barra-Jiménez A, Aronen TS, Toribio JAM (2015) Cryopreservation of embryogenic tissues from mature holm oak trees. Cryobiology 70(3):217–225
Beier MP, Jinno C, Noda N, Nakamura K, Sugano S, Suzuki Y, Fujita T (2023) ABA signaling converts stem cell fate by substantiating a tradeoff between cell polarity, growth and cell cycle progression and abiotic stress responses in the moss *Physcomitrium patens*. Front Plant Sci 14:1303195
Benelli C (2021) Plant cryopreservation: a look at the present and the future. Plan Theory 10(12):2744
Benelli C, De Carlo A, Engelmann F (2013) Recent advances in the cryopreservation of shoot-derived germplasm of economically important fruit trees of *Actinidia*, *Diospyros*, *Malus*, *Olea*, *Prunus*, *Pyrus* and *Vitis*. Biotechnol Adv 31:175–185
Benelli C, Tarraf W, Izgu T, De Carlo A (2022) *In vitro* conservation through slow growth storage technique of fruit species: an overview of the last 10 years. Plan Theory 11:3188
Bhandari K, Nayyar H (2014) Low temperature stress in plants: an overview of roles of cryoprotectants in defense. In: Ahmad P, Wani M (eds) Physiological mechanisms and adaptation strategies in plants under changing environment. Springer, New York
Bilavcik A, Faltus M, Zamecnik J (2021) The survival of pear dormant buds at ultra-low temperatures. Plan Theory 10(11):2502
Bradaï F, Almagro-Bastante J, Sánchez-Romero C (2023) Effect of sucrose preculture and culture inoculum density on cryopreservation of olive somatic embryos. Sci Hortic 322:112385
Brennan AN, Pence VC, Taylor MD, Trader BW, Westwood M (2017) Tissue culture using mature material for the conservation of oaks. HortTechnol 27(5):644–649
Capuana M, Di Lonardo S (2013) *In vitro* conservation of chestnut (*Castanea sativa*) by slow growth. In vitro Cell Dev Biol-Plant 49:605–610
Carrero C, Beckman Bruns E, Frances A, Jerome D, Knapp W, Meyer A, Mims R, Pivorunas D, Speed D, Treher Eberly A, Westwood M (2022) Data sharing for conservation: a standardized checklist of US native tree species and threat assessments to prioritize and coordinate action. Plants People Planet 5:600–614
Carvalho MDJDSD, Souza ADS, Santos EB, Filho WDSS, Ledo CADS, Costa EMR, Souza FVD (2022) *In vitro* conservation of 'Florida Rough' lemon plants. Ciência Rural 52:1–9
Cazzolla Gatti R, Reich PB, Gamarra JGP, Crowther T, Hui C, Morera A, Bastin JF et al (2022) The number of tree species on earth. Proc Natl Acad Sci USA 119(6):e2115329119. https://doi.org/10.1073/pnas.2115329119. Erratum in: Proc Natl Acad Sci USA. 119(13):e2202784119
CBD (2025) Convention of biological diversity. https://www.cbd.int
Cha-um S, Kirdmanee C (2007) Minimal growth *in vitro* culture for preservation of plant species. Fruit Veg Cereal Sci Biotechnol 1:13–25
Chávez-Salgado LP, Vandenbossche V, Vilarem G (2022) *Tectona grandis* Linn. f. secondary metabolites and their bioactive potential: a review. iForest 15:112–120
Chmielarz P, Kotlarski S, Kalemba EM, Martins JPR, Michalak M (2023) Successful *in vitro* shoot multiplication of *Quercus robur* L. trees aged up to 800 years. Plan Theory 12(12):2230
Dawson IK, Guariguata MR, Loo J, Weber JC, Lengkeek A, Bush D, Cornelius J, Guarino L, Kindt R, Orwa C, Russell J, Jamnadass R (2013) What is the relevance of smallholders' agroforestry systems for conserving tropical tree species and genetic diversity *in circa situm*, *in situ* and *ex situ* settings? A review. Biodivers Conserv 22:301–324
de Vries SMG, Alan M, Bozzano M, Burianek V, Collin E, Cottrell J, Ivankovic M, Kelleher CT, Koskela J, Rotach P, Vietto L, Yrjänä L (2015) Pan-European strategy for genetic conservation of forest trees and establishment of a core network of dynamic conservation units. European Forest Genetic Resources Programme (EUFORGEN). Bioversity International, Rome, Italy, p xii + 40
Dosmann MS (2006) Research in the garden: averting the collections crisis. Bot Rev 72:207–234

Engelmann F (1990) Use of atmospheres with low oxygen contents for the storage of oil palm (*Elaeis guineensis* Jacq.) somatic embryo cultures. c.R. Acad Sei Paris 310(3):679–684
Fahy GM, Wowk B (2015) Principles of cryopreservation by vitrification. Methods Mol Biol 1257:21–82
FAO (2018) The state of the world's forests. https://www.fao.org/interactive/state-of-forests/2018/en/
FAO (2025) The global plan of action for forest genetic resources. https://www.fao.org/forest-genetic-resources/global-plan-of-action/en/
Farhadi-Tooli S, Ghanbari A, Kermani MJ, Zeinalabedini M, Bettoni JC, Naji AM, Kazemi N (2022) Droplet-vitrification cryotherapy and thermotherapy as efficient tools for the eradication of apple chlorotic leaf spot virus and apple stem grooving virus from virus-infected quince *in vitro* cultures. Eur J Plant Pathol 162:31–43
Farrant JM, Hilhorst H (2022) Crops for dry environments. Curr Opinion Biotechnol 74:84–91
Feng J, Dan X, Cui Y, Gong Y, Peng M, Sang Y, Ingvarsson PK, Wang J (2024) Integrating evolutionary genomics of forest trees to inform future tree breeding amid rapid climate change. Plant Commun 5(10):101044
Fenning T, O'Donnell M, Preedy K, Bézanger A, Kenyon D, Lopez G (2022) The rooting ability of *in vitro* shoot cultures established from a UK collection of the common ash (*Fraxinus excelsior* L.) and their *ex vitro* survival. Ann Forest Sci 79:30
Fernández A, León-Lobos P, Contreras S, Ovalle JF, Sershen, van der Walt K, Ballesteros D (2023) The potential impacts of climate change on ex situ conservation options for recalcitrant-seeded species. Front For Glob Change 6:1110431
Fki L, Bouaziz N, Chkir O, Benjemaa-Masmoudi R, Rival A, Swennen R, Drira N, Panis B (2013) Cold hardening and sucrose treatment improve cryopreservation of date palm meristems. Biol Plant 57:375–379
Foster JA, Walsh SK, Havens K, Kramer AT, Fant JB (2022) Supporting long-term sustainability of *ex situ* collections using a pedigree-based population management approach. Appl Plant Sci 10(5):e11491
Funnekotter B, Bunn E, Mancera R (2016) Cryo-mesh: a simple alternative technique for cryopreservation. Cryobiology 73(3):441
Gantait S, Kundu S, Wani SH, Das PK (2016) Cryopreservation of forest tree seeds: a mini-review. J Forest Environ Sci 32(3):311–322
GGBN (2025) Global Genome Biodiversity Network. https://www.ggbn.org
Gianní S, Sottile F (2015) *In vitro* storage of plum germplasm by slow growth. Hortic Sci 42:61–69
Gómez-Lama Cabanás C, Fernández-González AJ, Cardoni M, Valverde-Corredor A, López-Cepero J, Fernández-López M, Mercado-Blanco J (2021) The banana root endophytome: differences between mother plants and suckers and evaluation of selected bacteria to control *Fusarium oxysporum* f.sp. *cubense*. J Fungi 7(3):194
Hamann A, Smets P, Yanchuk AD, Aitken SN (2005) An ecogeographic framework for *in situ* conservation of forest trees in British Columbia. Canad J Forest Res 35(11):2553–2561
Hanson J, Ellis RH (2020) Progress and challenges in *ex situ* conservation of forage germplasm: grasses, herbaceous legumes and fodder trees. Plan Theory 9:446
Harris J, van Zonneveld M, Achigan-Dako EG, Bajwa B, Brouwer ID, Choudhury D, de Jager I, de Steenhuijsen Piters B, Dulloo ME, Guarino L, Kindt R, Mayes S, McMullin S, Quintero M, Schreinemachers P (2022) Fruit and vegetable biodiversity for nutritionally diverse diets: challenges, opportunities, and knowledge gaps. Glob Food Sec 33:100618
Hazubska-Przybył T, Wawrzyniak MK, Obarska A, Salaj T (2024) Cryopreservation of *Abies alba* × *A. numidica* and *Pinus nigra* embryogenic tissues by stepwise dehydration method. Plant Met 20:10
Hodkinson TR, Waldren S, Parnell JAN, Kelleher CT, Salamin K, Salamin N (2007) DNA banking for plant breeding, biotechnology and biodiversity evaluation. J Plant Res 120:17–29
Hughes CE, Styles BT (1984) Exploration and seed collection of multiple purpose dry zone trees in Central America. Int Tree Crops J 3(1):1–31

Ikhlaq M, Hafiz IA, Micheli M, Ahmad T, Abbasi NA, Standardi A (2010) *In vitro* storage of synthetic seeds: effect of different storage conditions and intervals on their conversion ability. Afr J Biotechnol 9:5712–5721

Irish BM, Goenaga R, Reed BM (2013) Amending storage vessel and media increase transfer interval of Musa spp. tissue culture plantlets. J Agric Univ Puerto Rico 97(1–2):1–14

Jactel H, Bauhus J, Boberg J, Bonal D, Castagneyrol B, Gardiner B, Gonzalez-Olabarria JR, Koricheva J, Meurisse N, Brockerhoff EG (2017) Tree diversity drives forest stand resistance to natural disturbances. Curr Forestr Rep 3:223–243

Jung YY, Gi AL, Sok YL, Jong WC, Il SS (2013) Cryopreservation of winter-dormant apple buds using two-step freezing. Plant Breed Biotech 1(3):283–289

Kader A, Sinha SN, Ghosh P (2022) A strategy for development of genetically stable synseeds of *Azadirachta indica* A. Juss. (Neem) suitable for *in vitro* storage. Plant Cell Tissue Organ Cult 151:47–58

Kasso M, Balakrishnan M (2013) *Ex situ* conservation of biodiversity with particular emphasis to Ethiopia. Int Scholarly Res Notices 2013:985037

Kaviani B, Kulus D (2022) Cryopreservation of endangered ornamental plants and fruit crops from tropical and subtropical regions. Biology 11(6):847

Kew (2019) Living Collections Strategy. https://www.kew.org/

Koç I, Akdemir H, Onay A, Çiftçi YÖ (2014) Cold-induced genetic instability in micropropagated *Pistacia lentiscus* L. plantlets. Acta Physiol Plant 36:2373–2384

Koroleva OV, Molkanova OI, Vysotskaya ON (2023) Development of cryopreservation technique for meristems of *Syringa vulgaris* L. cultivars. Int J Plant Biol 14(3):625–637

Kovalchuk I, Lyudvikova Y, Volgina M, Reed BM (2009) Medium, container and genotype all influence in vitro cold storage of apple germplasm. Plant Cell Tissue Organ Cult 96:127–136

Krishna H, Sairam R, Singh S, Patel V, Sharma R, Grover M, Nain L, Sachdev A (2008) Mango explant browning: effect of ontogenic age, mycorrhization and pre-treatments. Sci Hortic 118:132–138

Kulus D (2019a) Application of synthetic seeds in propagation, storage, and preservation of Asteraceae plant species. In: Faisal M, Alatar A (eds) Synthetic seeds. Springer, Cham

Kulus D (2019b) Managing plant genetic resources using low and ultra-low temperature storage: a case study of tomato. Biodivers Conserv 28:1003–1027

Kundu S, Sutradhar M, Salma U (2019) Synthetic seed technology in forest trees: a promising technology for conservation and germplasm exchange. In: Faisal M, Alatar A (eds) Synthetic seeds. Springer, Cham

Lambardi M, Shaarawi S (2017) Importance of *in vitro* culture for developing cryopreservation strategies of woody plants. Acta Hortic 1187:177–188

Leach WD (2006) Public involvement in USDA forest service policymaking: a literature review. J Forestr 104(1):43–49

Lennon JT, den Hollander F, Wilke-Berenguer M, Blath J (2021) Principles of seed banks and the emergence of complexity from dormancy. Nat Commun 12:4807

Leva AR, Petruccelli R, Rinaldi LMR (2012) Somaclonal variation in tissue culture: a case study with olive. In: Recent advances in plant *In vitro* culture. InTech, Croatia, Balkans

Li M, Fu Z, Yu G, Ma Z (2025) Effects of different cryoprotectants on cryopreservation of sperm from Noble Scallop *Mimachlamys nobilis*. J Marine Sci Eng 13(3):607

Li JW, Ozudogru EA, Li J, Wang M-R, Bi W-L, Lambardi M, Wang Q-C (2018) Cryobiotechnology of forest trees: recent advances and future prospects. Biodivers Conserv 27:795–814

Liu Y, Huang T, Lin S, Jiang Y (2023) Programmed cooling and vitrification method applied to the germplasm preservation of *Eriobotrya* plants. HortSci 58(2):164–169

Madhu GS, Harish D, Mir JI, Nabi SU, Sharma OC (2024) Conservation and use of temperate fruit and nut genetic resources. In: Rajasekharan PE, Rao VR (eds) Fruit and nut crops. Handbooks of crop diversity: conservation and use of plant genetic resources. Springer, Singapore

Maheswarappa V, Vasudeva R, Hegde R (2022) *Circa situm* conservation of coffee agroforests: farmer's perception in Kodagu landscape of Karnataka, India. Trop Ecol 63:104–112

Malabadi RB, Nataraja K (2006) Cryopreservation and plant regeneration via somatic embryogenesis using shoot apical domes of mature *Pinus roxburghii* sarg, trees. In vitro Cell Dev Biol-Plant 42:152–159
Marcantonini G, Bartolini D, Zatini L, Costa S, Passerini M, Rende M, Luca G, Basta G, Murdolo G, Calafiore R, Galli F (2022) Natural cryoprotective and cytoprotective agents in cryopreservation: a focus on melatonin. Molecules 27:3254
Martinez P, Serpe M, Barron R, Buerki S (2023) Acclimation and hardening of a slow-growing woody species emblematic to western North America from *in vitro* plantlets. Appl Plant Sci 11(2):e11515
Mascarello C, Sacco E, Di Silvestro D, Pamato M, Ruffoni B (2017) Reduction of the subculture frequency in three Mediterranean species during micropropagation. Acta Hortic 1155:461–466
McCown BH, Lloyd G (1981) Woody plant medium (WPM) - a mineral nutrient formulation for microculture of woody plant species. HortSci 16:453–453
Miler N, Kulus D, Woźny A, Rymarz D, Hajzer M, Wierzbowski K, Nelke R, Szeffs L (2019) Application of wide-spectrum light-emitting diodes in micropropagation of popular ornamental plant species: a study on plant quality and cost reduction. In vitro Cell Dev Biol-Plant 55:99–108
Miller DC, Muñoz-Mora JC, Rasmussen LV, Zezza A (2020) Do trees on farms improve household well-being? Evidence from national panel data in Uganda. Front For Glob Change 3:101
Monticelli S, Gentile A, Forni C, Caboni E (2020) Slow growth in "Pisana" apricot and in "Cerina" apple, two Italian cultivars. Acta Hortic 1285:117–124
Morrissey TM, Gustafson WA Jr (1990) The use of dormant-bud cryopreservation for long-term storage of pecan and walnut germplasm. HortSci 25(9):1142c–1142
Mudoi KD, Gogoi B, Borah G, Hussain M, Tasfia T, Borah K, Lekhak H, Saikia SP (2023) An assessment for *in vitro* propagation and genetic stability of *Phoebe goalparensis* Hutchinson, an endemic valuable timber tree of North East India. J Genet Eng Biotechnol 21:29
Murashige T, Skoog F (1962) A revised medium for rapid growth and bio assays with tobacco tissue cultures. Physiol Plant 15:473–497
Naqvi RZ, Siddiqui HA, Mahmood MA, Najeebullah S, Ehsan A, Azhar M, Farooq M, Amin I, Asad S, Mukhtar Z, Mansoor S, Asif M (2022) Smart breeding approaches in post-genomics era for developing climate-resilient food crops. Front Plant Sci 13:972164
Nguyen VA, Nguyen PT, Le MA, Bazrafshan A, Sisunandar S, Kalaipandian S, Adkins SW, Nguyen QT (2023) A practical framework for the cryopreservation of palm species. In vitro Cell Dev Biol-Plant 59:425–445
Niino T, Sakai T (1992) Cryopreservation of alginate-coated in vitro-grown shoot tips of apple, pear and mulberry. Plant Sci 87(2):199–206
Nin S, Benelli C, Petrucci WA, Turchi A, Pecchioli S, Gori M, Giordani E (2019) *In vitro* propagation and conservation of wild bilberry (*Vaccinium myrtillus* L.) genotypes collected in the Tuscan Apennines (Italy). J Berry Res 9:411–430
NordGen (2025). https://www.nordgen.org/our-work/forest/
Normah MN, Sulong NA, Reed BM (2019) Cryopreservation of shoot tips of recalcitrant and tropical species: advances and strategies. Cryobiology 87:1–14
O'Brien C, Hiti-Bandaralage J, Folgado R, Hayward A, Lahmeyer S, Folsom J, Mitter N (2021) Cryopreservation of woody crops: the avocado case. Plan Theory 10(5):934
Ochatt S, Lambardi M, Panis B, Pathirana R, Revilla MA, Wang Q-C (2021) Cryopreservation and *in vitro* banking: a cool subject – Preface from the editors. Plant Cell Tissue Organ Cult 144:1–5
Oldfield SF (2009) Botanic gardens and the conservation of tree species. Trends Plant Sci 14(11):81–583
Oldfield S, Newton AC (2012) Integrated conservation of tree species by botanic gardens: a reference manual. Botanic Gardens Conservation International, Richmond
Pachas ANA, Sakanphet S, Midgley S, Dieters M (2019) Teak (*Tectona grandis*) silviculture and research: applications for smallholders in Lao PDR. Aust For 82(sup1):94–105
Panis B, Piette B, Swennen R (2005) Droplet vitrification of apical meristems: a cryopreservation protocol applicable to all Musaceae. Plant Sci 168(1):45–55

Parab AR, Lynn CB, Subramaniam S (2021) Assessment of genetic stability on *in vitro* and *ex vitro* plants of *Ficus carica* var. black jack using ISSR and DAMD markers. Mol Biol Rep 48(11):7223–7231

Paul H, Daigny G, Sangwan-Norreel B (2000) Cryopreservation of apple (*Malus* × *domestica* Borkh.) shoot tips following encapsulation-dehydration or encapsulation-vitrification. Plant Cell Rep 19:768–774

Pawłowska B, Ptak A (2023) Application of *in vitro* culture and biotechnology for the protection of endangered national plant species in Poland. Acta Soc Bot Polon 92(1):1–13

Pence VC, Chaiken MF (2021) Shoot tip cryopreservation as a conservation tool for species of *Quercus*: effects of species and environment on recovery. Cryo-Letters 42(3):159–167

Pence VC, Ballesteros D, Walters C, Reed BM, Philpott M, Dixon KW, Pritchard HW, Culley TM, Vanhove A-C (2020) Cryobiotechnologies: tools for expanding long-term *ex situ* conservation to all plant species. Biol Conserv 250:108736

Pence VC, Meyer A, Linsky J, Gratzfeld J, Pritchard HW, Westwood M, Beckman Bruns E (2022) Defining exceptional species - a conceptual framework to expand and advance *ex situ* conservation of plant diversity beyond conventional seed banking. Biol Conserv 266:109440

Philpott M, Pence VC, Bassüner B, Clayton AS, Coffey EED, Downing JL, Edwards CE, Folgado R, Ligon JJ, Powell C, Ree JF, Seglias AE, Sugii N, Zale PJ, Zeldin J (2022a) Harnessing the power of botanical gardens: evaluating the costs and resources needed for exceptional plant conservation. Appl Plant Sci 10(5):e11495

Philpott M, Pence VC, Coffey EED (2022b) Building capacity in the conservation of exceptional plant species. Appl Plant Sci 10(5):e11498

Pokorná E, Faltus M, Máchová P, Zámečník J, Fulín M (2020) Grey poplar explant acclimation to improve the dehydration tolerance and cryopreservation. Biol Plant 64:119–128

Popova E, Shukla M, Kim HH, Saxena PK (2021) Root cryobanking: an important tool in plant cryopreservation. Plant Cell Tissue Organ Cult 144:49–66

Popova E, Kulichenko I, Kim H-H (2023) Critical role of regrowth conditions in *post*-cryopreservation of *in vitro* plant germplasm. Biol 12(4):42

Potter KM, Jetton RM, Bower A, Jacobs DF, Man G, Hipkins VD, Westwood M (2017) Banking on the future: Progress, challenges and opportunities for the genetic conservation of forest trees. New For 48:153–180

Pritchard HW, Moat JF, Ferraz JBS, Marks TR, Camargo JLC, Nadarajan J, Ferraz IDK (2014) Innovative approaches to the preservation of forest trees. Forest Ecol Manag 333:88–98

Quoirin M, Lepoivre P (1977) Etudes de milieux adaptés aux cultures *in vitro* de *Prunus*. Acta Hort 78:437–442

Reed BM, DeNoma J (2016) Medium-term *in vitro* storage of pear as a complementary germplasm preservation technique. Acta Hortic 1113:251–256

Restrepo-Osorio C, Gil-Correal A, Chamorro-Gutiérrez L, Ramírez-Ríos V, Álvarez JC, Villanueva-Mejía D (2020) Efficient direct shoot organogenesis and genetic stability in micropropagated sacha inchi (*Plukenetia volubilis* L.). BMC Res Notes 13:414

Ribeiro de Souza R, Oliveira Paiva PD, Freitas RT, Corrêa da Silva DP, dos Reis MV, Nery FC, Rufato RP (2023) Cryopreservation of *Genipa americana* seeds. Rev Ciênc Agron 54:e20228531

Righetti PG, Esteve C, D'Amato A, Fasoli E, Marina ML, García MC (2023) Proteomics of some "recalcitrant" tropical fruits: avocado, banana and mango. World J Agric Soil Sci 8(4):1–7

Rittenhouse CD, Rissman AR (2005) Forest cover, carbon sequestration, and wildlife habitat: policy review and modeling of tradeoffs among land-use change scenarios. Environ Sci Pol 21:94–105

Rodrigues PHV, Oliveira EL, Demetrio CA, Ambrosano GB, Piedade SMS (2022) Effects of different light spectra on the slow-grown *in vitro* storage and quality of banana plantlets cv. Prata Catarina (AAB). Plant Cell Tissue Organ Cult 150:479–485

Roque-Borda CA, Kulus D, Vacaro de Souza A, Kaviani B, Vicente EF (2021) Cryopreservation of agronomic plant germplasm using vitrification-based methods: an overview of selected case studies. Int J Mol Sci 22(11):6157

Ryynänen L, Häggman H (2001) Recovery of cryopreserved silver birch shoot tips is affected by the pre-freezing age of the cultures and ammonium substitution. Plant Cell Rep 20:354–360
Sakai A (1956) Survival of plant tissue of super-low temperatures. Low Temp Sci B 14:17–23
Sakai A (1960) Survival of the twigs of woody plants at −196°C. Nature 185:393–394
Sakai A, Kobayashi S, Oiyama I (1990) Cryopreservation of nucellar cells of navel orange (*Citrus sinensis* Osb. var. *brasiliensis* Tanaka) by vitrification. Plant Cell Rep 9:30–33
Salgotra RK, Chauhan BS (2023) Genetic diversity, conservation, and utilization of plant genetic resources. Genes 14:174
San José MC, Janeiro LV, Corredoira E (2015) Simple strategy for the *in vitro* conservation of *Alnus glutinosa* (L.) Gaertn. germplasm. Trees 29:539–549
Scottez C, Chevreau E, Godard N, Arnaud Y, Duron M, Dereuddre J (1992) Cryobiology 29(6):691–700
Sedlak J, Zidova P, Paprstein F (2019) Slow growth *in vitro* conservation of fruit crops. Acta Hortic 1234:119–124
Sharma S, Shahzad A, Mahmood S, Saeed T (2015) High-frequency clonal propagation, encapsulation of nodal segments for short-term storage and germplasm exchange of *Ficus carica* L. Trees 29:345–353
Silva SV, Andermann T, Zizka A, Kozlowski G, Silvestro D (2022) Global estimation and mapping of the conservation status of tree species using artificial intelligence. Front Plant Sci 13:839792
Sota V, Kongjika E (2014a) Slow growth *in vitro* conservation of *Zizyphus jujuba* Mill. Poljopr I Sumar 60:27–37
Sota V, Kongjika E (2014b) The effect of nutrient media in micropropagation and *in vitro* conservation of wild population of mahaleb cherry (*Prunus mahaleb* L.). J Microbiol Biotech Food Sci 3:453–456
Sota V, Benelli C, Myrselaj M, Kongjika E, Gruda NS (2023) Short-term conservation of *Juglans regia* L. via synthetic seed technology. Horticulturae 9(5):559
Souza JA, Bogo A, Bettoni JC, Dalla Costa M, da Silva FN, Casa RT, Rufato L (2020) Droplet-vitrification cryotherapy for eradication of apple stem grooving virus and apple stem pitting virus from "Marubakaido" apple rootstock. Trop Plant Pathol 45:148–152
Srivastava LM (2002) Seed development and maturation. In: Srivastava LM (ed) Plant growth and development. Academic Press, pp 431–446
Staszak AM, Pawlowski TA (2012) Forest tree research in post genomic era. Introduction to systems biology of broadleaves. Dendrobiology 68:113–123
Storme V, Vanden Broeck A, Ivens B, Halfmaerten D, Van Slycken J, Castiglione S, Grassi F, Fossati T et al (2004) *Ex-situ* conservation of Black poplar in Europe: genetic diversity in nine gene bank collections and their value for nature development. Theor Appl Genet 108:969–981
Tailor A, Joshi A (2024) The role of synthetic seeds in germplasm conservation of forest tree species. In: Thomas TD, Razdan MK, Kumar A (eds) Biotechnological approaches for sustaining forest trees and their products. Springer, Singapore
Tambe S, Rawat GS, Krishen P, Ranjitsinh MK, Ghosh N, Rawat AS, Takpa J (2022) Compensatory afforestation policy in India: an analysis using an ecorestoration lens. Internat Forestr Rev 24(4):607–618
Tarraf W, De Carlo A (2024) *In vitro* biotechnology for conservation and sustainable use of plant genetic resources. Plan Theory 13:1897
Thomas G, Sucher R, Wyatt A, Jiménez I (2022) *Ex situ* species conservation: predicting plant survival in botanic gardens based on climatic provenance. Biol Conserv 265:109410
Tsai CJ, Hubscher SL (2004) Cryopreservation in *Populus* functional genomics. New Phytol 164(1):73–81
Turdiyev TT, Kovalchuk IY, Kabylbekova BZ, Chukanova NI, Frolov SN (2020) *In vitro* germplasm cold storage of fruit and berry plants of Kazakhstan. Eurasian J Biosci 1219:1213–1219
Turner-Skoff JB, Paist S, Byrne A, Westwood M (2021) ArbNet: 10 years of fostering collaborations, furthering professionalism, and advancing the planting and conservation of trees through the global network of arboreta. Plants People Planet 4:128–135

Tweddle JC, Dickie JB, CC, Baskin JM (2003) Baskin ecological aspects of seed desiccation sensitivity. J Ecol 91:294–304
VianiS RAG, Holl KD, Padovezi A, Strassburg BBN, Farah FT, Garcia LC, Chaves RB, Rodrigues RR, Brancalion PHS (2020) Protocol for monitoring tropical forest restoration: perspectives from the Atlantic forest restoration pact in Brazil. Trop Conserv Sci 10(1):1–8
Vidyagina EO, Kharchenko NN, Shestibratov KA (2021) Efficient cryopreservation of *Populus tremula* by *in vitro*-grown axillary buds and genetic stability of recovered plants. Plan Theory 10(1):77
Visscher AM, Pritchard HW, Neri G, Ballesteros D (2023) How do we transport plant species with desiccation-sensitive germplasm in space? Life Sci Space Res 36:135–137
Volk GM, Bonnart R, Araújo de Oliveira AC, Henk AD (2022) Minimizing the deleterious effects of endophytes in plant shoot tip cryopreservation. Appl Plant Sci 10(5):e11489
Vujović T, Ružić Đ, Cerović R (2021) Cryopreservation of apple shoot tips by vitrification and subsequent plant regeneration. Acta Hortic 1308:33–40
Vujović T, Anđelić T, Vasilijević B, Jevremović D, Engelmann F (2023) Cryopreservation of indigenous plums and monitoring of multiplication and rooting capacity of shoots obtained from cryopreserved specimens. Plan Theory 12(17):3108
Walters C (2015) Orthodoxy, recalcitrance and in-between: describing variation in seed storage characteristics using threshold responses to water loss. Planta 242:397–406
Wang MR, Chen L, Teixeira da Silva JA, Volk GM, Wang QC (2018) Cryobiotechnology of apple (*Malus* spp.): development, progress and future prospects. Plant Cell Rep 37:689–709
Wang Q, Batuman Ö, Li P, Bar-Joseph M, Gafny R (2002a) Cryopreservation of *in vitro*-grown shoot tips of 'Troyer' citrange [*Poncirus trifoliata* (L.) Raf. × *Citrus sinensis* (L.) Osbeck.] by encapsulation-dehydration. Plant Cell Rep 20:901–906
Wang Q, Batuman Ö, Li P, Bar-Joseph M, Gafny R (2002b) A simple and efficient cryopreservation of *in vitro*-grown shoot tips of 'Troyer' citrange [*Poncirus trifoliata* (L.) Raf.× *Citrus sinensis* (L.) Osbeck.] by encapsulation-vitrification. Euphytica 128:135–142
Wang M-R, Bao J-H, Ma X-Y, Yan Z-H, Cui Z-H, Zhu L-Y, Zhang D, Wang Q-C (2024) Vitrification cryo-foil method for shoot tip cryopreservation and virus eradication in apple. Cryobiology 117:104957
Winkeljohn M, Pence VC, Culley TM (2022) Improving culture initiation of mature oak shoots through use of silver thiosulfate. Appl Plant Sci 10(5):e11497
Wu JY, Zhao YH, Engelmann F, Zhou MD, Zhang DM, Chen SY (2001) Cryopreservation of apple dormant buds and shoot tips. CryoLett 22(6):375–380
Yang R-C, Yeh FC (1992) Genetic consequences of *in situ* and *ex situ* conservation of forest trees. For Chron 68(6):720–729
Yi J, Liang XM, Zhao G, He X (2014) An improved model for nucleation-limited ice formation in living cells during freezing. PLoS ONE 9(5):e98132
Zegeye H (2017) *In situ* and *ex situ* conservation: complementary approaches for maintaining biodiversity. Int J Res Environ Stud 4:1–12
Zhang XH, Yu XZ, Yue DM (2016) Phytotoxicity of dimethyl sulfoxide (DMSO) to rice seedlings. Int J Environ Sci Technol 13:607–614
Zhang A-L, Wang M-R, Li Z, Panis B, Bettoni JC, Vollmer R, Xu L, Wang Q-C (2023) Overcoming challenges for shoot tip cryopreservation of root and tuber crops. Agronomy 13(1):219
Zheng L, Lu J, Liu H, Chen X, Yesou H (2023) Evidence of vegetation greening benefitting from the afforestation initiatives in China. Geo-Spatial Inf Sci 27(3):683–702

9 Biotechnological Approaches for In Vitro Conservation of *Petiveria alliacea* Germplasm

Rachel Fatima Gagliardi, Jamine de Almeida Pettinelli, Bianka de Oliveira Soares, and Flavio Jose da Silva Dantas

Abstract

In vitro conservation of germplasm is considered the best alternative for preserving medicinal plants, enabling the multiplication and use of specific genotypes and contributing to the reduction of genetic erosion caused by anthropogenic or environmental impacts. These plants constitute an invaluable reservoir of substances with pharmacological activities and are, for this reason, overexploited, generally in an unsustainable manner, whether for popular, commercial, or research uses. Although the pharmaceutical industry produces many drugs, the chemical synthesis of complex molecules is not always possible. This chapter addresses the in vitro conservation of *Petiveria alliacea*, a tropical herbaceous species traditionally used as an analgesic, sedative, or abortifacient in folk medicine, as well as in religious rituals, owing to its neurological effects, indicating the great potential for the pharmacological use of bioactive substances synthesized in the plant against various human and animal pathologies, including neoplasms. This species has received much attention from researchers because of its biological activities, generating very complete knowledge about its phytochemical profile and the mechanisms of action of its main components, but conservation actions are scarce, being carried out, generally, at home in small gardens or

R. F. Gagliardi (✉) · J. de Almeida Pettinelli
Núcleo de Biotecnologia Vegetal, Universidade do Estado do Rio de Janeiro, Rio de Janeiro, RJ, Brazil
e-mail: gagliard@uerj.br

B. de Oliveira Soares
Instituto Federal de Educação, Ciência e Tecnologia do Rio de Janeiro, Rio de Janeiro, RJ, Brazil

F. J. da Silva Dantas
Laboratório de Rádio e Fotobiologia, Universidade do Estado do Rio de Janeiro, Rio de Janeiro, RJ, Brazil

S. Gantait, P. Chmielarz (eds.), *Conservation of Plant Genetic Resources*,
https://doi.org/10.1007/978-981-96-8331-4_9

in pots, by those who use or sell them. Research focused on in vitro conservation includes the establishment and standardization of micropropagation and cryopreservation protocols, as well as efficient systems for genetic monitoring of preserved material in addition to phytochemical studies and the biological effects of plants in vitro. Providing appropriate methodologies for the production and long-term conservation of the genetic resources of this species can contribute to the recovery of impacted natural populations, in addition to providing material for phytochemical and pharmacological research.

Keywords

Cryopreservation · Micropropagation · Somatic embryogenesis · Histological analysis · Phytochemical profile · Genetic stability · Cytotoxicity · Ethnobotanic

9.1 Introduction

Medicinal plants are those that contain special chemical components with therapeutic biological activities. Some of these components are volatile and aromatic and are also used by the cosmetics and food industries. These substances are produced by secondary metabolism throughout the evolution of plants in variable environments and ecosystems and can cause variations in the expression of genetic potential. Consequently, the phytochemical composition of plants collected in the field is not uniform.

Furthermore, the impacts caused by civilization and the predatory collection of native species have caused genetic erosion in natural populations, a situation as serious as extinction, when the possible loss of genes responsible for the biosynthesis of bioactive substances is considered (Ebert and Engels 2020). The vast majority of species with medicinal potential remain unknown and are at risk of disappearing, depriving society of resources that could be useful in the treatment of many pathologies for which there are still no effective therapeutic options. In this context, medicinal and aromatic plants are seriously threatened throughout the world.

The species *Petiveria alliacea*, which is the focus of this chapter, produces several metabolites of medicinal importance. However, it is a semidomesticated species that is not produced agronomically, making it difficult to obtain the compounds of interest, which are heterogeneous from natural sources and insufficient for large-scale production. Although many bioactive substances are synthesized chemically, this process is quite complex and requires the establishment of alternative production systems. In this context, plant cell and tissue culture techniques constitute promising systems for the production of chemical substances of medicinal interest because of the possibility of manipulating culture conditions, leading to different metabolic pathways.

Despite the importance of medicinal species, their conservation has been carried out only at home in small gardens or in pots by those who use or sell them. Therefore, more consistent conservation actions for these natural resources are a high priority.

Among the available strategies, in situ conservation is the best way to conserve species in their habitats, which coevolves as part of the ecosystem (CBD 1992). However, exposure to climate disasters, pests, and pathogens, in addition to fluctuating government policies, has made this type of conservation difficult, and complementary ex situ conservation is recommended (Chen et al. 2016).

Moreover, ex situ conservation in the field outside of natural sites is problematic, as large areas are needed for the conservation of genetic diversity, and plants are also subject to climate variations. Furthermore, endemic species do not adapt to other environments. In seed banks, many species cannot be stored due to the physiological characteristics of their seeds or their absence. Therefore, among the ex situ conservation strategies available, in vitro culture associated with cryopreservation has proven to be the most suitable alternative for medicinal species because of the guarantee of material quality and safety in the storage of selected genotypes, in addition to the possibility of conserving intraspecific variability in much smaller spaces compared with other methods (Reed et al. 2011).

9.2 In Vitro Conservation of Medicinal Plants

In vitro conservation is based on plant tissue culture, which is used to produce and multiply plants under controlled conditions. This method involves the cultivation of explants (segments) from a mother plant, which are inoculated in aseptic and appropriate culture media with controlled light, temperature, and humidity conditions. Clones are generated vegetatively from preexisting meristems or by adventitious regeneration induced by phytoregulators.

The use of tissue culture methods enables the multiplication of specific genotypes, offering an additional guarantee of germplasm availability, which is essential to enable the conduct of phytochemical and pharmacological studies with these species, whose raw material has become scarce owing to the increase in extractivism. Plant cell and/or tissue cultures are useful tools for modulating the biosynthesis and accumulation of secondary metabolites, constituting indispensable tools in research preceding the development of new drugs. Furthermore, large-scale culture in bioreactors is already technically feasible for many species (Murthy et al. 2021). Unfortunately, in vitro systems, in general, cannot be maintained for long periods of time, as they may present a loss of propagation potential or limitations related to genetic instability, with the risk of alterations in metabolism (Karp 1989). The genetic variation induced by in vitro culture steps is called somaclonal (Larkin and Sowcroft 1981) and constitutes a restrictive factor for the application of in vitro methodologies in programs that require the maintenance of genetic fidelity. In addition to the regeneration pathway, many factors affect the nature and frequency of these variations, including culture time, medium composition, genotype, and others, and there is a need to optimize each new species studied. Reducing or eliminating favorable conditions for the emergence of somaclonal variations is therefore highly important and should be considered when establishing cultures aimed at preserving germplasm or producing metabolites.

In vitro conservation by slow growth is based on reducing tissue growth rates and increasing the time intervals between subcultures. In this type of preservation, germplasm can be maintained for medium periods, and for this purpose, several techniques are used separately or in combination, such as the addition of high levels of osmotic or hormonal agents, in addition to adjustments to the physical environment, such as lowering the temperature and incubation in the dark, with the aim of establishing the minimum conditions for growth (Engelmann 1997). However, cryopreservation is the most reliable method for long-term in vitro preservation of germplasm due to the genetic stability of the material (Mazur 1970; Reed et al. 2011) and is therefore the ideal method for preserving medicinal species. Several techniques are available, such as encapsulation–dehydration, vitrification, encapsulation–vitrification, and drop vitrification, allowing the preservation of many plant species (Engelmann 2011a, b). However, research on native medicinal species is still needed to establish specific protocols (Kundu et al. 2018). There is great interest in the cryopreservation of systems such as suspension cells, roots (adventitious or transformed), and somatic embryos, aimed at the production and storage of secondary metabolites and other pharmacological products.

9.3 *Petiveria alliacea*

This chapter presents the in vitro germplasm conservation strategies developed for the species *Petiveria alliacea*, an herbaceous plant with enormous medicinal potential that has been used for centuries in folk medicine in several countries and is used as a raw material by the pharmaceutical industry. Its medicinal properties generate high demand, and the species appears to be one of the most commercialized in the Brazilian market (Rodriguez 2023).

9.3.1 Characterization and Distribution

The species is a perennial dicotyledonous plant that can reach up to 1 meter in height and is characterized by erect branches; smooth, alternate, elliptical leaves; and tiny white flowers in terminal inflorescences (Cronquist 1981; Camargo 2007). The plant exudes a strong garlic smell (Ramirez 2003), and histochemical analyses of the aerial part indicated the presence of phytoliths with mineral inclusions, which are useful for species identification in herbal compositions (Duarte and Lopes 2005).

The genus *Petiveria* belongs to the Phytolaccaceae family and is monotypic (Udulutsch et al. 2007); however, the plants may present morphological differences that are not easily distinguishable on the basis of ovarian morphology, and four varieties or subspecies have been described (*alliacea*, *tetrandra*, *grandifolia*, and *octandra*) (Tropicos.org 2025). On the basis of this parameter, the existence of two varieties of *P. alliacea* (var. *alliacea* and var. *tetrandra*) was reported in populations occurring in Rio de Janeiro, Brazil (Soares et al. 2013). The literature also indicates many synonymies, among which the most relevant are *P. corrientina* Rojas Acosta,

P. hexandria Sessé & Moc., *P. octandra* L., and *P. paraguayensis* D. Parodi (Marchioretto 2020).

The species is native to the Amazon rainforest, but its present geographic distribution is wide, being found from Central America, including the Caribbean, to South America as well as in some countries on the African and Asian continents with a tropical climate. It can be found in its natural state throughout the territorial extension of Brazil, with a greater population concentration in rainforest regions (Luz et al. 2016; Marchioretto 2020).

9.3.2 Ethnobotany

P. alliacea receives many popular names depending on the place of occurrence; in English-speaking countries, it is called Guinea-Hen weed; in Spanish-speaking countries, it receives names such as Anamú, Zorrillo, and Ipicina; in the Amazon rainforest, it is known as Mucuracaá; in Tupi-Guarani; and in other regions of Brazil, Pipi, Guiné-pipi, Tipi, Amansa-senhor, Guiné, and others (Camargo 2007). In Africa, among Yoruba speakers, it is called Awogba or Ojusaju; in Hausa, it is known as kanunfari; and in Igbo, it is called Akwa-Ose (Olomieja et al. 2021; Lawal et al. 2024). Among the Brazilian names, "amansa-senhor" was coined by slaves during the colonial period (1500–1822), who added powdered roots to food destined for their masters in retaliation for the mist treatment they suffered. The effect appeared over time in the form of convulsive crises followed by lethal apathy (Camargo 2007; Luz et al. 2016). Research has confirmed that the root of a plant contains organic sulfides, nitrates, alkaloids, which cause hyperexcitation, hallucinations, convulsions, apathy, and brain damage (Matos et al. 2011).

The first reports of the medicinal use of this species include antispasmodic, diuretic, emmenagogue, antirheumatic, anthelmintic, abortifacient, analgesic, antitumor, and sedative effects, among others. The root is used as an anesthetic to relieve toothache, and crushed leaves, in the form of compresses and poultices, are used to treat headaches and rheumatic pain (Camargo 2007). In general, the internal use of the ethanolic extract is based on its antispasmodic, diuretic, abortifacient, hypoglycemic, anti-inflammatory, and anticancer activities (Conceição et al. 2023), in addition to effects on the central nervous system, such as anxiolytics, in the treatment of dementia and nervous disorders (Salomón et al. 2022). Aqueous infusions or decoctions (teas) of various parts of the plant (roots, leaves, or whole plants) are also used to treat numerous pathologies, such as arthritis, asthma, cancer, colds, coughs, fever, headache, inflammation, nervousness, spasms, pain, anemia, snake bites, urinary insufficiency, venereal disease, worms, malaria and aphrodisiac, sedatives, and inducers of labor and abortion (Taylor 2005; Picking et al. 2015).

The mystical use of this species is very present in religious rituals. In Brazil, the leaves are used in the composition of the "seven herbs" incense (Garcia et al. 2016), and in smokers they are used in smoke rituals to harmonize the environment, in addition to being used in baths to treat spiritual diseases (Alves et al. 2019). In Haiti, they are used in the composition of the zombie potion used in voodoo rituals

(Albuquerque et al. 2012). Knowledge of the relationships between traditional communities and plants can provide good indications of their pharmacological potential, and ethnobotanical studies constitute the most promising initial strategy for the development and discovery of new phytopharmaceuticals. Most current medicines are derived from plants that were originally discovered through traditional medicine and from the popular knowledge of native peoples.

9.3.3 Phytochemistry

Phytochemical analysis of *Petiveria alliacea* has already been widely performed, which can be seen from the numerous records in the literature on the chemical variety produced by the secondary metabolism of this species (Delle-Monache and Suarez 1992; Delle-Monache et al. 1996; De Sousa et al. 1990; Kubec and Musah 2001; Kubec et al. 2002, 2003; Ramirez 2003; Kubec and Musah 2005; Castellar et al. 2014). Several chemical compounds have been characterized in *P. alliacea*, with varied distributions in different parts of the plant. Some of the compounds found preferentially in the roots are benzaldehyde, dibenzyl disulfide, cis- and trans-stilbene, dibenzyl trisulfide, and coumarin (De Sousa et al. 1990; Kubec et al. 2002). Flavonoids such as 6-C-formyl and 6-C-hydroxymethyl were identified in the leaves, in addition to other compounds such as triterpene saponin glycosides, isoarborinol, isoarborinol acetate, isoarborinol cinnamate, steroids, alkaloids, and tannins (De Sousa et al. 1990; Delle-Monache and Suarez 1992; Delle-Monache et al. 1996).

Like the metabolism of allicin in *Allium* species (Kubec and Musah 2001; Ramirez 2003), the organosulfur compounds that confer the odor of garlic in *P. alliacea*, known as petiverin (cysteine sulfoxides), are thiosulfinates that are stored in plastids in the roots and undergo enzymatic cleavage (C-S lyase) after tissue damage, yielding several lipophilic, nonpolar, volatile compounds (e.g., dibenzyl sulfides, benzaldehyde, trithiolaniacin, stilbenes, benzyl mercaptan, benzoic acid) that are not found in intact plant tissues (De Sousa et al. 1990; Kubec and Musah 2001; Kubec et al. 2002).

9.3.4 Biological Effects

Medicinal plants contain a complex mixture of substances that can exert biological actions in isolation or synergistically (Silva et al. 2016). In general, the chemical composition of crude plant extracts is not fully known since the extracts constitute complex mixtures of chemical substances with diverse biological actions (Zhang 2004). On the other hand, the fractionation of extracts for analysis and definition of the active principle is not always effective, since certain effects are only observed with a combination of the substances present in the crude extract, and the fractionation method may not reveal these effects. Research on the biological effects of *P. alliacea* has increased interest in the effects of the compounds present in this species for therapeutic and prophylactic use, and the literature is rich in works confirming the activities described in folk medicine through the evaluation of crude extracts and fractions in different experimental models (Conceição et al. 2023).

Various activities have been proven with crude extracts, such as anticoagulant (Villar et al. 1997), nematicide and acaricide (Ponte et al. 1996), anti-inflammatory and analgesic (Di Stasi et al. 1998; Sertie et al. 1995a, b; Martins et al. 2002), abortifacient (Feng et al. 1964; Oluwole and Bolarinwa 1998), anti-snake venom (Trevisan et al. 2022), antimitotic (Mallpezzi et al. 1994), fungicidal (Benevides et al. 2001), and repellent (Akintan and Akinneye 2020) activities. An allelopathic effect was also observed through the inhibition of seed germination in sorghum (Ayala et al. 1994) and other species (Leal et al. 2005). In addition, antioxidant activity has been confirmed via biochemical tests (Olomieja et al. 2021), animal models (Gunawan et al. 2020), and in vitro assays with plasmid DNA (Soares et al. 2014; Maciel 2022). The modulation of the immune response by increasing the activity of killer cells (Queiroz et al. 2000) and the action against pathogenic agents, *such as Trypanosoma cruzi* (Cáceres et al. 1998) and bovine diarrhea virus (Ruffa et al. 2002b), has also been described. Bioassay-guided fractionation has shown narcotic (Zoghbi et al. 2002) and depressant activity (Gomes et al. 2008).

Antitumor activity has been studied for decades (Mata-Greenwood et al. 2001; Ruffa et al. 2002a; Taylor 2005), and the antineoplastic effect is attributed to dibenzyl trisulfide (DTS), which is considered the main lipophilic compound of the species and has been identified as an immunomodulatory component that also presents neurotoxic and cytostatic properties (De Sousa et al. 1990; Williams et al. 1997; Kinghorn et al. 1998; Rosner et al. 2001). This substance affects the microtubule-dependent cellular events of tumor cells, which are altered, inhibiting the proliferation of several types of cancer (Ruffa et al. 2002a; Williams et al. 2007). Commercial DTS (Sigma-Aldrich) can kill breast adenocarcinoma cells in culture via several mechanisms, such as cell cycle arrest in the G2 phase, the induction of cytoskeletal reorganization, and DNA fragmentation (Maciel 2022). Furthermore, DTS has been reported to have inhibitory activity against key cytochrome P450 (CYP) enzymes, which are responsible for the metabolism of foreign chemicals (including drugs and pollutants), generating free radicals that can bind to DNA, causing mutations that initiate carcinogenesis (Murray et al. 2016). Therefore, CYP inhibitors may have a chemopreventive effect on this cancer initiation process (Clarke and Irvine 2022).

Pharmacology The possibility of chemical transformations and interconversions has been extensively explored, and several synthetic products derived from *P. alliacea* polysulfides have been successfully obtained. Among these, sulfonic anhydride has shown increased biological activity, with potential for use in controlling nematode infestations and pathogenic microorganisms in crops (Williams et al. 2003).

On the basis of research on its biological activities, DTS is considered a very promising agent for the development of antitumor drugs. This class of agents has the potential to suppress the carcinogenesis process and is indicated for chemotherapy (Kim et al. 2006; An et al. 2006; Murray et al. 2016; Clarke and Irvine 2022).

The chemical synthesis of DTS was defined from the precursors but resulted in a mixture of di-/tri-/tetrasulfides, with DTS being obtained at relatively low rates. A simplified protocol, starting from a commercially available sulfur dichloride solution, allowed the synthesis of new derivatives with excellent yield (An et al. 2006), and their chemical synthesis and form of antineoplastic use were patented (Willian and Levy 2007). The isolation of this substance allowed modifications in the chemical structure, which generated new products that showed high antimicrotubule cytotoxic activity, such as fluorapacin, whose therapeutic potential stimulated the validation of an analytical method for an experimental drug (An et al. 2006; Bao et al. 2008). Recently, new DTS analogs were designed in silico with the aim of increasing specificity of the inhibition of cytochrome P450 (Salomón et al. 2022). After synthesis and in vitro validation of their inhibitory properties, these compounds may be useful for preventing cancer (Clarke and Irvine 2022).

Cytotoxicity *P. alliacea* is one of the most widely sold medicinal herbs in fairs and popular markets in the form of a dried or fresh herbal drug (Melo et al. 2009). There are dozens of herbal medicine companies selling this herb in several countries for the preparation of infusions, decoctions, or use in baths or in the form of capsules, which are easily accessible on the Internet, with various indications for use, on the basis of proven biological activities. However, herbal medicines and herbal drugs, without adequate toxicity assessment by competent authorities, can pose a risk to the health of the consumer. Research has confirmed the toxic effects on the CNS observed in popular use due to organic sulfides present in the plant's roots (Matos et al. 2011). The dose-dependent cytotoxic effect was defined by a reduction in the survival of bacteria (Maciel 2017) and yeasts (Cal et al. 2022). Furthermore, the rate of mutagenesis induced by the fractions in yeast cultures was significantly higher than the rate of spontaneous mutagenesis in these organisms (Cal et al. 2022).

The implementation of molecular studies on the cytotoxic and mutagenic potential of the extracts constitutes one of the prerequisites for the release of therapeutic extracts from this species, since beneficial and harmful effects are reported in the literature in different systems. The performance of in vitro and in vivo tests with crude extracts or substances isolated from a plant allows the evaluation of their toxicity at the cellular or molecular level. These tests can be based on the evaluation of cell survival, membrane integrity, and the presence of mitochondria or DNA and have contributed to the development and discovery of new medical and pharmaceutical products. Dose-dependent genotoxic activity has been reported on the basis of the analysis of cytogenetic alterations (Hoyos et al. 1992). The genotoxic effects of the hydroalcoholic extract were evaluated through an in vitro comet assay, which revealed significant damage to the DNA of human lymphocytes at low concentrations (50–100 mg/ml) (Silva et al. 2016). However, studies on the specific effects of *P. alliacea* extracts on DNA are lacking. Thus, experimental models using plasmid DNA have proven to be very useful in evaluating the genotoxicity or antigenotoxicity caused by various chemical and physical agents, including plant extracts, since the ability to induce DNA breaks can be assessed by changing the structural conformation of the plasmid through the electrophoretic profile compared with a control (Dantas et al. 1999; De Mattos et al. 2000). The crude ethanolic extract and its fractions (ethyl acetate and butanol) produced breaks in plasmid DNA by this model, varying directly as a function of concentration (Cal et al. 2022). Biomonitoring, after phytochemical screening, may provide support for a better understanding of the genotoxicity caused by extracts and greater reliability in the consumption of this plant, in addition to serving as a reference for new scientific studies.

9.4 Plant Tissue Culture

The process of obtaining plants in vitro is based on the totipotency of plant cells, which enables the formation of a new individual who is genetically identical to the parent plant. To this end, segments of any part of the plant, called explants, are excised, subjected to decontamination and cultivated under appropriate conditions, including phytohormones that can induce dedifferentiation and redifferentiation in the explant cells, resulting in the formation of new tissues, organs, or the entire organism. The morphogenic capacity of plant cells is characterized by competence, which is dependent on the degree of differentiation, to express receptors for different chemical and physical signals, such as growth regulators and environmental factors. On the other hand, meristematic cells are undifferentiated and naturally competent, remaining in constant division. In practice, in vitro morphogenesis modulation from plant tissues can be performed through variations in the basal composition (macronutrients, micronutrients), supplementation (growth regulators, nitrogen supplements, carbon sources, and vitamins), and consistency of the culture media (solid or liquid, kept in static conditions or under agitation) for the growth and development of specific structures. The balance of phytoregulators used can affect the differentiation and multiplication of cells, inducing certain regeneration pathways.

The plant cultures were incubated in culture chambers with automated control of the light intensity and photoperiod. Micropropagation allows vegetative propagation with high multiplication rates under aseptic conditions and in reduced spaces throughout the year without interference from seasonal variations. In addition to large-scale propagation for commercial purposes, micropropagation is also used to multiply endangered species and genotypes of interest, such as plants with agricultural, ornamental, or medicinal potential (Rout et al. 2000).

P. alliacea was propagated in vitro for the first time through the cultivation of nodal segments (Castellar et al. 2011), producing a single plant per explant. Subsequently, clonal multiplication was established through multiple shoots, and different varieties were characterized for use as matrices (Soares 2011; Soares et al. 2013). Adventitial regeneration was established through somatic embryogenesis (Cantelmo et al. 2013; Soares 2016). Undifferentiated cultures, such as callogenetic and suspension cell cultures (Castellar et al. 2011), have also been established, and the volatile components from essential oils produced in vitro and ex vitro have been characterized, revealing qualitative and quantitative differences, depending on the plant structure analyzed, confirming the usefulness of in vitro culture as a tool for the modulation of substances of pharmacological interest (Rout et al. 2000; Castellar et al. 2014). The tissue culture techniques used with the species from preexisting meristems and adventitious regeneration are summarized below.

9.4.1 Micropropagation from Preexisting Buds

The culture of isolated apical meristems requires a specialized technique for explant removal because of their small size (0.5–1 m). However, this method is the best way to produce germ-free plants because of the anatomical isolation of meristems and is very suitable for establishing field plants for in vitro culture. In general, propagation from preexisting meristems reduces the risk of somaclonal variation (Sharma et al. 2009), since regeneration occurs through the route closest to natural ontogeny, enabling rapid and efficient micropropagation, which is widely used in commercial-scale propagation.

Propagation by the apex culture technique has been applied very successfully to several medicinal species; in *Phyllanthus fraternus*, propagation rates of more than 20 shoots/explants were obtained in response to BAP and KIN (Das et al. 2011). However, culture conditions, such as the types and combinations of cytokinins with auxins (Sasidharan and Jayachitra 2017), the concentrations of mineral salts in the culture medium (Kim and Jang 2004), and the use of silver nitrate in association with growth regulators (Ozudogru et al. 2005), must be specifically evaluated. Once proliferation is induced, shoots must be individualized and recultivated with a view to elongation and subsequent rooting, a step that requires transfer to another medium, generally containing auxins. Individualization and subculturing of axillary shoots formed from primary explants continues the proliferation process. The greatest advantage of obtaining plants from preexisting meristems is the genetic fidelity between the clones and the mother plant. Shoots develop from isolated meristems apical or nodal segments, generally without the need for plant growth regulators, but can be induced to proliferate axillary buds, often by supplementing the culture medium with cytokinins, which act to break the apical dominance in nodal segments, generating multishooting.

P. alliacea plants were cloned from nodal segments (Castellar et al. 2011). However, shoot growth after cultivation in MS media (Murashige and Skoog 1962) without plant growth regulators (MS0) resulted in only one plant for each inoculated explant. However, the established micropropagation protocol, although suitable for in vitro conservation, proved to be insufficient in the short term to obtain material in sufficient quantity for phytochemical and pharmacological evaluations, which require relatively high amounts of material. Therefore, increasing the production of aerial parts through cytokinin-induced multishoots from shoot apex explants of clones is necessary. The combination of 4.4 μM BAP (benzyl amino purine) with 0.54 μM NAA (naphthalene acetic acid) resulted in approximately eight shoots per explant, depending on the origin of the matrix (Fig. 9.1a). Individual shoots

Fig. 9.1 In vitro culture of *P. alliacea*: (**a**) multishoots in response to 4.4 μM BAP and 0.54 μM NAA; (**b**) rooting of shoots in the presence of 0.6 μM IAA; (**c**) somatic embryogenesis from leaf explants; (**d**) somatic embryogenesis from root explants. Bar = 1 cm. (Source: unpublished photos of the authors)

subcultured in MS supplemented with 0.6 μM IAA (indole acetic acid) gave rise to complete plants (Fig. 9.1b). On the other hand, in the presence of the best concentration of TDZ (2.27 μM), friable callus formation occurred at the base of the shoots, making their individualization difficult (Soares 2011).

9.4.2 Adventitial Regeneration

In vitro-*generated* plants regenerated via adventitious pathways are valuable biotechnological products. The acquisition of new skills by specialized cells must be preceded by a process of dedifferentiation, followed by genetic and metabolic reprogramming that can be induced by growth regulators, among other forms, such as stress factors, gene expression regulatory proteins, and gene transcription factors. The newly formed cells subsequently take a new route, becoming committed to a specific regeneration pathway, embryogenic or organogenic (Fehér et al. 2003; Schwarz et al. 2004).

When new plants are formed by adventitious regeneration, they can arise from a direct process, from the explant cells, or indirectly, from the callus cells, which can eventually form buds or somatic embryos in response to the induction conditions. Several medicinal plant species have been propagated via adventitious regeneration aimed at the conservation of germplasm or the production of secondary metabolites through direct (Lata et al. 2013; Piątczak et al. 2014; Martín et al. 2015), indirect

(Jiang et al. 2012; Jafari et al. 2017; Hesami and Daneshvar 2018), and direct or indirect SE. In addition to micropropagation, in vitro manipulation of plant material also allows the establishment of disorganized and organized cultures, which constitute the most widely used systems with medicinal species, owing to the possibility of modulating secondary metabolism and, therefore, can be applied to *P. alliacea*, the results of which are discussed below.

9.4.3 Calli and Cell Suspensions

The establishment of nonmorphogenic calli and suspension cell cultures are systems that have proven valuable in obtaining different molecules (Bourgaud et al. 2001; Vanisree et al. 2004; Jeong et al. 2006; Balbuena et al. 2009). In vitro callus induction occurs through hormonal stimuli, mainly the presence of auxins, which modify the quiescent state of the specialized tissue to a state of intense cell proliferation that gives rise to a mass of undifferentiated or poorly differentiated cells. Calli vary in competence and morphogenetic and metabolic potential and may present different responses to different induction techniques and conditions. In addition, calli may differ in appearance, color, and degree of compaction, being classified as compact when they have densely aggregated cells or friable when their cells are loosely associated with each other, indicating their ability to initiate suspension cell cultures. Some explants or specific cell types do not form friable calli, making it impossible to culture cells in suspension, which can eventually be overcome by reducing the concentration of the solidifying agent in the medium or by successive subcultures (Slater et al. 2008).

In *P. alliacea*, the development of two types of friable calli was observed: (i) calluses formed at the basal end of shoots grown in MS0, which showed continuous growth when subcultured, and (ii) calluses formed on the surface of leaf explants grown in the presence of 22.6 μM 2,4D (2,4-dichlorophenoxyacetic acid) or 20.0 μM PIC (picloram), which presented a similar pattern, with the initial formation of small, compact white protuberances on the central vein of the leaf segments, which expanded over the surface of the explants, followed by the development of friable, nonmorphogenic white calluses throughout the subcultures (Castellar et al. 2011). Callogenesis was influenced by the type and concentration of the growth regulator and by the temperature, and among the various auxins tested, the best callus formation responses were induced by PIC and 2,4D at approximate frequencies (90–100%); however, the cultures maintained at 22.6 μM 2,4D were oxidized after some subcultures. As PIC produces more friable calluses without oxidation, it was the phytoregulator selected for the most in-depth evaluation of the callogenesis process from leaf explants (see item 3.2.3). Many pharmacologically important chemical compounds have been obtained from callus cultures (Khurshid et al. 2018).

Suspension cell cultures are established from friable, nonmorphogenic calluses, through their isolated cells or in small clusters, dispersed in a liquid medium under agitation so that the cells multiply and produce substances, which can be excreted into the medium, facilitating chemical extraction. In these cultures, agitation minimizes the formation of aggregates and sedimentation of cells, uniformizing the distribution of nutrients and diluting toxic substances excreted by the cells. Suspension cell cultures were developed in the 1990s and aroused great interest for the production of metabolites because of the possibility of large-scale production in a short period of time. They present some limitations related to genetic instability and decreased biosynthetic capacity throughout subcultures (Bajaj 1995; Patnaik et al. 1999) that have been overcome with new technologies, such as elicitation, genetic transformation, metabolic engineering, and cryopreservation, which have enabled the production and conservation of cell lines with high rates of metabolite production for the production of bioactive substances (Bourgaud et al. 2001; Aijaz et al. 2011; Valdes et al. 2020). Many advances have been achieved in recent decades from cell suspensions in bioreactors in association with cryopreservation, enabling the storage of cells and pharmacological products (Hashimoto and Azechi 1988; Schmale et al. 2006; Valdes et al. 2020).

However, in *P. alliacea*, the suspension cell culture was initiated from the two types of friable calli in liquid media, but with the same composition of salts and phytoregulators that induced the calli (MS0, MS supplemented with PIC or 2,4D), which still requires optimization before phytochemical evaluation. Initially, the cultures in the presence of PIC and 2,4D presented several aggregates (0.5–1.0 mm in diameter), which increased in quantity after subculture. The growth kinetics of the cell suspensions from the different types of calli revealed that the MS0 cultures decreased in growth after 3 weeks, reflecting the accumulation of biomass. The cells grown in the presence of PIC and 2,4D presented a lag phase of 4–5 weeks, with a moderate increase in biomass at the end of the fifth week and a stationary phase that was maintained until the end of the observation period (60 days). Cultures maintained at 22.6 μM 2,4D gradually turned brown, indicating the 2,4D-induced oxidative process previously observed in callus cultures (Castellar et al. 2011).

9.4.4 Adventitious Root Cultures

Root production is due to the differentiation of xylem cells induced by IAA, an auxin that is responsible for modulating the activity of peroxidases involved in this process (Rout et al. 2000). The mechanism of adventitious rhizogenesis was clarified in several studies with different species, which demonstrated that the increase in IAA catabolism, mediated by peroxidase, was initiated by means of H2O2 generated in the metabolism of putrescine, a very common polyamine in plants derived from arginine or ornithine (Hausman 1993; Rout et al. 2000).

The production of adventitious roots from root explants of *P. alliacea* in response to auxins at different concentrations was explored. However, the response of the increase in biomass did not occur, and individualized, elongated, and poorly branched roots were observed only in the presence of the lowest concentration of IBA (indole butyric acid) (1.02 μM). In the presence of the other plant growth regulators, regardless of the concentration tested, the roots curled intensely, generating globular structures that were dispersed in the liquid medium and were visible to the naked eye after 120 days. Direct SE was confirmed by histological analysis of the clusters (Soares 2016).

Compared with field plants, adventitious root culture systems have been researched for years (Bourgaud et al. 2001; Murthy et al. 2016) for the production of bioactive substances due to greater control of product quality (Murthy et al. 2018). However, commercial-scale production has been limited due to genetic and biochemical control problems (Murthy et al. 2021). Attempts to overcome these difficulties have been successfully carried out through genetic transformation techniques in hairy roots (Valdes et al. 2020). However, production processes closer to natural ones are preferable because of their use as food or medicine and elicitation techniques (Aijaz et al. 2011), and mass production in bioreactors shows positive results for some species (Murthy et al. 2021).

9.4.5 Somatic Embryogenesis

Somatic embryogenesis is a regeneration process that consists of the formation of embryos from somatic cells. SEs are bipolar structures with a development pattern similar to that of zygotic embryos, but they do not have suspensor tissue or endosperm, do not form teguments, do not enter dormancy, and have no connections with the tissue that originated them (Zimmerman 1993). SEs can form directly on the surface of the explant or from callus cells at different stages of differentiation (George 2008), and SE can occur through two pathways: unicellular and multicellular (Figueroa et al. 2006). Embryogenic cells have characteristics common to meristematic cells, such as small size, a high division rate, dense cytoplasm with starch grains, large nuclei, small vacuoles, and a thick cell wall (Nonohay et al. 1999; Figueroa et al. 2006). The plant genotype is the main factor in defining the potential for somatic embryogenesis, and the process by which somatic cells acquire embryogenic competence is related to reprogramming and gene expression in response to induction conditions such as osmotic stress, pH changes, thermal shocks, and phytoregulators, mainly auxins (Fehér et al. 2003; Fehér 2006).

Primary somatic embryos can initiate a repetitive multiplication process, giving rise to an unlimited number of secondary embryos generated cyclically from a single primary embryo culture. This process presents a high multiplicative rate, in addition to the independence of a new source of explants to initiate the culture. On the other hand, the conversion rate to plants tends to be insufficient for large-scale clonal production (Zheng et al. 1996; Karami et al. 2007). Knowledge of the critical factors that interfere with the process of obtaining normal seedlings is highly important in mass propagation. However, somatic embryos produced from several medicinal species can be considered model systems when conversion is dispensable, and the objective is the extraction of plant metabolites of interest, which are often produced in greater quantities than the original plant material (Szypuła et al. 2005; Murthy et al. 2023).

In *P. alliacea*, SEs are produced from the leaves (Cantelmo et al. 2013) (Fig. 9.1c) and roots (Soares 2016) of in vitro-*grown* plants (Fig. 9.1d) in response to different auxins, including PIC, 2,4D, and IBA, which induce a process of direct SE, evolving into secondary embryogenesis or conversion to plants. Regardless of the explant of origin, the somatic embryos differentiated from cells peripheral to the vascular system associated with the phloem. Most available studies on somatic embryo formation report that embryogenic responses usually originate from epidermal, subepidermal, or procambial cells (Yang et al. 2010; Rocha et al. 2016). A comparison of SE processes in this species revealed that SE production from roots occurred more quickly, and 60 days after the start of culture, SE clusters were visible to the naked eye (Soares 2016), whereas 120 days were visible from leaves (Cantelmo et al. 2013); however, embryogenic development followed a similar path. Although the initiation processes of SE present some differences, embryo formation and development involve the same genes, regardless of activation signals (Fehér et al. 2016).

With respect to conversion, the literature indicates that subcultures should be grown in media free of plant growth regulators, with low conversion rates often attributed to poor formation of apical or root meristems of the embryonic axis (Sivanesan and Jeong 2016). In fact, the conversion of SEs of *P. alliacea* from leaves and roots was very low or did not occur in MS0 media solidified with 0.2% phytagel. However, replacing the solidifying agent with 0.7% agar and adding 0.6 μM IAA promoted conversion rates ranging from 90% to 100% (Paiva 2011; Soares 2016). Efficient SE protocols with a good conversion rate to plants constitute an interesting alternative for the large-scale propagation and conservation of endangered medicinal species, such as *Wedelia calendulacea* (Sharmin et al. 2014) and *Crinum brachynema* (Kaur et al. 2024).

On the other hand, the multiplication of SEs from leaves or roots, directly in medium supplemented with PIC in the culture, established a process of increasing secondary embryogenesis, up to the fourth subculture, which decreased after this period and led to a degenerative process in the semisolid medium. Monitoring of the subcultures revealed continuity of the secondary embryogenesis process over 120 days (Cantelmo et al. 2013; Soares 2016). The protocols established for the direct and repetitive multiplication of SEs are very useful for phytochemical studies, as they make the process genetically more stable (Sivanesan and Jeong 2016), and the phytochemical analysis of embryogenic cultures of *P. alliacea* indicates the great potential of this system for the production of polysulfides (see item 7).

9.4.6 Acclimatization

In most plant germplasm research and conservation laboratories, in vitro plant production involves a series of artificial conditions in the macro- and microenvironments of the cultures, which often results in malformations that make it difficult to transfer the seedlings to ex vitro conditions. The humidity inside the flask is high, and the irradiance is very low compared with that in the natural environment. The closure of the flasks limits gas exchange, preventing the entry of CO2 and the exit of gases produced by plant metabolism, which accumulate. In addition, the composition of the culture media normally used is not specifically optimized and includes high concentrations of sugars and phytoregulators. All these factors represent stresses that can be overcome, depending on the adaptations of each species and the adjustments made to the protocols (Pospisilova et al. 1999).

In vitro, shoots and seedlings are mixotrophic in their mode of nutrition, that is, autotrophic and/or heterotrophic, depending on the moment. However, under in vitro conditions, even those with a normal phenotypic appearance do not perform photosynthesis, since the exogenous supply of sucrose and low light intensity induce heterotrophic metabolism, compromising the development of the photosynthetic apparatus. Therefore, tissue culture-generated plants are poor in chlorophyll content, and the enzymes responsible for photosynthesis are inactive or absent, resulting in thicker roots, hairs, and an underdeveloped vascular system (Hazarika et al. 2006).

Acclimatization protocols should be developed with the implementation of specific steps, aiming to minimize water stress in the initial ex vitro stages. Some substances, such as glycol, paraffin, or grease, can be used for leaf coverage. Other strategies can be implemented, still in an in vitro growth chamber, to stimulate plant hardening before transplanting, among which treatment with ABA aims to induce tolerance to dehydration, while increasing aeration, light intensity, and CO2 concentration are measures that promote an increase in the photosynthetic rate and have already resulted in impressive increases in the growth of many ornamental species (Kitaya et al. 2005; Hazarika et al. 2006). With respect to medicinal species, in vitro root proliferation is a critical step for transferring the culture to greenhouses. For large-scale production, in vitro rooting of shoots is a laborious process, but it can be induced ex vitro after treatment with auxins (Rout et al. 2000; Phulwaria et al. 2013).

The acclimatization of *P. alliacea* plants produced via micropropagation via different regenerative processes was performed with 90–100% success. The plants were rooted in vitro and selected according to the size of the roots (>1 cm). The most efficient protocol included washing the roots and keeping them in water for 10 days in the growth chamber. The flasks covered with PVC plastic film were gradually perforated during this period (Fig. 9.2a) to balance the air humidity and allow the establishment of morphophysiological adaptations. The plants were subsequently transplanted into pots containing a mixture of sand and humus (1:1) at pH 5.5 (Fig. 9.2b) and returned normally until flowering, which indicates the use of

Fig. 9.2 Acclimatization: (**a**) washing and keeping the roots in water; (**b**) transplanting to the ex vitro environment. Bar = 1 cm. (Source: unpublished photos of the authors)

this protocol for all plants of this species, which are produced in tissue culture, regardless of the regeneration route (Paiva 2011; Soares 2016).

9.5 Cryopreservation Techniques

The need to preserve biotechnological products has led to the development of specific cryopreservation protocols. The integration of cryogenic conservation with plant biotechnology programs has many advantages, particularly given the need to maintain genetic stability, totipotency, and the potential for the biosynthesis of secondary metabolites. Therefore, the establishment of cryopreservation techniques is essential, whether for the conservation of germplasms of rare and endangered species or for the in vitro storage of clones with increased production of metabolites of medicinal and pharmaceutical interest. For this reason, germplasm collections maintained through cryopreservation assume an important role in the future, providing genetic material for chemical characterization, improvement, and knowledge of secondary metabolism (Adriaansen et al. 2022). Cryopreservation is considered the only reliable technique for long-term preservation of plant germplasms because of the suspension of cell division and metabolic activities under storage conditions of the material in liquid nitrogen (LN −150 to −196 °C) (Mazur 1970). The biological materials indicated for cryopreservation are propagules, which can regenerate new organisms through tissue culture, such as cell suspensions, shoot tips, calluses, somatic or zygotic embryos, and cell suspensions, which contain large amounts of intracellular water and must be artificially dehydrated to induce tolerance to damage caused by the crystallization of intracellular water (Mazur 1984).

Success will depend on achieving a level of dehydration that results in water contents low enough to prevent crystal formation without making it impossible for cells to survive after the LN is removed (Engelmann 2011b). During storage, the plant material remains immersed in LN, taking up little space, in small volumes protected from contamination, and requiring little maintenance for a theoretically infinite period of time. Sample dehydration can be induced during temperature reduction, as in the first established (classical) techniques, by slow freezing of the material. This technique depends on the use of programmable freezers, which automatically control the temperature reduction from 1 to 10 °C/hour down to −40 °C, when the dehydrated material can be immersed in LN (Sakai 2004). Under optimal conditions, most of the freezable water will have left the cells during evaporative dehydration, making the cellular compartment highly concentrated by the accumulation of metabolic products such as sugars and amino acids, establishing amorphous or glassy conditions (Fahy et al. 1984). This technique is used for the cryopreservation of cold-climate species, which are naturally tolerant (Engelmann 2011a). Tropical species, however, do not express these mechanisms, and tolerance must be artificially induced via the use of cryoprotective substances that promote osmotic dehydration, establishing the vitreous state in samples rapidly immersed in LN. Thus, many agronomic species have been successfully cryopreserved through these modern and contemporary techniques, which are based on the vitrification process (Gonzales-Arnao et al. 2008; Lambardi et al. 2008).

The basic techniques have been optimized (desiccation, vitrification, encapsulation–dehydration and encapsulation–vitrification, drop vitrification, and cryoplate), and some innovations, such as vacuum infiltration (Nadarajan and Pritchard 2014) and cryo-mesh (Funnekotter et al. 2017), have been evaluated, increasingly aiming at practicality and speed in the simultaneous manipulation of many samples. Some of these techniques have already been successfully used in endemic species (Coelho et al. 2020). However, studies on medicinal species, especially wild ones, are lacking (Kundu et al. 2018; Gowthami et al. 2024). The main results of the cryopreservation of SEs of *P. alliacea* are described below.

9.5.1 Desiccation

Desiccation is the simplest technique, requiring no expensive equipment or cryoprotectant solutions. This method is based on physical dehydration (evaporation) by exposing the explants to laminar flow or silica gel for varying lengths of time until they reach an optimal water content. For postfreeze recovery, samples should reach a water content below 25–30% on the basis of their fresh weight (FW) (Fang et al. 2009). The optimal water content for cryopreservation, however, varies considerably depending on the species and type of material and must be determined experimentally (Popova et al. 2010).

9.5.2 Preculture Dehydration

Osmotic dehydration based on exposure to high concentrations of soluble sugars, mainly sucrose, has been used to induce the vitreous state and increase dehydration tolerance in somatic embryos prior to immersion in LN (Hoekstra et al. 2001). This method has been described for the cryopreservation of embryogenic cultures of many agronomic species, such as *Elaeis guineensis* (Dumet et al. 1993), *Daucus carota* (Tessereau et al. 1994), and *Coriandrum sativum* (Popova et al. 2010), and has been evaluated in a few medicinal species, such as the cell suspensions of *Ginkgo biloba* (Lu et al. 2009) and *Petiveria alliacea* (Pettinelli et al. 2017).

9.5.3 Vitrification

In the vitrification technique, plant material is treated with highly concentrated cryoprotectant solutions for short periods. Among the various compositions of the PVS series vitrification solutions, the most widely used is PVS2 (15% ethylene glycol + 30% glycerol + 15% DMSO + 13.7% sucrose) (Sakai et al. 1990). The standard protocol comprises several steps: pretreatment, preconditioning, preculture, osmoprotection, dehydration, cooling, heating, dilution, and recovery, which must be optimized, depending on the material, to the ideal point for vitrification during immersion in NL and which allows recovery. The vitrification technique was standardized for several explants of some medicinal plant species already used in the phytotherapeutic or pharmaceutical industry, such as *Rauvolfia serpentina* (Ray and Bhattacharya 2008), *Hypericum perforatum* (Skyba et al. 2010), *Tetraclinis articulata* (Martínez and Casas 2011), *Artemisia herba-alba* (Shatnawi 2011), *Wasabia japonica* (Maki et al. 2015), *Dioscorea opposita* (Li et al. 2009), *Catharanthus roseus* (Fatima et al. 2009), *Dendrobium nobile* (Vendrame and Faria 2011), *Anemarrhena asphodeloides* (Rong and Hua 2012a), few wild species, *Cleome rosea* (Cordeiro et al. 2015), and *Petiveria alliacea* (Pettinelli 2013).

Thus, the ESs of *P. alliacea* were treated with a loading solution (2 M glycerol and 0.4 M sucrose) (Matsumoto et al. 1994), which initiated conditioning, and then exposed to PVS2 (Sakai et al. 1990) or PVS3 (Nishizawa et al. 1993) for different periods of time. In the next step, the tubes containing the samples were immersed directly in NL. After removal, reheating was performed at 37 °C, and the vitrification solutions were replaced with an unloading solution (0.8–1.2 M sucrose) for progressive rehydration of the samples. Next, the explants were inoculated in a medium appropriate for the species and incubated in the dark for 1 week. The plants were then transferred to standard culture conditions for recovery and regeneration. Despite the lower toxicity of PVS3, which does not contain DMSO in its composition, the use of PVS2 for 15 min promoted the highest survival and recovery rates, which were between 80% and 86%; however, callus formation occurred during recovery, indicating the need for optimization (Pettinelli et al. 2017).

9.5.4 Encapsulation-Dehydration

The encapsulation of explants is a successful adaptation based on the technique of artificial seed production (Redenbaugh et al. 1991). The capsule is formed by the polymerization of calcium alginate around the explant, which is then precultured in liquid medium enriched with sucrose for 1–7 days or exposed to silica gel or laminar flow until a water content of approximately 20% (fresh weight) is reached, followed by direct immersion in NL. The capsule physically protects the explant against excessive dehydration (Engelmann 2011b), survival is usually high, and growth recovery is generally rapid and direct, without the formation of calluses. This technique has been applied to shoot tips, as well as to cell suspensions and somatic embryos of several agronomic species of temperate or tropical origin (Gonzalez-Arnao and Engelmann 2006; Engelmann et al. 2008), and is one of the most widely used techniques with medicinal plants in general, with reports of more than 60 cryopreserved species (Kundu et al. 2018).

Several studies have shown that preculture conditions, such as relatively high sucrose concentrations or cold hardening, strongly influence tolerance to freezing or dehydration. SEs of *P. alliacea* were encapsulated and dehydrated by laminar flow air for different exposure times (0–15 min), and recovery rates of up to 100% were observed (Pettinelli et al. 2017) (Fig. 9.3a, b).

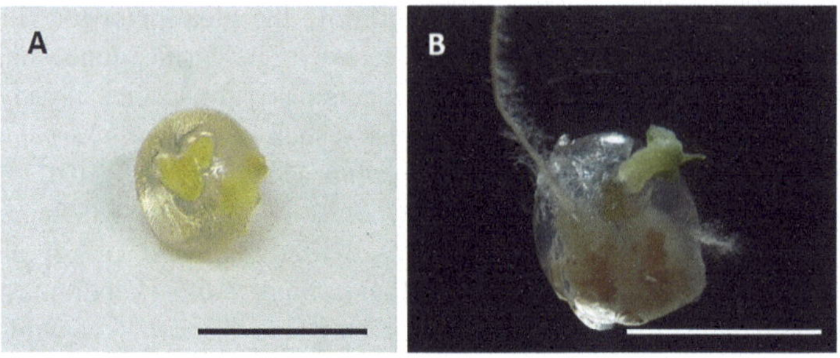

Fig. 9.3 Encapsulation–dehydration: (**a**) encapsulated before dehydration; (**b**) recovery after NL. Bar = 0.5 cm. (Source: unpublished photos of the authors)

9.5.5 Encapsulation–Vitrification

The combination of encapsulation–dehydration with vitrification has been tested to reduce the toxicity of cryoprotective solutions since explants are osmoprotected by a gradient without direct contact with vitrification solutions. The samples encapsulated in calcium alginate were subjected to the standard vitrification protocol described above. This technique has been applied to the apices of several temperate and tropical species (Sakai and Engelmann 2007). Among the medicinal species, several have been cryopreserved by these methods, such as *Dendrobium candidum* Wall. ex Lindl. (Yin and Hong 2009), *Dioscorea bulbifera* (Yin and Hong 2010), *Anemarrhena asphodeloides* (Rong and Hua 2012b), *Dioscorea opposita* (Zhao et al. 2016), and others. SEs of *P. alliacea* were encapsulated and treated with PVS2 for different periods of time (0–60 minutes), after which the capsules were immersed in LN. Embryo recovery was possible at all treatment times, reaching high regeneration rates (76–100%) (Pettinelli et al. 2017).

9.5.6 Cryoplate

This technique is adapted from drop vitrification (Panis et al. 2005), in which the explants are dripped, together with the vitrification solution, onto strips of aluminum foil directly immersed in LN, ensuring high-speed cooling and reheating (Sakai and Engelmann 2007). In the cryoplate technique (Sekizawa et al. 2011; Niino et al. 2013), the explants are placed in wells containing aluminum cryoplates, where a calcium alginate solution is dripped, promoting explant adhesion. Thus, the plate can be immersed in the loading solution and then exposed to laminar flow (D-cryoplate) for physical dehydration or immersed in vitrification solution (V-cryoplate) for osmotic dehydration before immersion in NL. The main advantage of these techniques is their speed, which is due to the ease of handling several explants at the same time (Fig. 9.4).

The D-cryoplate technique has already been tested on different explants for the conservation of germplasms of several agronomic and ornamental species, such as

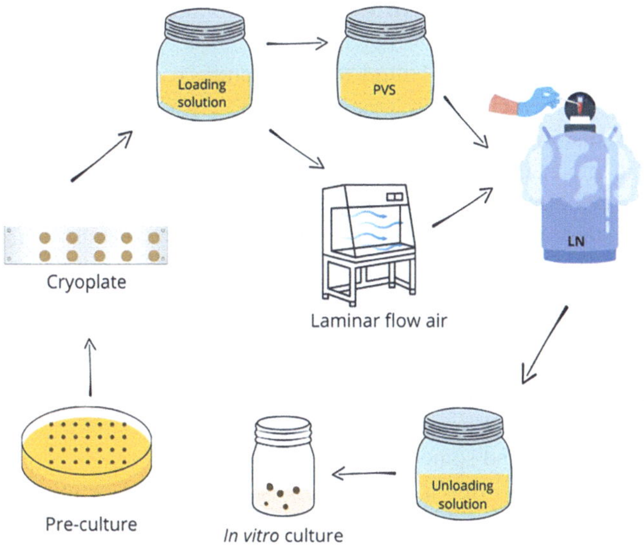

Fig. 9.4 Cryopreservation of somatic embryos of *Petiveria alliacea*, using cryoplate techniques. (Source: unpublished diagrams of authors)

Solanum tuberosum (Yamamoto et al. 2015), *Diospyros kaki* Thunb. (Matsumoto et al. 2015) and *Phoenix dactylifera* (Salma et al. 2014), medicinalis *Lomandra longifolia* (Niino et al. 2013), *Juncus decipiens* (Niino et al. 2014), and *Petiveria alliacea* (Pettinelli 2018; Goncalves et al. 2021). Similarly, the V-cryoplate technique has already been successfully applied to *Tanacetum cinerariifolium* (Yamamoto et al. 2011), *Dianthus caryophyllus* L. (Sekizawa et al. 2011), *Fragaria* × *ananassa* Duch. (Yamamoto et al. 2012a), *Saccharum officinarum* (Rafique et al. 2015), *Solanum tuberosum* (Yamamoto et al. 2015), *Phoenix dactylifera* (Salma et al. 2014), and a few medicinal species, such as *Mentha spicata* (Yamamoto et al. 2012b), *Cleome rosea* (Cordeiro et al. 2015), *Lomandra longifolia* (Niino et al. 2013), and *Petiveria alliacea* (Pettinelli et al. 2020).

Some species were cryopreserved via both cryoplate techniques, such as *P. alliacea*, whose SEs cryopreserved by D-cryoplate had high viability, regardless of the original explant (leaf or root). Dehydration under laminar flow was evaluated for various time periods (0–140 min), achieving 85–93% SE survival, with recovery of the multiplicative potential, generating 19 SEs/SE after 90–120 min of exposure (Pettinelli et al. 2020; Goncalves et al. 2021). With V-cryoplate, root ES were treated with loading solution (20 min), followed by PVS2 at different exposure times (0–60 min), with the highest viability after 15 min (95–100%). This technique promoted an increase in the recovery of SEs after cryopreservation, ranging from approximately 13 SEs/SE in classical vitrification (Pettinelli et al. 2017) to approximately 22 SEs/SE, in addition to multiplication by secondary embryogenesis in culture (Pettinelli 2018; Pettinelli et al. 2020).

9.5.7 Cryotolerance Assessment

The ability of plants to withstand ultralow temperatures essentially depends on their genetic makeup, which confers tolerance permanently, or in response to environmental stimuli that induce acclimatization through anatomical, physiological, biochemical, and molecular processes. At the cellular level, when the temperature decreases, the cell freezes from the outside up to the intracellular medium, where it forms ice crystals, which damage the membrane. The constituent polar lipids undergo a transition from a fluid phase to a gel phase, which causes the extravasation of solutes, interrupting the organization of the membrane (Pearce 2004).

In species that have tolerance mechanisms, ultralow temperatures also cool the cell from the outside, but the cytoplasm solidifies without the formation of crystals. This phenomenon, known as vitrification, occurs during dehydration caused by a decrease in temperature. When water is lost, the cytoplasm becomes concentrated by the solutes present, and some species even increase the synthesis of sugars and amino acids, making the cytoplasm so viscous that it solidifies without forming ice crystals. This process occurs in many cold-climate woody species but not in tropical or subtropical herbaceous plants, which constitute a large part of the agronomic and most medicinal species, in which the cellular dehydration induced by extracellular freezing is the main cause of death (Pearce 2004).

The need to preserve the germplasm of species of economic interest has driven research aimed at preserving valuable genetic resources for the development of society, and many cryopreservation protocols have been established in recent decades on the basis of artificial vitrification. These protocols require optimization on the basis of differences in the water content of the materials. The recovery of the explants, on the basis of the evaluation of the lesions caused by the dehydration stages, will define the necessary adjustments. The quantification of plasmolysis during these stages has been applied to different explants of cryopreserved agronomic species, such as *Dioscorea alata* (Barraco et al. 2014), *Rubia akane* (Salma et al. 2014), *Passiflora pohlii* (Simão et al. 2018), and many reports about the effects of osmotic stress, specifically in somatic embryos, such as *Citrus* spp. (González-Arnao et al. 2003), *Musa* spp. (Georget et al. 2009), *Picea* spp. (Vondráková et al. 2010), *Theobroma cacao* (Fang and Wetten 2011), *Bactris gasipaes* (Heringer et al. 2013), and *Olea europaea* (Bradaï et al. 2017).

In *P. alliacea*, histological evaluation of SEs subjected to cryopreservation via the V-cryoplate technique revealed the effects of successive steps of the protocol compared with those of an untreated embryo (Fig. 9.5a), especially the occurrence of plasmolysis after treatment with loading solutions (Fig. 9.5b) and PVS2 (Fig. 9.5c), since in the case of extensive dehydration, generalized plasmolysis occurs in different cell types and can impair recovery (Pettinelli et al. 2020). The process of cytoplasmic vacuolization after osmotic dehydration has been described in other species, such as *Rubia akane* (Yil et al. 2012) and *Bactris gasipaes* (Heringer et al. 2013), and can generate severe dehydration, as reported in *Artocarpus heterophyllus* (Wesley-Smith et al. 2001).

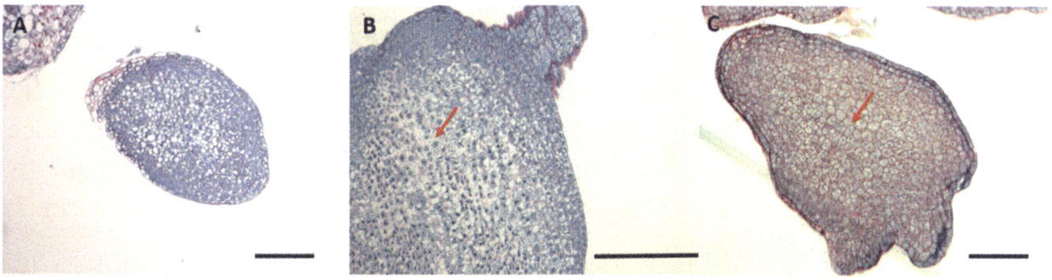

Fig. 9.5 Histological evaluation of SEs cryopreserved by V-cryoplate: (**a**) root ES, untreated; (**b**) ES after dehydration by loading solution, red arrow indicating the plasmolyzed cell; (**c**) ES recovered after dehydration and immersion in NL, red arrow indicating the recovered cell. Bar = 100 μm. (Source: unpublished photos of the authors)

The starch accumulation observed in dehydrated cells may be related to the energy reserves required for the development of SE (Carvalho et al. 2013) or to the increase in the intracellular sucrose concentration during the osmotic dehydration process (Godbole et al. 2004; Zhu et al. 2006). However, high survival rates were detected soon after rewarming, although low levels of plasmolysis were still detected after 30 days in culture. The cellular integrity observed in the different cell types throughout the recovery process demonstrated the importance of the vitrification solution to protect and induce dehydration tolerance, which is indispensable for developing an efficient protocol. Compared with conventional techniques, the use of the V-cryoplate technique made it possible to optimize the recovery of *P. alliacea* SEs. Furthermore, the use of cryoprotectant solutions did not cause irreversible damage to the cells constituting the SE, as demonstrated by the histological evaluation performed in this study (Pettinelli et al. 2020).

9.6 Monitoring Genetic Stability

In vitro plant conservation has been considered the best alternative for preserving medicinal species, enabling the multiplication and use of specific genotypes. However, in vitro culture protocols can generate genetic variations capable of modifying the phytochemical profile; in addition, monitoring variations can help identify new variants useful in biotechnological processes (Mosa et al. 2023; Cornescu et al. 2023). Furthermore, genetic instability due to cellular dedifferentiation can generate somaclonal variations (Larkin and Sowcroft 1981) that can restrict the use of regenerated plants. With respect to cryopreservation, the possibility of genetic variations occurring during the storage period in LN is considered null (Sakai 2004); however, plant recovery depends on tissue culture, which could generate variability (Matsumoto 2017). Therefore, to identify the genetic stability of medicinal plant clones obtained after propagation or in vitro conservation of a genotype, methods to evaluate somaclonal variation are essential, and the use of several different techniques to evaluate the phenotypic, cytogenetic, phytochemical, and genetic/epigenetic profiles of in vitro-regenerated plants is recommended (Cornescu et al. 2023).

However, the choice of monitoring method depends on several factors related to the objective, available time, laboratory infrastructure, and assessment of the advantages and limitations of each technique. In the case of medicinal plants, interest lies in the production of bioactive secondary metabolites; therefore, phytochemical analysis of the regenerants is essential and was performed to complement the in vitro production and conservation protocols of *Petiveria alliacea* (see item 7), together with molecular analysis.

Molecular techniques allow for greater genome tracking and are based on the detection of altered DNA segments, also called polymorphisms, which are characterized as molecular markers. Many types of molecular markers have been developed in recent decades; however, PCR-based methods with arbitrary primers, such as random amplified polymorphic DNA (RAPD) (Karp et al. 1997), are the most common. ISSR (intersimple sequence repeat) and SCOT (start codon targeted) methods have been the most commonly used methods for genetic monitoring of clones produced or preserved in vitro.

Instability monitoring in *P. alliacea* clones was based on RAPD, a highly polymorphic marker that allows the detection of minimal (punctual) variations between clones through visual comparison of the presence or absence of bands between samples after agarose gel electrophoresis. Although several reports indicate a low level of polymorphism detection (Cornescu et al. 2023), most studies that indicate a monomorphic profile compare clones regenerated from preexisting meristems with each other and with the mother plant to verify genetic stability with optimized micropropagation protocols, generally without the use of phytoregulators, or in the presence of low concentrations. However, RAPD has proven to be a useful tool for monitoring genetic stability among plants originating from adventitious regeneration, maintained for many subculture cycles, or subjected to high concentrations of phytoregulators (Rumińska and Mellem 2017; Bradaï et al. 2017; Mamdouh et al. 2021; Kanjanaphachoat et al. 2020; Tikendra et al. 2021). In addition to its practicality, speed, and ease of automation, the protocol requires minimal amounts of DNA, which becomes a fundamental factor when a single seedling corresponds to a DNA sample.

The optimization of the RAPD technique depends on the prior identification of polymorphic primers, which are selected from the numerous sequences available in the literature, and a broad genome evaluation, which varies depending on the number of loci accessed, showing consistent patterns that are independent of the age of the plant. For *P. alliacea*, the technique was used successfully. The ability of meristematic explants of *P. alliacea* to produce shoots, even in the absence of phytoregulators, is particularly important for maintaining the genetic stability of the material propagated and preserved in vitro (Castellar et al. 2011). However, as described for other medicinal species produced biotechnologically (Chinnadurai et al. 2019; Murthy et al. 2021), the need to increase the amount of clonal materials for phytochemical studies and biotechnological production indicates the use of adventitious regeneration protocols with high multiplication rates, which require the use of phytoregulators.

Therefore, the genetic stability of *P. alliacea* plants obtained in vitro was evaluated via RAPD markers via a screening of 89 primers sequenced from the literature, some of which have been used in phylogenetic studies of the Phytolaccaceae family (Semagn et al. 2001). A standard amplification program with predetermined cycling was adopted (Mançano et al. 2019), and the products were analyzed via visual comparison with DNA ladder (1 kb) after electrophoresis in a 1.4% agarose gel and fluorescence staining. Polymorphisms were determined by the change in the banding pattern (presence or absence), and the reproducibility of the method was confirmed with two or three replicates (Soares 2016).

Clones from multibudding versus cytokinins, as well as clones maintained in the absence of phytoregulators but originating from SE, were evaluated among themselves (Fig. 9.6a) and by comparison with the corresponding mother plants (Fig. 9.6b). The selected markers were suitable for monitoring clones from multibudding and somatic embryogenesis, in addition to being used in monitoring materials from cryopreservation. A high polymorphism index (22–57%) (Table 9.1) was observed among clones produced in the presence of phytoregulators, but no polymorphisms were detected among plants originating from somatic embryos before (Fig. 9.6c) or after cryopreservation (Fig. 9.6d). Thus, the use of RAPD markers was appropriate, constituting a practical tool for controlling the level of variability induced by in vitro culture.

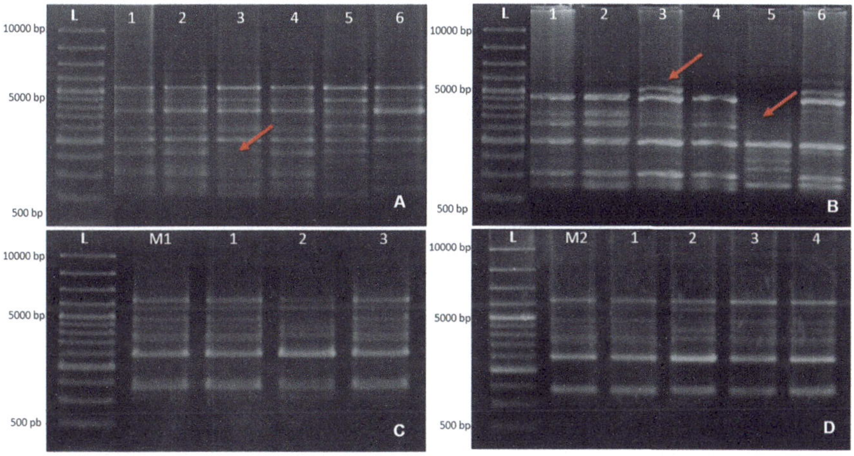

Fig. 9.6 RAPD amplification pattern in *P. alliacea* clones: (**a**) Primer 59 in BAP-induced multishoots; (**b**) Primer 49 in BAP-induced multishoots; (**c**) Primer 88 in SEs-converted clones; (**d**) Primer 88 in SEs-converted clones. L = ladder 1 kb. M = mother plant. Arrows indicate polymorphisms. (Source: unpublished photos of the authors)

Table 9.1 RAPD banding pattern in *P. alliacea* clones in vitro with selected primers

NBV[a] code	Nucleotide sequence	Bands (N°)	Polimosphyc bands (N°)	Polimosphysm (%)
4	GAGCTC GTGT	13	0	–
5	GCGGCTGGAG	6	2	33
9	CGACCAGAGC	10	4	40
11	CTCGGGTGGG	7	4	57
23	AGCCAGGCTG	7	3	43
41	ACCGCGAAGG	14	0	–
43	CCCGGCATAA	11	0	–
45	ACGCCCAGGT	15	0	–
46	CCACACTACC	11	0	–
49	CAGCCTACCA	12	3	25
51	GAGAGCCAAC	10	0	–
53	GTCAGAGTCC	10	0	–
55	ACGGGCCAGT	11	6	54
59	ACAGCCCCCA	9	2	22
61	AATCGGGCTG	10	0	–
64	ACCCCCGAAG	10	0	–
65	GGCACCCTTAC	10	0	–
68	ACGGAAGTGG	12	0	–
69	CAAAGGGCGG	11	0	–
72	AACGGGCGTC	14	0	–
76	TGCCAAGAGG	12	0	–
77	CTGAAGCGCA	12	0	–
78	GAGAGGCTCC	10	0	–
79	CTGGTGCTGA	11	0	–
80	GACAGTCCCT	10	0	–
88	GGGTAACGCC	9	2	22
89	GTAGACCCGT	9	4	44

[a]NBV = Plant Biotechnology Nucleus

9.7 Phytochemical Evaluation

The phytochemical profile was used to evaluate the composition of the essential oils of the clones produced and subcultured in vitro without phytoregulators. The oil was extracted by SDE (simultaneous distillation and extraction) and SPME (solid-phase microextraction), and a comparative analysis between field and in vitro plants via gas chromatography–mass spectrometry (GC–MS) revealed that the essential oil was composed mainly of benzaldehyde (48.3%) and dibenzyl disulfide (23.3%). By SDE, the only substance found in large quantities and in all in vitro and ex vitro structures was benzaldehyde. By SPME, volatile sulfur compounds responsible for garlic odor, such as dimethylsulfide, ethylenesulfide, 2,3-dimethyl-thiirane, benzyl-methyldisulfide, 2,3-dimethyl-thiirane, and benzylmethyldisulfide, were identified in the essential oils of in vitro and ex vitro plants. The in vitro-generated plants presented lower diversity; however, substances exclusive to this organ (p-cymene and trans-vertrocitral C) were detected in the leaves, whereas other substances were detected only in the roots of the in vitro-*generated* plants (dimethylsulfide, cymenene, and menthol). The extraction techniques used allowed characterization of small samples, such as those obtained in tissue culture, and the comparative analysis of essential oils extracted from leaves and roots by SPME confirmed greater chemical diversity in the roots (Castellar et al. 2014) (Table 9.2). In addition to the evaluation of essential oils, the phytochemical profile of crude hydroalcoholic extracts of fresh leaves was also comparatively evaluated via GC–MS (Table 9.3).

By comparing the profiles obtained with ethanolic extracts of leaves and roots in vitro with those obtained with three commercial standards, namely, MDS, DDS, and DTS (Sigma-Aldrich), through HPLC-DAD-UV-254 nm (high-performance liquid chromatography coupled with an iodine detector), plants that were cloned and propagated in response to cytokines were evaluated only for polysulfide production. Three different types of polysulfides were identified (MDS, DDS, DTS) in the leaf extracts of the plants that multiplied in response to BAP, but in the root samples, the DMS and DDS were identified, and only the DDS was identified in all the samples.

Table 9.2 Substances identified in the essential oil from in vitro and ex vitro plants of *P. alliacea*

	Leaves		Root	
Substance	In vitro	Ex vitro	In vitro	Ex vitro
Benzaldehyde	12.6	12.8	34.0	55.1
2-Pentylfurane	–	–	t	–
Decane	–	–	t	–
Benzyl alcohol	3.0	–	t	t
Undecane	13.8	t	3.2	5.1
Dodecane	2.4	2.8	7.0	1.3
Tridecane	1.7	t	4.4	1.7
Guaiacol < *p*-vinyl->	13.6	24.3	–	–
β-Damascenone	–	t	–	–
Tetradecane	t	t	1.3	t
Pentadecane	–	–	29.4	7.6
Elemol	–	–	–	t
Spathulenol	–	–	–	5.0
Caryophyllene oxide	–	–	–	2.7
Hexadecene	–	–	–	–
epi-α-Muurolol	–	–	–	t
α-Muurolol	–	–	–	t
α-Cadinol	–	–	–	2.4
14-Hydroxy-α-humulene	–	–	–	t
Isobicyclogermacrenal	–	–	–	t
β-Acoradienol	–	–	–	t
β-Bisabolen-12-ol	–	–	–	t
Heneicosane	–	32.4	–	–
Tetracosane	–	–	1.5	–

Table 9.3 Comparison of substances identified after GC-MS of the crude extract of plant leaves in vitro with the field matrix of *P. alliacea*

Substances	T.R. (min)	Leaves	
		In vitro	Ex vitro
Pentanoic acid	2.816	+	–
Propane-1,2,3-trio	2.832	+	+
Monoglyceride acid	2.841	+	+
Butanoic acid	3.576	+	–
Furan	4.210	+	–
Butyric anhydride	7.835	+	–
Tetradecanoic acid	13.827	+	–
Hexadecanoic acid	13.846	+	–
Benzaldehyde	15.264	–	+
Phytol	15.620	+	+
Hexadecanoic acid	19.771	+	–
Octadecanoic acid	21.483	+	+
Skeleton	22.652	+	+
Vitamin E	25.276	+	+
Stigmasterol	26.566	+	–
Cholestanol	27.294	+	+

OBS. Hydroalcoholic extracts from fresh leaves

The phytochemical profiles of embryogenic culture systems were previously evaluated through GC–MS, which identified the presence of DTS in embryogenic calli (Webster et al. 2008). Therefore, ethanolic extracts of somatic root embryos were also monitored via HPLC-DAD-UV 254 nm to determine the presence of polysulfides using the commercial markers mentioned above. Signals suggestive of DMS and DDS were detected in the extracts of leaf and root embryos. In addition, the analyses were performed before and after the cryopreservation process and revealed a signal compatible with DTS in the extracts of only dehydrated embryos (Fig. 9.7a) and signals compatible with DDS and DTS in the extracts of cryopreserved embryos (Fig. 9.7b) (Pettinelli 2018). However, the accurate identification of these polysulfides requires confirmation by HPLC-MS, since the concentration of the constituents in the sample can alter the absorption spectra at different wavelengths, although the retention times coincide with the standards, by the established methods. Polysulfide research has also been performed on plants converted from cryopreserved somatic embryos, and DTS was identified after changing the UV from 250 nm to 300 and 340 nm, indicating that the wavelength strongly influences the detection of DTS (Soares 2016).

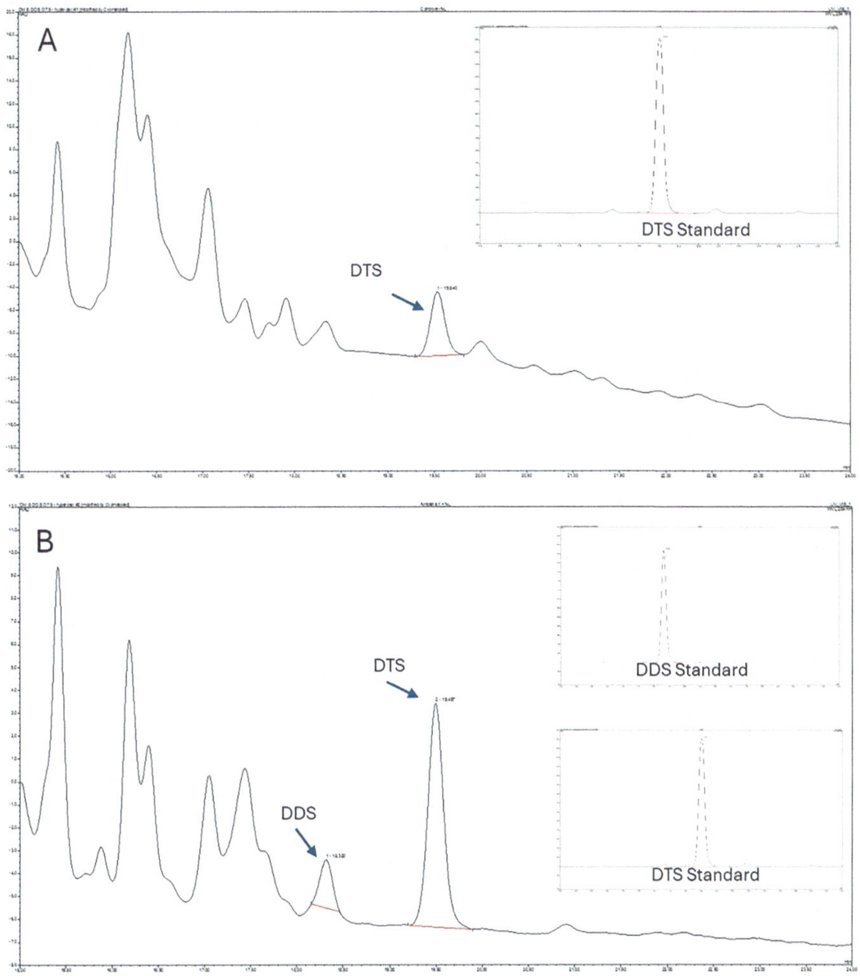

Fig. 9.7 Chromatogram obtained by HPLC-DAD-UV (254 nm) of the extract of somatic embryos dehydrated by the D-cryoplate technique. (**a**) not immersed in LN identification of DTS; (**b**) immersed in LN with identification of DDS and DTS extract obtained by absolute alcohol (100%-v/v). (Source: unpublished photos of the authors)

9.8 Progress and Prospects

The in vitro conservation of *P. alliacea* genetic resources was carried out through academic research on the basis of the biological nature and economic importance of the species, depending on the availability of the available infrastructure and the practicality of the chosen methods. Thus, the conservation program included the botanical characterization of different varieties, the micropropagation of different matrices through different regenerative pathways, the cryopreservation of ES, and phytochemical and cytotoxicity evaluations of field and in vitro plants, in addition to the genetic monitoring of plants produced and preserved in vitro. However, each stage opened questions for new avenues of investigation, mainly in relation to the increase in the scale of production and conservation of in vitro plants and the production and quantification of DTS, from in vitro cultivation, in addition to the deepening of the basic requirements for the production of other substances of interest, which can help to significantly advance the knowledge about this species with enormous pharmacological potential for the treatment of numerous human and animal pathological conditions.

Competing Interests The authors have no relevant financial or nonfinancial interests to disclose.

References

Adriaansen J, Stanton J, Schaut W, Bowden R (2022) Compliance and cost control for cryopreservation of cellular starting materials: An industry perspective. Cryotherapy 24:750–753

Aijaz A, Jain S, Hariharan AG (2011) Effect of elicitation on the production of phyto-constituents through plant tissue culture technique—a review. Int J Drug Discov Herb Res 1(2):84–90

Akintan MO, Akinneye JO (2020) Fumigant toxicity and phytochemical analysis of *Petiveria alliacea* (Linneaus) leaf and root bark oil on adult *Culex quinquefasciatus*. Bull Nat Res Centre 44:129

Albuquerque UP, Melo JG, Medeiros MF, Menezes IR, Moura GJ, El-Deir ACA, Alves RRN, Medeiros PM, Araujo TAS, Ramos MA, Silva RR, Almeida AL, Almeida CFC (2012) Natural Products from Ethno directed Studies: revisiting the ethnobiology of the Zombie Poison. Evid-Based Compl Alt Med 2012 202508

Alves TC, Rodrigues E, Lago JHG, Prado CM, Girardi CEN, Hipólide DC (2019) *Petiveria alliacea*, a plant used in Afro-Brazilian smoke rituals, triggers pulmonary inflammation in rats. Rev Bras 29:656–664

An H, Zhu J, Wang X, Xu X (2006) Synthesis and anti-tumor evaluation of new trisulfide derivatives. Bioorg Med Chem Lett 16(18):4826–4829

Ayala JR, Cruz AM, Miranda Z (1994) Effects of the aqueous extract of *Canavalia ensiformis*, *Bursera graveolens*, *Petiveria alliacea* and Parthenium hysterophorus on *Sorghum* germination – Technical Note - Cuban. J Agri Sci 28:371–373

Bajaj YPS (1995) Cryopreservation of plants germplasm I. In: Bajaj YPS (ed) Biotechnology in agriculture and forestry, vol 32. Springer-Verlag, Berlin Heidelberg, pp 419–434

Balbuena TS, Santa-Catarina C, Silveira V, Kato MJ, Floh EIS (2009) *In vitro* morphogenesis and cell suspension culture establishment in *Piper solmsianum* C. DC. (Piperaceae). Acta Bot Bras 23(1):274–281

Bao Y, He Y, Xu X, Mo X, Xu X, Wang X, An H (2008) Quantitative determination of anti-tumor agent bis(4-fluorobenzyl)trisulfide, fluorapacin and its pharmaceutical preparation by high-performance liquid chromatography. J Pharm Biomed Anal 46:206–210

Barraco G, Sylvestre S, Colin M, Escoute J, Lartaud M, Verdeil JL (2014) Histocytological analysis of yam (*Dioscorea alata*) shoot tips cryopreserved by encapsulation-dehydration. Protoplasma 251:177–189

Benevides PJC, Young MCM, Giesbrecht AM, Roque NF, Bolzani VDS (2001) Antifungal polysulphides from *Petiveria alliacea* L. Phytochemistry 57:743–747

Bourgaud F, Gravot A, Miseli E, Gontie E (2001) Production of plant secondary metabolites: a historical perspective. Plant Sci 161:839–851

Bradaï F, Almagro-Bastante J, Sánchez-Romero C (2017) Cryopreservation of olive somatic embryos using the droplet-vitrification method: The importance of explant culture conditions. Sci Hortic 218:14–22

Cáceres A, Lopes B, Gonzales S, Berger I, Tada I, Maki J (1998) Plants used in Guatemala for the treatment of the protozoal infections. I. Screening of activity to bacteria, fungi and American trypanosomes of 13 native plants. J Ethnopharmacol 62:195–202

Cal BBF, Araújo LBN, Nunes BM, Silva CR, Oliveira MBN, Soares BO, Leitão AAC, Pádula M, Nascimento D, Chaves DSA, Gagliardi RF, Dantas FJS (2022) Cytotoxicity of extracts from *Petiveria alliacea* leaves on Yeast. Plan Theory 11:3263

Camargo MT (2007) The weapon of the slaves against their masters. Rev Pós Ciênc Soc 4:31–42

Cantelmo L, Soares BO, Rocha LP, Pettinelli JA, Calado CH, Mansur E, Castellar A, Gagliardi RF (2013) Repetitive somatic embryogenesis from leaves of the medicinal plant *Petiveria alliacea* L. Plant Cell Tissue Organ Cult 115:385–393

Carvalho MA, Paiva R, Alves E, Nogueira RC, Stein VC, Castro EM, Paiva PDO, Vargas DP (2013) Morphogenetic potential of native passion fruit (*Passiflora gibertii* N. E. Brown.) calli. Braz J Bot 36:141–151

Castellar A, Gagliardi RF, Mansur E (2011) *In vitro* propagation and establishment of callus and cell suspension cultures of *Petiveria alliacea* L., a valuable medicinal plant. J Med Plant Res 5(7):1113–1120

Castellar A, Gagliardi RF, Mansur E, Bizzo H, Souza AM, Leitão SG (2014) Volatile constituents from *in vitro* and *ex vitro* plants of *Petiveria alliacea* L. J Essent Oil Res 26:19–23

Chen SL, Yu H, Luo HM, Wu Q, Li CF, Steinmetz A (2016) Conservation and sustainable use of medicinal plants: problems, progress, and prospects. Chin Med 11:37

Chinnadurai V, Viswanathan P, Kalimuthu K, Vanitha A, Ranjitha V, Pugazhendhi A (2019) Comparative studies of phytochemical analysis and pharmacological activities of wild and micropropagated plant ethanol extracts of *Manihot esculenta*. Biocatal Agric Biotechnol 19:101166

Clarke N, Irvine W (2022) In silico design and SAR study of dibenzyl trisulfide analogues for improved CYP1A1 Inhibition. Chemistryopen 11(5):e202200016

Coelho N, Gonçalves S, Romano A (2020) Endemic plant species conservation: biotechnological approaches. Plan Theory 9(3):345

Conceição BC, Silva TA, Pantoja LVPS, Luz DA, Cardoso EKS, Reis LDS, Silva MCR, Kussler MS, Maia CSF, Júnior EAF (2023) Amazonian plants: a global bibliometric approach to *Petiveria alliacea* L. pharmacological and toxicological properties. Plan Theory 12:3343

Convention on Biological Diversity, Rio de Janeiro (1992). Available at https://www.cbd.int/convention/text

Cordeiro LS, Simões-Gurgel C, Albarello N (2015) Multiplication and cryopreservation of adventitious roots of *Cleome rosea*. In Vitro Cell Dev Biol - Plant 51:249–257

Cornescu GD, Constantin N, Pojoga DM, Nicuta D, Simon-Gruita A (2023) Somaclonal variation—advantage or disadvantage in micropropagation of the medicinal plants. Int J Mol Sci 24(1):838

Cronquist A (1981) An integrated system of classification of flowering plants. Columbia University, New York, p 1262

Dantas FJS, Moraes MO, Mattos JCP, Bezerra RJAC, Carvalho EF, Bernardo-Filho M, Caldeira-de-Araujo A (1999) Stannous chloride mediates single strand breaks in plasmid DNA through reactive oxygen species formation. Toxicol Lett 110:129–136

Das M, Mishra PK, Mohapatra D (2011) Mass propagation via shoot tip culture and detection of genetic variability of *Saccharum officinarum* clones using biochemical markers. Asian J Biotechnol 3:378–387

De Mattos JCP, Dantas FJS, Bezerra RJAC, Bernardo-Filho M, Cabral-Neto JB, Lage CAS, Leitão AC, Caldeira-de-Araujo A (2000) Damage induced by stannous chloride in plasmid DNA. Toxicol Lett 116:159–163

De Sousa JR, Demuner AJ, Pinheiro JA, Bretmair E, Cassels BK (1990) Dibenzyl trisulphide and trans-N-methyl-4-methoxyproline from *Petiveria alliacea*. Phytochemistry 29:3653–3655

Delle-Monache F, Suarez LEC (1992) 6-c-formil and 6-c-hydroxymethyl flavanones from *Petiveria alliacea*. Phytochemistry 31:2481–2482

Delle-Monache F, Menichini F, Suarez LEC (1996) From *Petiveria alliacea*: II further flavonoids and triterpenes. Gazz Chim Ital 126:275–278

Di Stasi LC, Costa M, Mendacolli LJ, Kirizawa M, Gomes C, Trolin G (1998) Screening in mice of some medicinal plants used for analgesic purposes in the state of São Paulo. J Ethnopharmacol 2(3):205–208

Duarte MR, Lopes JF (2005) Leaf and stem morphoanatomy of *Petiveria alliacea*. Fitoterapia 76:599–607

Dumet D, Engelmann F, Chabrillange N, Duval Y (1993) Cryopreservation of oil palm (*Elaeis guineensis* Jacq.) somatic embryos involving a desiccation step. Plant Cell Rep 12:352–355

Ebert AW, Engels MMJ (2020) Plant biodiversity and genetic resources matter! Plan Theory 9:1706

Engelmann F (1997) In vitro conservation methods. In: Biotechnology in agriculture series. CAB International, Wallingford, pp 119–162

Engelmann F (2011a) Use of biotechnologies for the conservation of plant biodiversity. In Vitro Cell Dev Biol - Plant 47:5–16

Engelmann F (2011b) Encapsulation-dehydration for cryopreservation: past, present and future. Acta Hortic 908:165–171

Engelmann F, Arnao MTG, Wu Y, Escobar R (2008) Development of encapsulation dehydration. In: Plant cryopreservation: a practical guide. Springer Science, Corvallis, pp 59–68

Fahy GM, MacFarlane DR, Angell CA, Meryman HT (1984) Vitrification as an approach to cryopreservation. Cryobiology 21:407–426

Fang JY, Wetten A (2011) Importance of structural integrity of somatic embryos for long-term cryopreservation of cocoa (*Theobroma cacao* L.) germplasm. Afr J Agric Res 6:3954–3961

Fang JY, Sacandé M, Pritchard H, Wetten A (2009) Influence of freezable/non-freezable water and sucrose on the viability of *Theobroma cacao* somatic embryos following desiccation and freezing. Plant Cell Rep 28:883–889

Fatima S, Mujib A, Nasim SA, Siddiqui ZH (2009) Cryopreservation of embryogenic cell suspensions of *Catharanthus roseus* L. (G) Don. Plant Cell Tissue Organ Cult 98:1–9

Fehér A (2006) Why somatic plant cells start to form embryos? In: Mujib A, Šamaj J (eds) Somatic embryogenesis Serie plant cell monographs. Springer, Berlin/Heidelberg, pp 85–101

Fehér A, Pasternak TP, Dudits D (2003) Transition of somatic plant cells to an embryogenic state. Plant Cell Tissue Organ Cult 74:201–228

Fehér A, Bernula D, Gémes K (2016) The many ways of somatic embryo initiation. In: Loyola-Vargas V, Ochoa-Alejo N (eds) Somatic embryogenesis: fundamental aspects and applications. Springer, Cham, pp 23–37

Feng PC, Haynes LJ, Magnus KE, Plimmer JR (1964) Further pharmacological screening of some West Indian medicinal plants. J Pharm Pharmacol 16:115

Figueroa FRQ, Herrera RR, Avalos RMG, Vargas VML (2006) Embryo production through somatic embryogenesis can be used to study cell differentiation in plants. Plant Cell Tissue Organ Cult 86:285–301

Funnekotter B, Bunn E, Mancera RL (2017) Cryo-mesh: a simple alternative cryopreservation protocol. Cryo-Lett 38:155–159

Garcia D, Ribeiro C, Santos JFL, Soares JAR, A RL, Santos TSD, Rodrigues E (2016) Smokers with possible anxiolytic effects used in the center of umbanda Caboclo Ubirajara and Exú Ventania, Diadema, SP: an ethnopharmacological study. Ethnos 1:1–12

George EF (2008) Plant tissue culture procedure – background. In: Goerge EF, Hal MA, Klerk GJD (eds) Plant propagation by tissue culture, vol 1, 3rd edn. Springer, The Netherlands, pp 1–28

Georget F, Engelmann F, Domergue R, Côte F (2009) Morpho-histological study of banana (*musa* spp. cv. Grande Naine [AAA]) cell suspensions during cryopreservation and regeneration. Cryo-Lett 30:398–407

Godbole S, Sood A, Sharma M, Nagar PK, Ahuja PS (2004) Starch deposition and amylase accumulation during somatic embryogenesis in bamboo (*Dendrocalamus hamiltonii*). J Plant Physiol 161:245–248

Gomes PB, Noronha EC, Melo CTV, Bezerra JNS, Neto MA, Lino CS, Vasconcelos SMM, Viana GSB, Sousa FCF (2008) Central effects of isolated fraction from the root of *Petiveria alliacea* L. (tipi) in mice. J Ethnopharmacol 120:209–214

Goncalves AS, Pettinelli JA, Soares BO, Mansur E, Engelmann F, Gagliardi RF (2021) Evaluation of cryopreservation of *Petiveria alliacea* somatic embryos based on stress caused for the method used. Braz J Dev 7:74759–74774

Gonzales-Arnao MTG, Panta A, Roca WM, Escobar RH, Engelmann F (2008) Development and large-scale application of cryopreservation techniques for shoot and somatic embryo cultures of tropical crops. Plant Cell Tissue Organ Cult 92:1–13

Gonzalez-Arnao MT, Engelmann F (2006) Cryopreservation of plant germplasm using the encapsulation-dehydration technique: review and case study on sugarcane. Cryo-Lett 27(3):155–168

González-Arnao MT, Ortega JC, Navarro L, Duran-Vila N (2003) Cryopreservation of ovules and somatic embryos of Citrus using the encapsulation-dehydration technique. Cryo-Lett 24:85–94

Gowthami R, Chander S, Gupta S, Srivastava V, Puneeth GM, Deepa DA, Singh K (2024) Cryogenic conservation of medicinal plants. In: Tissue culture techniques and medicinal plants: enhancing propagation and production. CRC Press. https://doi.org/10.1201/b23374

Gunawan VA, Soetjipto H, Mustika A (2020) Hypoglicemic and antioxidant activity of *Petiveria alliacea* in diabetic rat models. Biomol Health Sci J 3(1). https://doi.org/10.20473/bhsj.v3i1.19129

Hashimoto T, Azechi S (1988) Bioreactors for the large-scale culture of plant cells. In: Biotechnology in agriculture and forestry, vol 4. Medicinal and aromatic plants I. Springer, Berlin/Heidelberg/New York, pp 104–122

Hausman JF (1993) Changes in peroxidase activity, auxin level and ethylene production during root formation by poplar shoots raised *in vitro*. Plant Growth Regul 13:263–268

Hazarika BN, Silva JAT, Talukdar A (2006) Effective acclimatization of *in vitro* cultured plants: methods, physiology and genetics. In: Floriculture, ornamental and plant biotechnology volume II. Global Science Books, UK, pp 427–438

Heringer AS, Steinmacher DA, Schmidt EC, Bouzon ZL, Guerra MP (2013) Survival and ultrastructural features of peach palm (*Bactris gasipaes* Kunth) somatic embryos submitted to cryopreservation through vitrification. Protoplasma 250:1185–1193

Hesami M, Daneshvar MH (2018) In vitro adventitious shoot regeneration through direct and indirect organogenesis from seedling-derived hypocotyl segments of *Ficus religiosa* L.: an important medicinal plant. HortScience 53(1):55–61

Hoekstra FA, Golovina EA, Tetteroo FA, Wolkers WF (2001) Induction of desiccation tolerance in plant somatic embryos: how exclusive is the protective role of sugars? Cryobiology 43(2):140–150

Hoyos LS, Au WW, Heo MY, Morris DL, Legator MS (1992) Evaluation of the genotoxic effects of a folk medicine, *Petiveria alliacea* (anamu). Mutat Res 280:29–34

Jafari S, Daneshvar MH, Salmi MS, Abad ALJ (2017) Indirect organogenesis and plant regeneration in common Sage (*Salvia officinalis* L.): An important medicinal plant of Iran. Mod Appl Sci 11(5):22–29

Jeong C, Chakrabarty D, Hahn E, Lee H, Paek K (2006) Effects of oxygen, carbon dioxide and ethylene on growth and bioactive compound production in bioreactor culture of ginseng adventitious roots. Biochem Eng J 27:252–263

Jiang W, Chen L, Pan Q, Qiu Y, Shen Y, Fu C (2012) An efficient regeneration system via direct and indirect organogenesis for the medicinal plant *Dysosma versipellis* (Hance) M. Cheng and its potential as a podophyllotoxin source. Acta Physiol Plant 34:631–639

Kanjanaphachoat P, Topoonyanont N, Jantawichai P (2020) RAPD marker analysis of genetic variation in Kluai Nam Wa (Musa ABB group) banana plants regenerated after 8 and 9 subcultures. Maejo Int J Sci Technol 14(2):130–140

Karami O, Deljou A, Pour AM (2007) Repetitive somatic embryogenesis in carnation on picloram supplemented media. Plant Growth Regul 51:33–39

Karp A (1989) Can genetic instability be controlled in plant tissue cultures? IAPTC Newslett 58:2–11
Karp A, Kresovich S, Bhat KV, Ayad WG, Hodgkin T (1997) Molecular tools in plant genetic resources conservation: a guide to the technologies. IPGRI Technical Bulletin n°2, 47p. International Plant Genetic Resources Institute, Rome, Italy
Kaur H, Lekhak MM, Ochatt SJ, Kumar V (2024) Somatic embryogenesis and genetic homogeneity assessment of regenerated plants of *Crinum brachynema* (Amaryllidaceae): an endemic critically endangered medicinal plant. Plant Cell Tissue Organ Cult 156:33
Khurshid R, Khan T, Zaeem A, Laurine Garros L, Hano C, Abbasi BH (2018) Biosynthesis of precious metabolites in callus cultures of *Eclipta alba*. Plant Cell Tissue Organ Cult 135:287
Kim K-S, Jang G-W (2004) Micropropagation of Drosera peltata, a tuberous sundew, by shoot tip culture. Plant Cell Tissue Organ Cult 77:211–214
Kim S, Kubec R, Musha RA (2006) Antibacterial and antifungal activity of sulphur-containing compounds from *Petiveria alliacea* L. J Ethnopharmacol 104:188–192
Kinghorn AD, Ito A, Kennelly EJ, Kim NC, Westenburg HE (1998) Studies on some edible and medicinal plants of Mesoamerica. Proc West Pharmacol Soc 41:253–258
Kitaya Y, Ohmura Y, Kubota C, Kozai T (2005) Manipulation of the culture environment on *in vitro* air movement and its impact on plantlets photosynthesis. Plant Cell Tissue Organ Cult 83:251–257
Kubec R, Musah RA (2001) Cysteine sulfoxide derivatives in *Petiveria alliacea*. Phytochemistry 58:981–985
Kubec R, Musah RA (2005) c-Glutamyl dipeptides in *Petiveria alliacea*. Phytochemistry 66:2494–2497
Kubec R, Kim S, Musah RA (2002) S-Substituted cysteine derivatives and thiosulfinate formation in *Petiveria alliacea*—part II. Phytochemistry 61:675–680
Kubec R, Kim S, Musah RA (2003) The lachrymatory principle of *Petiveria alliacea*. Phytochemistry 63:37–40
Kundu S, Salma U, Gantait S (2018) Cryopreservation of medicinal herbs: major breakthroughs, hurdles and future. In: Biotechnological approaches for medicinal and aromatic plants: conservation, genetic improvement and utilization. Springer, Singapore, pp 353–381
Lambardi M, Ozudogru EA, Benelli C (2008) Cryopreservation of embryogenic cultures. In: Plant cryopreservation: a practical guide. Springer, Corvallis, pp 177–210
Larkin P, Sowcroft W (1981) Somaclonal variation - a novel source of variability from cell cultures for plant improvement. Theor Appl Genet 60:197–214
Lata H, Chandr S, Wang YH, Raman V, Khan IA (2013) TDZ-induced high frequency plant regeneration through direct shoot organogenesis in *Stevia rebaudiana* Bertoni: an important medicinal plant and a natural sweetener. Am J Plant Sci 4(1):27643
Lawal IO, Sowunmi LI, Olaniyi MB, Jegede AE, Akanni FO, Akinwumi IA, Babalola OY (2024) Ethnobotany, chemistry and toxicity of *Petiveria alliacea*: a review. Funct Food Sci 4(2):69–82
Leal RP, Mateos MRG, Rojas TRV, Leon TCC (2005) Allelopathic potential of *Petiveria alliacea* L. Agron Sustain Dev 25:177–182
Li M, Zhao X, Hong S, Zhang X, Li P, Liu J, Xie C (2009) Cryopreservation of plantlet nodes of *Dioscorea opposita* Thunb. using a vitrification method. Cryo-Lett 30:19–28
Lu ZW, Popova EV, Wu CH, Lee EJ, Hahn EJ, Paek KY (2009) Cryopreservation of *Ginkgo biloba* cell culture: effect of pretreatment with sucrose and ABA. Cryo-Lett 30(3):232–243
Luz DA, Pinheiro AM, Silva ML, Monteiro MC, Prediger RD, Maia CSF, Júnior EAF (2016) Ethnobotany, phytochemistry and neuropharmacological effects of *Petiveria alliacea* L. (Phytolaccaceae): a review. J Ethnopharmacol 185:182–201
Maciel VB (2017) Evaluation of toxicity and mutagenicity of ethanolic extracts of *Petiveria alliacea* L. leaves in Escherichia coli and plasmid DNA. Dissertation. Universidade do Estado do Rio de Janeiro
Maciel VB (2022) Evaluation of the toxicity of polysulfides from *Petiveria alliacea* L. in association with ionizing radiation, on plasmid DNA and tumor cells. Thesis. Universidade do Estado do Rio de Janeiro

Maki S, Hirai Y, Niino T, Matsumoto T (2015) Assessment of molecular genetic stability between long-term cryopreserved and tissue cultured wasabi (*Wasabia japonica*) plants. Cryo-Lett 36:318–324

Mallpezzi ELA, Davino SC, Costa LV, Freitas JC, Giesbrecht AM, Roque NF (1994) Antimitotic action of extracts of *Petiveria alliacea* on sea urchin egg development. Braz J Med Biol Res 27:749–754

Mamdouh D, Mahgoub HA, Gabr AM, Ewais EA, Smetanska I (2021) Genetic stability, phenolic, flavonoid, ferulic acid contents, and antioxidant activity of micropropagated *Lycium schweinfurthii* plants. Plan Theory 10(10):2089

Mançano AP, Oliveira SB, Garcia RO, Mansur EA, Gagliardi RF (2019) Genetic stability among *in vitro* eggplant clones induced by different plant growth regulators. Biotecnol Veg 19:75–83

Marchioretto MS (2020) Phytolaccaceae in Flora do Brasil 2020. Jardim Botânico do Rio de Janeiro. Available at https://floradobrasil2020.jbrj.gov.br/FB24173

Martín R, Pérez BC, Alonso NP (2015) Organogénesis in vitro en el género *Digitalis*. Biotecnol Veg 15:195–206

Martínez FS, Casas JL (2011) Cryopreservation of *Tetraclinis articulata* (Vahl.) masters. Cryo-Lett 32:248–255

Martins RABL, Pegoraro DH, Woisky R, Penna SC, Serti JAA (2002) The anti-inflammatory and analgesic effects of a crude extract of *Petiveria alliacea* L. (Phytolaccaceae). Phytomedicine 9:245–248

Mata-Greenwood E, Ito A, Westenburg H, Cui BL, Mehta RG, Kinghorn AD, Pezzuto N (2001) Discover of novel inducers of cellular differentiation using HL-60 promyelocytic cells. Anticancer Res 21:1763–1770

Matos FJA, Lorenzi H, Santos LFL, Silva MGV, Sousa MP (2011) Plantas Tóxicas: Estudo da fitotoxicologia química de plantas brasileiras. Plantarum, São Paulo, p 256

Matsumoto T (2017) Cryopreservation of plant genetic resources: conventional and new methods. Review methods. Rev Agric Sci 5:13–20

Matsumoto T, Sakai A, Yamada K (1994) Cryopreservation of in vitro-grown apical meristems of wasabi (*Wasabia japonica*) by vitrification and subsequent high plant regeneration. Plant Cell Rep 13(8):442–446

Matsumoto T, Yamamoto SI, Fukui K, Rafique T, Engelmann F, Niino T (2015) Cryopreservation of persimmon shoot tips from dormant buds using the D cryo-plate technique. Horticult J 84(2):106–110

Mazur P (1970) Cryobiology: The freezing of biological systems: The responses of living cells to ice formation are of theoretical interest and practical concern. Science 168:939–949

Mazur P (1984) Freezing of living cells: mechanisms and applications. Am J Physiol Cell Physiol 247(3):125–142

Melo JG, Amorim ELC, Albuquerque UP (2009) Native medicinal plants commercialized in Brazil – priorities for conservation. Environ Monit Assess 156:567–580

Mosa KA, Ahme AE, Hazem Kanawati IS, Abdullah A, Sori LH, Ali MA, Vendrame W (2023) Insights into cryopreservation, recovery and genetic stability of medicinal plant tissues. Fitoterapia 169:105555

Murashige T, Skoog F (1962) A revised medium for rapid growth and bioassay with tobacco tissue cultures. Physiol Plant 15:473–497

Murray J, Picking D, Lamm A, McKenzie J, Hartley S, Watson C, Williams L, Lowe H, Delgod R (2016) Significant inhibitory impact of dibenzyl trisulfide and extracts of *Petiveria alliacea* on the activities of major drug-metabolizing enzymes *in vitro*: an assessment of the potential for medicinal plant drug interactions. Fitoterapia

Murthy HN, Dandin VS, Paek KY (2016) Tools for biotechnological production of useful phytochemicals from adventitious root cultures. Phytochem Rev 15:129–145

Murthy HN, Dandin VS, Park SY, Paek KY (2018) Quality, safety and efficacy profiling of ginseng adventitious root produced in vitro. Appl Microbiol Biotechnol 102:7309–7317

Murthy HN, Dalawai D, Bhat MA, Dandin VS, Paek KY, Park SO (2021) Biotechnological production of useful phytochemicals from adventitious root cultures. In: Plant cell and tissue differentiation and secondary metabolites fundamentals and applications. Springer Nature, Cham, pp 469–486

Murthy HN, Joseph KS, Hahn JE et al (2023) Suspension culture of somatic embryos for the production of high-value secondary metabolites. Physiol Mol Biol Plants 29:1153–1177

Nadarajan J, Pritchard HW (2014) Biophysical characteristics of successful oilseed embryo cryoprotection and cryopreservation using vacuum infiltration vitrification: An innovation in plant cell preservation. PLoS One 9(5):e96169

Niino T, Yamamoto S, Fukui K, Martínez CRC, Arizaga MV, Matsumoto T, Engelmann F (2013) Dehydration improves cryopreservation of mat rush (*Juncus decipiens* Nakai) basal stem buds on cryoplates. Cryo-Lett 34:549–560

Niino T, Watanabe K, Nohara N, Rafique T, Yamamoto SI, Fukui K, Engelmann F (2014) Cryopreservation of mat rush lateral buds by air dehydration using aluminum cryo-plate. Plant Biotechnol 31(3):281–287

Nishizawa S, Sakai A, Amano Y, Matsuzawa T (1993) Cryopreservation of Asparagus (*Asparagus officinalis* L.) embryogenic suspension cells and subsequent plant regeneration by vitrification. Plant Sci 91:67–73

Nonohay JS, Mariath JEA, Winge H (1999) Histological analysis of somatic embryogenesis in Brazilian cultivars of barley, *Hordeum vulgare vulgare*, Poaceae. Plant Cell Rep 18:929–934

Olomieja AO, Olanrewaju IO, Ayo-Ajayi JI, Jolayemi GE, Daniel UO, Mordi RC (2021) Antimicrobial and antioxidant properties of *Petiveria alliacea*. IOP Conf Ser Earth Environ Sci 655:012015

Oluwole FS, Bolarinwa AF (1998) The uterine contractile effects of *Petiveria alliacea* seeds. Fitoterapia 69:3–6

Ozudogru EA, Ozden-Tokatli Y, Akcin A (2005) Effect of silver nitrate on multiple shoot formation of Virginia-type peanut through shoot tip culture. In Vitro Cell Dev Biol-Plant 41:151–156

Paiva LC (2011) Culture of dedifferentiated tissues and somatic embryogenesis of *Petiveria alliacea* aiming at the production of bioactive compounds. Dissertation. Universidade do Estado do Rio de Janeiro

Panis B, Piette BMAG, Swennen R (2005) Droplet vitrification of apical meristems: a cryopreservation protocol applicable to all Musaceae. Plant Sci 168(1):45–55

Patnaik J, Sahoo S, Debata BK (1999) Somaclonal variation in cell suspension culture derived regenerants of *Cymbopogon martini* (Roxb.) wats var. *motia*. Plant Breed 118:351–354

Pearce RS (2004) Adaptation of higher plants to freezing. In: Fuller BJ, Lane N, Benson EE (eds) Life in the frozen state. CRC Press LLC, Boca Raton, pp 171–194

Pettinelli JA (2013) In vitro conservation of *Petiveria alliacea* L. germplasm using tissue culture and cryopreservation. Dissertation. Universidade do Estado do Rio de Janeiro

Pettinelli JA (2018) *In vitro* production and conservation of somatic embryos and roots of *Petiveria alliacea* L., aiming to obtain polysulfides. Thesis. Universidade do Estado do Rio de Janeiro

Pettinelli JA, Soares BO, Cantelmo L, Garcia RO, Mansur E, Engelmann F, Gagliardi RF, Gagliardi RF (2017) Cryopreservation of somatic embryos from *Petiveria alliacea* L. by different techniques based on vitrification. In Vitro Cell Dev Biol-Plant. https://doi.org/10.1007/s11627

Pettinelli JA, Soares BO, Collin M, Mansur E, Engelmann F, Gagliardi RF (2020) Cryotolerance of somatic embryos of Guinea (*Petiveria alliacea*) to V-cryoplate technique and histological analysis of their structural integrity. Acta Physiol Plant 42:40–50

Phulwaria NSM, Shekhawat, Rathore JS, Singh RP (2013) An efficient *in vitro* regeneration and *ex vitro* rooting of *Ceropegia bulbosa* Roxb. - a threatened and pharmaceutical important plant of Indian Thar Desert. Ind Crop Prod 42:25–29

Piątczak E, Grzegorczyk-Karolak I, Wysokińska H (2014) Micropropagation of *Rehmannia glutinosa* Libosch.: production of phenolics and flavonoids and evaluation of antioxidant activity. Acta Physiol Plant 36:1693–1702

Picking D, Delgoda R, Younger N, Germosén-Robineau L, Boulogne I, Mitchell S (2015) TRAMIL ethnomedicinal survey in Jamaica. J Ethnopharmacol 169:314–327

Ponte JJ, Franco A, Silveira-Filho J (1996) Preliminary investigation on the nematocide potentiality of Guine's plants (*Petiveria alliacea*). Fitopatol Venez 9(1):14–15

Popova E, Kim HH, Paek KY (2010) Cryopreservation of coriander (Coriandrum sativum L.) somatic embryos using sucrose preculture and air desiccation. Sci Hortic 124(4):522–528

Pospisilova J, Ticha I, Kadlecek P, Haisel D, Plzakova S (1999) Acclimatization of micropropagated plants to *ex vitro* conditions. Biol Plant 42(4):481–497

Queiroz MS, Quadros MR, Santos LMB (2000) Cytokine profile and natural killer cell activity in *Listeria monocytogenes* infected mice treated orally with *Petiveria alliacea* extract. Immunopharm Immunot 22(3):501–518

Rafique T, Yamamoto SI, Fukui K, Mahmood Z, Niino T (2015) Cryopreservation of sugarcane using the V cryo-plate technique. Cryo-Lett 36(1):51–59

Ramirez EC (2003) Alliinases. Food Sci Technol 122:1043–1050

Ray A, Bhattacharya S (2008) Cryopreservation of *in vitro* grown nodal segments of *Rauvolfia serpentina* by PVS2 vitrification. Cryo-Lett 29:321–328

Redenbaugh K, Fuji JA, Slade D, Viss P, Kossler M (1991) Artificial seeds: encapsulated somatic embryos. In: Biotechnology in agriculture and forestry, vol 17. Springer-Verlag, Berlin, pp 395–414

Reed BM, Sarasan V, Kane M, Bunn E, Pence VC (2011) Biodiversity conservation and conservation biotechnology tools. In Vitro Cell Dev Biol - Plant 47:1–4

Rocha DI, Kurczyńska E, Potocka I, Steinmacher DA, Otoni WC (2016) Histology and histochemistry of somatic embryogenesis. In: Somatic embryogenesis: fundamental aspects and applications. Springer, Cham, pp 471–473

Rodriguez MA (2023) Tracing plant histories: linking botanical collections, peoples, and illustrations in seventeenth century Dutch Brazil. https://hdl.handle.net/1887/3640541

Rong HS, Hua YM (2012a) A simple and efficient protocol for cryopreservation of embryogenic calli of the medicinal plant *Anemarrhena asphodeloides* Bunge by vitrification. Plant Cell Tissue Org Cult 109:287–296

Rong HS, Hua YM (2012b) High-efficiency encapsulation-vitrification protocols for cryopreservation of embryogenic calli of the oriental medicinal plant *Anemarrhena asphodeloides* Bunge. Cryo-Lett 33:190–200

Rosner H, Williams LAD, Jung A, Kraus W (2001) Disassembly of microtubules and inhibition of neurite outgrowth, neuroblastoma cell proliferation, and MAP-kinase tyrosine dephosphorylation by dibenzyl trisulphide. BBA-Mol Cell Res 1540:166–177

Rout GR, Samantaray S, Das P (2000) *In vitro* manipulation and propagation of medicinal plants. Biotechnol Adv 18:91–120

Ruffa MJ, Ferraro G, Wagner ML, Calcagno ML, Campos RH, Cavallaro L (2002a) Cytotoxic effect of Argentine medicinal plant extracts on human hepatocellular carcinoma cell line. J Ethnopharmacol 79:335–339

Ruffa MJ, Perusina M, Alfonso V, Wagner ML, Suriano M, Vicente C, Cavallaro L (2002b) Antiviral activity of *Petiveria alliacea* against the bovine viral diarrhea virus. Chemotherapy 48(3):144–147

Rumińska JL, Mellem A (2017) Genetic diversity of chrysanthemum plants derived via somatic embryogenesis using rapd markers. Acta Sci Pol-Hortoru Cultus 16(6):149–156. https://doi.org/10.24326/asphc.2017.6.13

Sakai A (2004) Plant cryopreservation. In: Life in the frozen state. CRC Press, Boca Raton, pp 329–341

Sakai A, Engelmann F (2007) Vitrification, encapsulation-vitrification and droplet-vitrification: a review. Cryo-Lett 28(3):151–172

Sakai A, Kobayashi S, Oiyama I (1990) Cryopreservation of nucellar cells of navel orange (*Citrus sinensis* Osb. var. brasiliensis Tanaka) by vitrification. Plant Cell Rep 9:30–33

Salma M, Engelmann-Sylvestre I, Collin M, Escoute J, Lartaud M, YIJ, Kim H, Verdeil J, Engelmann F (2014) Effect of the successive steps of a cryopreservation protocol on the structural integrity of *Rubia akane* Nakai hairy roots. Protoplasma 251:649–659

Salomón KCC, Rodríguez RIC, Juárez JVE, Salomón AC, Aranda AB, Lau NR, Valdiviezo VMR (2022) In vivo and in silico study of the antinociceptive and toxicological effect of the extracts *of Petiveria alliacea* L. leaves. Pharmaceuticals 15:943

Sasidharan P, Jayachitra A (2017) Direct shoot bud regeneration from shoot tip explants of *Enicostema axillare*: an important medicinal plant. Agrofor Syst 91:471–477

Schmale K, Rademacher T, Fischer R, Hellwig S (2006) Towards industrial usefulness – cryo-cell-banking of transgenic BY-2 cell cultures. J Biotechnol 124:302–311

Schwarz OJ, Sharma A, Beaty RM (2004) Propagation from nonmeristematic tissues: organogenesis. In: Plant development and biotechnology. CRC Press, Boca Raton, pp 159–172

Sekizawa K, Yamamoto SI, Rafique T, Fukui K, Niino T (2011) Cryopreservation of in vitro-grown shoot tips of carnation (*Dianthus caryophyllus* L.) by vitrification method using aluminium cryo-plates. Plant Biotechnol 28(4):401–405

Semagn K, Stedje B, Bjornstad A (2001) Analysis of genetic diversity and structure in Ethiopian populations of *Phytolacca dodecandra* using RAPD. Hereditas 135:50–51

Sertie JAA, Germano DHP, Sudo LS, Hanada S (1995a) *Petiveria alliacea* - analgesic effect and preclinical toxicity. J Dent Res 74(3):793–793

Sertie JAA, Hanada S, Sudo LS, Germano DHP (1995b) *Petiveria-alliacea* - Antiinflammatory effect and gastric mucous protection. J Dent Res 74(3):793–793

Sharma M, Rai SK, Purshottam DK, Jain M, Chakrabarty D, Awasthi A, Nair KN, Sharma AK (2009) *In vitro* clonal propagation of *Clerodendrum serratum* (Linn.) Moon (barangi): a rare and threatened medicinal plant. Acta Physiol Plant 31:2379–2383

Sharmin SA, Alam J, Sheikh MI, Sarker KK, Khalekuzzaman M, Haque A, Alam MF, Alam I (2014) Somatic embryogenesis and plant regeneration in *Wedelia calendulacea* Less. An endangered medicinal plant. Braz Arch Biol Technol 57:394–401

Shatnawi MA (2011) Multiplication and cryogenic storage of *Artemisia herba-alba*: a medicinal plant. Food Agric Environ 9:340–344

Silva JPB, Oliveira FR, Paixão TP, Malcher NS, Santos PC, Baetas AC, Batittuci MCP, Monteiro MC, Andrade M (2016) A. *In vitro* and *in vivo* assessment of genotoxic activity of *Petiveria alliacea*. Pharm Pharmacol 10(34):718–727

Simão MJ, Collin M, Garcia RO, Mansur E, Pacheco G, Engelmann F (2018) Histological characterization of *Passiflora pohlii* Mast. root tips cryopreserved using the V-Cryo-plate technique. Protoplasma 255:741–750

Sivanesan I, Jeong BR (2016) Optimizing factors affecting somatic embryogenesis in *Cineraria*. In: Somatic embryogenesis in ornamentals and its applications. Springer, New Delhi, pp 55–65

Skyba M, Urbanová M, Kapchina-Toteva V, Košuth J, Harding K, Čellárová E (2010) Physiological, biochemical and molecular characteristics of cryopreserved *Hypericum perforatum* L. shoot tips. Cryo-Lett 31:249–260

Slater A, Scott N, Fowler M (2008) Plant tissue culture. In: Plant biotechnology: the genetic manipulation of plants. Oxford University Press, New York, pp 37–53

Soares BO (2011) Micropropagação e avaliação da potencialidade genotóxica de *Petiveria alliacea* L Dissertação. Universidade do Estado do Rio de Janeiro

Soares BO (2016) Tissue culture and assessment of genetic and metabolic stability of *Petiveria alliacea* L. produced and maintained *in vitro*. Thesis. Universidade do Estado do Rio de Janeiro

Soares BO, Fernandes D, Cantelmo L, Rocha LP, Pettinelli JA, Christo AG, Coelho MGP, Gagliardi RF (2013) Botanical characterization of *Petiveria alliacea* L. from Rio de Janeiro, Brazil: systematic and functional implications. Plant Biosyst 147(2):411–417

Soares BO, Oliveira MBN, Mansur E, Dantas FJS, De Mattos JCP, Caldeira-de-Araujo A, Gagliardi RF (2014) Effect of extracts from field and *in vitro* plants of *Petiveria alliacea* L. on plasmidial DNA. J Med Plant Res 8(35):1101–1109

Szypuła W, Pietrosiuk A, Suchocki P, Olszowska O, Furmanowa M, Kazimierska O (2005) Somatic embryogenesis and *in vitro* culture of *Huperzia selago* shoots as a potential source of huperzine A. Plant Sci 168:1443–1452

Taylor L (2005) The healing power of rainforest herbs: a guide to understanding and using herbal Medicinals. Square One Publishers, New York, pp 114–166

Tessereau H, Florin B, Meschine MC, Thierry C, Pétiard V (1994) Cryopreservation of somatic embryos: a tool for germplasm storage and commercial delivery of selected plants. Ann Bot 74(5):547–555

Tikendra L, Potshangbam AM, Dey A, Devi TR, Sahoo MR, Nongdam P (2021) RAPD, ISSR, and SCoT markers based genetic stability assessment of micropropagated *Dendrobium fimbriatum* Lindl. var. oculatum Hk. f.- an important endangered orchid. Physiol Mol Biol Plants 27:341–357

Trevisan M, Rodrigues PSM, Nunes AIS, Falcão MDS, Martins HC, Scapin E, Seibert CS (2022) Antiedematogenic activity of *Petiveria alliacea* L. in bothropic poisoning. J Med Plant Res 16(12):315–325

Tropicos.org. ©Missouri Botanical Garden - *Petiveria alliacea* – synonyms. Missouri. Available at https://tropicos.org Accessed 07 Jan 2025

Udulutsch RG, Pinheiro MHO, Tannus JLS, Dias P, Furlan A (2007) Phytolaccaceae. In: Flora Fanerogâmica do Estado de São Paulo, vol 5. FAPESP, São Paulo, p 494

Valdes NG, Häkkinen ST, Lemasson C, Guillet M, Caldentey KMO, Ritala A, Cardon F (2020) Hairy root cultures - a versatile tool with multiple applications. Front Plant Sci 11:33

Vanisree M, Lee CY, Lo SF, Nalawade SM, Lin CY, Tsay HS (2004) Studies on the production of some important secondary metabolites from medicinal plants by plant tissue cultures. Bot Bull Acad Sin 45:1–22

Vendrame WA, Faria RT (2011) Phloroglucinol enhances recovery and survival of cryopreserved *Dendrobium nobile* protocorms. Sci Hortic 128(2):131–135. https://doi.org/10.1016/j.scienta.2010.12.018

Villar R, Calleja JM, Morales C, Cáceres A (1997) Screening of 17 Guatemalan medicinal plants for platelet antiaggregant activity. Phytother Res 11(6):441–445

Vondráková Z, Cvikrová M, Eliášová K, Martincová O, Vágner M (2010) Cryotolerance in Norway spruce and its association with growth rates, anatomical features and polyamines of embryogenic cultures. Tree Physiol 30:1335–1348

Webster SA, Mitchell SA, Gallimore WA, Williams LAD, Ahmad MH (2008) Biosynthesis of dibenzyl trisulfide (DTS) from somatic embryos and rhizogenous embryogenic callus derived from Guinea hen weed (*Petiveria alliacea* L.) leaf explants. In Vitro Cell Dev Biol Plant 44:112–118

Wesley-Smith J, Pammenter NW, Berjak P, Walters C (2001) The effects of two drying rates on the desiccation tolerance of embryonic axes of recalcitrant Jackfruit (*Artocarpus heterophyllus* Lamk.) seeds. Ann Bot 88:653–664

Williams LAD, The TL, Gardner MT, Fletcher CK, Navarane A, Gibbs N, Fleishhacker R (1997) Immunomodulatory activities of *Petiveria alliacea* L. Phytother Res 11:252–253

Williams LAD, Vasquez E, Klaiber I, Kraus W, Rosner H (2003) A sulfonic anhydride derivative from dibenzyl trisulphide with agro-chemical activities. Chemosphere 51:701–706

Williams LAD, Rosner H, Levy HG, Barton EN (2007) A critical review of therapeutic potential of dibenzyl trisulphide isolated from *Petiveria alliacea* L. (Guinea heenweed, anamu). West Indian Med 56:13–17

Willian LAD, Levy HG (2007) Composition and method for treating Cancer. Patent US 2008 0070839 A1

Yamamoto SI, Rafique T, Priyantha WS, Fukui K, Matsumoto T, Niino T (2011) Development of a cryopreservation procedure using aluminium cryo-plates. Cryo-Lett 32(3):256–265

Yamamoto SI, Fukui K, Rafique T, Khan NI, Martinez CRC, Sekizawa K, Niino T (2012a) Cryopreservation of in vitro-grown shoot tips of strawberry by the vitrification method using aluminium cryo-plates. Plant Genet Resour 10(1):14–19

Yamamoto SI, Rafique T, Fukui K, Sekizawa K, Niino T (2012b) V-cryo-plate procedure as an effective protocol for cryobanks: case study of mint cryopreservation. Cryo-Lett 33(1):2–23

Yamamoto SI, Wunna RT, Arizaga MV, Fukui K, Gutierrez EJC, Niino T (2015) The aluminum cryo-plate increases efficiency of cryopreservation protocols for potato shoot tips. Am J Potato Res 92:250–257

Yang JL, Seong ES, Kim MJ, Ghimire BK, Kang WH, Yu CY, Li CH (2010) Direct somatic embryogenesis from pericycle cells of broccoli (*Brassica oleracea* L. var. italica) root explants. Plant Cell Tissue Organ Cult 100:49–58

Yil JY, Sylvestre I, Colin M, Salma M, Lee SY, Kim HH, Park HJ, Engelmann F (2012) Improved cryopreservation using droplet vitrification and histological changes associated with cryopreservation of madder (*Rubia akane* Nakai). Korean J Hortic Sci 30:79–84

Yin M, Hong S (2009) Cryopreservation of *Dendrobium candidum* Wall. ex Lindl. protocorm-like bodies by encapsulation-vitrification. Plant Cell Tissue Organ Cult 98:179–185

Yin M, Hong S (2010) A simple cryopreservation protocol of *Dioscorea bulbifera* L. embryogenic calli by encapsulation-vitrification. Plant Cell Tissue Organ Cult 101:349–358

Zhang ZJ (2004) Therapeutic effects of herbal extracts and constituents in animal models of psychiatric disorders. Life Sci 75:1659–1699

Zhao X, Li P, Jiang ZY, Song Z, Han L, Li M (2016) Effects of preculture on cryopreservation of nodes from *Dioscorea opposita* Thunb. cv. Tiegun plantlets by encapsulation-vitrification. Med Plant 7:22–26

Zheng Q, Dessai AP, Prakash CS (1996) Rapid and repetitive plant regeneration in sweet potato via somatic embryogenesis. Plant Cell Rep 15:381–385

Zhu GY, Geuns JMC, Dussert S, Swennen R, Panis B (2006) Change in sugar, sterol and fatty acid composition in banana meristems caused by sucrose-induced acclimation and its effects on cryopreservation. Physiol Plant 128:80–94. https://doi.org/10.1111/j.1399-3054.2006.00713

Zimmerman JL (1993) Somatic embryogenesis: a model for development in higher plant. Plant Cell 5:1411–1423. https://doi.org/10.1139/b83-074

Zoghbi MGB, Andrade EHA, Maia SJG (2002) Volatile constituents from *Adenocalymma alliaceum* Miers and *Petiveria alliacea* L., two medicinal herbs of the Amazon. Flavour Fragr J 17:133–135

Multiomics Approaches for Conservation of Endangered Orchids: Advances, Applications, and Future Prospects

10

Yavar Vafaee, Lisa Wallace, Amir Khoddamzadeh, and Ayoub Molaahmad Nalousi

Abstract

The conservation of endangered orchids presents unique challenges that require innovative biotechnological approaches. Recent advances in multiomics technologies, including genomics, transcriptomics, proteomics, and metabolomics, have revolutionized our understanding of orchid biology and conservation strategies. This chapter provides a comprehensive review of how integrated omics approaches are transforming orchid conservation efforts. We examine how genomic tools uncover genetic diversity patterns and population structure in endangered orchid species, while transcriptomic analyses reveal mechanisms of adaptation and stress response. The application of proteomics and metabolomics in understanding orchid physiology and secondary metabolism is discussed, with particular emphasis on their relevance to conservation strategies. Specific case studies demonstrating the successful integration of multiomics data in orchid conservation programs are presented, highlighting both ex situ and in situ conservation approaches. Additionally, we explore emerging technologies and future directions in omics-based conservation, including environmental DNA (eDNA) monitoring, CRISPR–Cas9 genome editing, and machine learning for predicting

Y. Vafaee (✉) · A. M. Nalousi
Department of Horticultural Sciences, Faculty of Agriculture, University of Kurdistan, Sanandaj, Iran

Research Center of Medicinal Plants Breeding and Development, University of Kurdistan, Sanandaj, Iran
e-mail: y.vafaee@uok.ac.ir

L. Wallace
Department of Biological Sciences, Old Dominion University, Norfolk, VA, USA

A. Khoddamzadeh
Department of Earth and Environment, Institute of Environment, Florida International University, Miami, FL, USA

S. Gantait, P. Chmielarz (eds.), *Conservation of Plant Genetic Resources*,
https://doi.org/10.1007/978-981-96-8331-4_10

modeling and data integration. This synthesis provides valuable insights for researchers and conservation practitioners working to preserve endangered orchid species through advanced biotechnological interventions.

Keywords

Biotechnology · Endangered species · Multiomics · Orchid conservation · Genomics · Transcriptomics

10.1 Introduction

Orchids, members of the family Orchidaceae, are among the most diverse and ecologically significant groups of flowering plants on Earth. Over 28,000 species account for nearly 8% of the global angiosperm diversity, revealing an extraordinary range of floral forms, colors, and ecological adaptations (Pérez-Escobar et al. 2024; Chase et al. 2015). Figure 10.1 shows the flower and inflorescence morphology of some terrestrial orchids, which are characterized by different shapes and colors. In addition to their aesthetic appeal, orchids play vital roles in ecosystems as bioindicators, hosts for pollinators, and contributors to biodiversity. Their cultural significance, from traditional medicine to horticulture, further underscores their importance. However, despite their ecological and cultural value, orchids face unprecedented threats that jeopardize their survival (Wang et al. 2024).

Today, approximately 56.5% of evaluated orchid species are listed as threatened on the IUCN Red List, with many classified as endangered or critically endangered (Fay 2018; Hinsley et al. 2018) (Fig. 10.2). The primary drivers of this decline include habitat loss due to urbanization, deforestation, and agricultural expansion; climate change, which disrupts delicate pollination and mycorrhizal relationships; overharvesting for horticulture and traditional medicine; and competition from invasive species (Phillips et al. 2020; Vitt et al. 2023). Orchids are particularly vulnerable because of their specialized ecological requirements (Vafaee et al. 2021). Many species depend on specific mycorrhizal fungi for seed germination and growth, and they often rely on highly specialized pollinators for reproduction. These intricate relationships make orchids exceptionally sensitive to environmental changes, rendering them among the early indicators of ecosystem disruption (Anghelescu et al. 2023; Swarts and Dixon 2009).

To address these challenges, scientists and conservationists are increasingly turning to omics technologies—genomics, transcriptomics, proteomics, and metabolomics—to elucidate the molecular, genetic, and biochemical mechanisms that underpin orchid biology. These cutting-edge approaches provide unprecedented insights into orchid diversity, adaptation, and resilience. For example, genomics has revealed patterns of genetic diversity and population structure, helping identify at-risk populations and adaptive traits (Givnish et al. 2015; Fay et al. 2018; Balao et al. 2022). Transcriptomics has shed light on how orchids respond to environmental stressors, such as drought and pathogens, whereas proteomics and metabolomics

Fig. 10.1 The morphology of flower and inflorescence of some terrestrial orchid species. (**a**) *Orchis mascula subsp. mascula* (L.) L.; (**b**) *Orchis mascula subsp. pinetorum* (Boiss. and Kotschy); (**c**) *Orchis simia* (Lam.); (**d**) *Anacamptis collina* (Banks and Sol.); (**e**) *Ophrys straussii* (H. Fleischm. and Bornm.); (**f**) *Ophrys schulzei* Bornm. and Fleischm.; (**g**) *Himantoglossum affine* (Boiss.) Schltr.; (**h**) *Steveniella satyrioides* (Spreng.) Schltr.; (**i**) *Gymnadenia conopsea* (L.); (**j**) *Dactylorhiza umbrosa* (Kar. and Kir.) Nevski. (Photos A-G, I-K © Yavar Vafaee; Photo H © Abdolbaset Ghorbani)

have illuminated the physiological and biochemical pathways that enable orchids to thrive in challenging environments (Zhang et al. 2021; Song et al. 2022b).

This chapter explores how the integration of multiomics technologies with traditional conservation strategies is transforming efforts to protect endangered orchids. We begin with an overview of the global conservation status of orchids and the major challenges they face. Next, we delve into the applications of genomics, transcriptomics, proteomics, and metabolomics in orchid research and conservation, supported by case studies that highlight successful applications of these approaches.

By synthesizing these advances, this chapter aims to provide a comprehensive framework for leveraging multiomics approaches to safeguard the extraordinary diversity of orchids. Through a combination of cutting-edge science and practical conservation strategies, we can address the ongoing threats to these remarkable plants and ensure their survival for future generations.

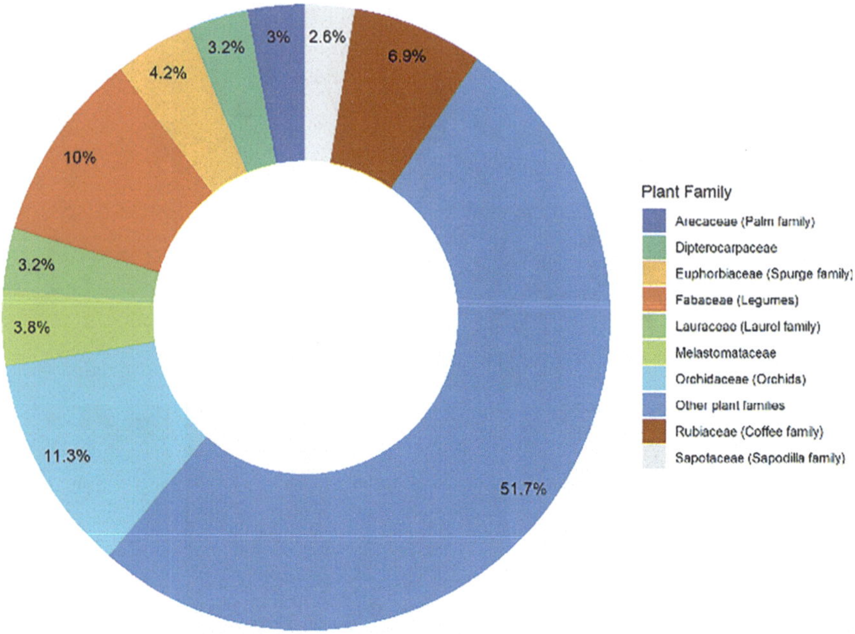

Fig. 10.2 Distribution of threatened plant species across major plant families based on IUCN Red List and global conservation assessments. Orchidaceae represents the largest proportion of threatened species (11.3%), followed by Fabaceae (10.0%) and Rubiaceae (6.9%). (Data compiled from the IUCN Red List of Threatened Species and conservation databases (2023–2024))

10.1.1 Global Status of Orchid Conservation

However, orchid conservation is becoming increasingly precarious. Approximately 56.5% of the evaluated orchid species are listed as threatened on the IUCN Red List, with many classified as endangered or critically endangered (Fay 2018; Hinsley et al. 2018). This alarming trend is driven by a combination of habitat loss, climate change, overharvesting, and invasive species, each of which poses unique challenges to orchid survival.

Threats to Orchids Habitat destruction is the foremost threat and is driven by urbanization, deforestation, and agricultural expansion. Orchids, particularly that endemic to biodiversity hotspots, are disproportionately affected (Kaki et al. 2020; Wallace 2006). For example, in China, human excavation and habitat destruction have pushed species such as Cypripedium macranthum and C. guttatum to the brink of extinction (Liu et al. 2021). In Mediterranean Europe, the vulnerable ghost orchid (*Epipogium aphyllum*) has experienced severe population decline due to forest fragmentation and changes in woodland management practices, with many historical populations disappearing across its range (Kull et al. 2016).

Climate change exacerbates these challenges by disrupting delicate ecological interactions. Rising temperatures and shifting precipitation patterns alter orchid–pollinator relationships, leading to reproductive failure. Species such as *Ophrys apifera* (bee orchid), which rely on specific pollinator interactions, are especially vulnerable to phenological mismatches, as climate change affects the emergence timing of both orchids and their insect pollinators (Robbirt et al. 2014; Kolanowska 2021). In the alpine regions of Europe, the rare terrestrial orchid *Nigritella rhellicani* (also known as *Gymnadenia rhellicani*) has undergone significant population decline and shift changes. These changes have contributed to reduced reproductive success and genetic diversity in Alpine populations, likely due to altered interactions with specialized pollinators, which threatens their long-term persistence as warming continues at high elevations (Geppert et al. 2020).

Overharvesting for horticulture, traditional medicine, and food further threatens orchids. Despite international regulations under the Convention on International Trade in Endangered Species (CITES), illegal trade persists, targeting high-demand species such as *Dendrobium officinale* for their medicinal properties (Hinsley et al. 2018; Phelps et al. 2016). In Iran and Turkey, terrestrial tuberous orchids, traditional drinks and ice cream ingredients, face severe pressure from collection for salep production. Species of *Orchis*, *Anacamptis*, and *Dactylorhiza* are particularly vulnerable, with an estimated 30 million orchid tubers harvested annually in Turkey alone. In northwestern Iran, populations of *D. umbrosa* and *O. coriophora* have declined by more than 60% due to intensive harvesting, whereas in Turkish Anatolia previously abundant *O. anatolica* populations have been decimated as collectors that target increasingly remote areas (Fatahi et al. 2022a, b; Ghorbani et al. 2014; Sezik 2022).

Invasive species compound these threats by outcompeting native orchids for resources and altering their habitats. In Florida, invasive plants such as *Lygodium microphyllum* have displaced native orchids, including ghost orchids (*Dendrophylax lindenii*), leading to local extinctions (Rodgers et al. 2014).

10.1.2 Conservation Strategies

Addressing these threats requires a multifaceted approach. Habitat preservation is critical, particularly in biodiversity hotspots where orchids are concentrated. Sustainable harvesting practices and stricter enforcement of antitrafficking laws are essential to curb illegal trade. Collaborative efforts between scientists, policymakers, and local communities have shown promise. For example, community-led conservation initiatives in Madagascar have increased populations of *Angraecum sesquipedale* through habitat restoration and sustainable harvesting (Cribb et al. 2007). Education and outreach are equally important. Programs engaging students in conservation science, such as those promoting native orchid propagation, foster public awareness and stewardship (Swarts and Dixon 2009).

Ex situ conservation techniques, such as seed banking and tissue culture, play a vital role in safeguarding genetic diversity. The successful propagation of endangered species such as *Himantoglossum affine* (Fatahi et al. 2022a), *Orchis simia* (Fatahi et al. 2022b), *Dactylorhiza umbrosa* (Fatahi et al. 2023), and *Ophrys straussii* (Ahmadzadeh et al. 2025) in controlled environments has been achieved through optimized cryopreservation protocols.

10.1.3 Evolution of Omics Technologies in Plant Conservation

The advent of omics technologies has revolutionized biological research, offering unprecedented insights into the molecular mechanisms that underpin plant biology. Historically, orchid conservation has relied on ecological observations and limited genetic markers, such as microsatellites. However, the emergence of genomics in the early twenty-first century marked a paradigm shift. The first orchid genome (*Phalaenopsis equestris*) was sequenced in 2015, revealing gene families linked to specialized adaptations such as crassulacean acid metabolism (CAM), a water-efficient photosynthetic pathway, and floral morphology (Cai et al. 2015). This breakthrough enabled researchers to decipher genetic blueprints of endangered orchids, such as *Dendrobium catenatum*, whose genome provides insights into polysaccharide synthase, floral development, and adaptive evolution (Zhang et al. 2016). Genomics has laid the foundation for subsequent omics methods, including transcriptomics, proteomics, and metabolomics, each contributing unique perspectives to orchid conservation.

Transcriptomics has emerged as a powerful tool for orchid conservation, providing unprecedented insights into gene expression dynamics under environmental stress conditions. By identifying differentially expressed genes and regulatory networks, transcriptomic analyses enable conservation biologists to understand adaptation mechanisms and resilience factors in endangered orchid populations. For example, RNA-seq analyses of *Phalaenopsis aphrodite* revealed complex transcriptional reprogramming during drought stress, with 1290 differentially expressed genes identified, including key transcription factors in the WRKY and MYB families that regulate stress responses. These findings have informed both in situ conservation strategies and ex situ breeding programs to increase resilience in vulnerable wild populations (Hsiao et al. 2011). RNA-seq analyses of *Dendrobium catenatum* revealed complex transcriptional reprogramming during drought stress, with numerous differentially expressed genes identified, including key transcription factors in the WRKY, MYB, and bZIP families that regulate stress responses. These findings have informed both in situ conservation strategies and ex situ breeding programs to increase resilience in vulnerable wild populations (Wan et al. 2019).

Proteomics, leveraging advances in mass spectrometry, has revealed crucial post-translational modifications in orchids that provide insights for conservation strategies. For example, Chen et al. (2021) identified 2331 proteins across different floral organs in *Cymbidium goeringii*, revealing distinct metabolic activities and hormone-related proteins that regulate floral development and morphogenesis. Their findings of organ-specific protein expression patterns, particularly phytohormone-related proteins and transcription factors, provide molecular markers that could help monitor developmental processes in threatened populations. On the other hand, Wang et al. (2017) identified a total of 54 DEP spots in two-dimensional electrophoresis (2DE) maps between self-pollinated (SP) and cross-pollinated (CP) pistils. Gene ontology analysis revealed an array of proteins belonging to the following different functional categories: metabolic process (8.94%), response to stimulus (5.69%), biosynthetic process (4.07%), protein folding (3.25%), and transport (3.25%). They concluded that DEP identification at the early response stage of pollination will hopefully provide new insights into the mechanism of the pollination response and help conserve orchid species. Sedeek et al. (2013) expanded this understanding by integrating transcriptome and proteome data to identify candidate genes responsible for pollinator attraction in sexually deceptive *Ophrys* orchids, such as those involved in hydrocarbon and anthocyanin biosynthesis critical for mimicry and reproductive isolation. Additionally, Valadares et al. (2014) revealed proteomic changes during symbiotic germination in *Oncidium sphacelatum*, offering critical insights into the mycorrhizal relationships essential for successful reintroduction programs. These proteomic studies provide valuable molecular markers and a biological understanding that can directly inform both in situ and ex situ conservation efforts for endangered orchid species, particularly in developing germination protocols and identifying resilient genotypes for restoration initiatives.

Metabolomics has further enriched this landscape through comprehensive profiling of secondary metabolites. Jie et al. (2023) conducted comprehensive chemical profiling of *Dendrobium chrysotoxum* and *Dendrobium officinale*, identifying active antihepatoma constituents such as chrysotoxine and dendrocrepine while linking metabolic diversity to ecological adaptation. Similarly, Zhan et al. (2020) revealed habitat-driven metabolite variations in *Dendrobium officinale* and demonstrated how phenylpropanoid and flavonoid biosynthesis pathways are modulated under environmental stress via integrated metabolomic–transcriptomic analyses. While these single-omics approaches have been transformative, studies such as Feng et al. (2017), which combined succinyl proteomics with metabolomic data to dissect glycolysis regulation in *Dendrobium officinale*, highlight the necessity of multiomics integration to unravel complex orchid biology and design targeted conservation strategies.

The mid-2010s witnessed the emergence of multiomics integration, propelled by computational advancements and systems biology frameworks. Combining genomic and transcriptomic data from *Apostasia shenzhenica* revealed evolutionary adaptations and identified gene networks crucial for orchid diversification (Zhang et al. 2017). Integrative approaches have elucidated complex traits such as orchid–mycorrhizal symbiosis. In this context, transcriptomics and proteomics have been combined to identify key genes (e.g., DOF transcription factors) and proteins involved in nutrient exchange during the symbiotic germination of *Dendrobium officinale* (Chen et al. 2017). Similarly, Valadares et al. (2014) documented proteomic shifts in *Oncidium sphacelatum* during mycorrhizal interactions, revealing enzymes linked to carbon metabolism and fungal communication. These multiomics studies collectively highlight how orchids dynamically regulate molecular pathways to establish and maintain symbiosis, although metabolomic insights into carbon–nitrogen exchange remain an open frontier. The integration of proteomics and metabolomics data in *Vanilla planifolia* provided comprehensive insights into flavonoid biosynthesis pathways crucial for both aromatic properties and stress adaptation (Palama et al. 2010). Tools like KEGG (Kyoto Encyclopedia of Genes and Genomes) and PlantCyc facilitated pathway enrichment analyses, bridging omics layers to model stress responses holistically (Kanehisa 2016; Tenenbaum 2016).

Multi-omics now underpins cutting-edge conservation strategies. Environmental DNA (eDNA) metabarcoding, coupled with genomic databases, enables noninvasive monitoring of orchid populations and their mycorrhizal partners (Thomsen and Willerslev 2015). Machine learning algorithms integrate multiomics datasets to predict habitat suitability, as demonstrated for rare orchid species in montane ecosystems (Wäldchen and Mäder 2018). CRISPR-Cas9 genome editing, guided by omics data, offers potential for crop improvement applicable to orchid conservation, though direct applications in endangered orchids remain theoretical. However, challenges persist, including data standardization (e.g., harmonizing proteomic and metabolomic datasets) and interdisciplinary collaboration (Hoban et al. 2022).

Emerging fields like single-cell omics and synthetic biology promise to further enhance conservation efforts. Single-cell RNA sequencing could unravel cell-type-specific responses in orchid protocorms during symbiotic germination, while synthetic biology might engineer microbial consortia to enhance symbiotic interactions (Selosse et al. 2017; de Vries et al. 2018).

10.2 Application of Genomics in Orchid Conservation

Orchids are pivotal models for studying plant evolution and adaptation, particularly due to their unique genetic innovations. The sequencing of the *Phalaenopsis equestris* genome marked a milestone in orchid genomics, revealing the genetic basis of crassulacean acid metabolism (CAM)—a water-efficient photosynthetic pathway critical for survival in arid and epiphytic habitats (Cai et al. 2015). This study identified gene families linked to CAM (e.g., *PPDK*, *PEPC*) and floral morphology, such as MADS-box transcription factors governing lip and column development. These genomic insights not only clarify orchid evolution but also inform conservation strategies for critically endangered species like *Cypripedium tibeticum*, where habitat-specific adaptations are genetically encoded and population structure analysis reveals distinct genetic clusters requiring targeted conservation approaches (Guo et al. 2015b).

High-throughput sequencing has revolutionized orchid genomics, enabling large-scale studies of genome structure, gene duplication, and chromosomal polyploidy. For instance, the genome of *Dendrobium catenatum* uncovered stress-responsive *WRKY* and *NAC* transcription factors, as well as polysaccharide synthases critical for drought tolerance (Zhang et al. 2016). Despite these advances, challenges persist in functional genomics. Orchids lack comprehensive gene knockout libraries and robust genetic transformation systems, hindering efforts to validate gene functions. Emerging CRISPR-Cas9 tools, however, are bridging this gap: For example, Tong et al. (2019) demonstrated high-efficiency editing of MADS-box genes (*PeSEP3*, *PeMADS6*) in *Phalaenopsis* orchids, enabling precise manipulation of floral organ identity—a breakthrough for tailoring floral morphology in ornamental or conservation contexts. Similarly, Kui et al. (2017) established a CRISPR toolkit for *Dendrobium officinale*, identifying constitutive promoters like *CaMV 35S* and *DnUbi* and successfully targeting *DoMYB1*, a regulator of anthocyanin biosynthesis, to enhance pigmentation and stress resilience. These advancements are complemented by protocorm-based transformation protocols (Hickerson and Samuel 2024), which leverage orchid protocorms as explants to edit symbiotic germination genes (e.g., *DOF* transcription factors), offering strategies to boost seedling survival in habitat restoration projects. Collectively, these studies underscore CRISPR's potential to overcome historical limitations in orchid genomics while accelerating trait optimization for conservation.

The OrchidBase database has become a cornerstone for comparative genomics, integrating genomic and transcriptomic data for various orchid species. OrchidBase 3.0 includes systemic transcriptomes of three *Cymbidium* species (*C. sinense*, *C. ensifolium*, *C. goeringii*), enabling researchers to analyze gene expression patterns and lineage-specific adaptations (Fu et al. 2011; Tsai et al. 2013). In this regard, long terminal repeat (LTR) retrotransposons have driven genome expansion in *Phalaenopsis equestris*, while plastid genome reductions in mycoheterotrophic orchids like *Epipogium aphyllum* reflect transitions to heterotrophy (Cai et al. 2015; Schelkunov et al. 2015).

Recent advances in phylogenomics and sequencing technologies have significantly enhanced our understanding of genetic diversity, evolutionary relationships, and adaptive potential in orchids, facilitating their conservation and management. High-throughput sequencing approaches, such as whole-genome sequencing, transcriptomics, and genotyping-by-sequencing, have provided unprecedented insights into orchid genome evolution, species diversification, and genetic adaptations to diverse ecological niches (Song et al. 2022a). Phylogenomic studies have further revealed that multiple drivers, including mycorrhizal associations, pollination strategies, and chromosomal rearrangements, have contributed to the extraordinary diversification of the Orchidaceae family (Givnish et al. 2015). For instance, an integrative analysis of genotypic and phenotypic diversity in the Eurasian genus *Neotinea* demonstrated significant intraspecific variation, highlighting the importance of genetic diversity studies for conservation planning (Trávníček et al. 2021). Similarly, genome-wide association studies (GWAS) in *Phalaenopsis* orchids have identified key genetic loci linked to aesthetic traits, which not only aid in breeding programs but also contribute to preserving genetic integrity in cultivated and wild populations (Hsu et al. 2022). Furthermore, double-digest restriction-site associated DNA sequencing (ddRAD-seq) has been employed to assess genetic variation among orchid species, providing valuable data for conservation strategies aimed at maintaining intraspecific diversity and preventing genetic erosion (Roy et al. 2017). The integration of phylogenomics with conservation genomics allows for the identification of evolutionarily significant units (ESUs), guiding reintroduction efforts and ensuring the long-term survival of endangered orchid taxa. As sequencing technologies continue to evolve, the application of genomic approaches in orchid conservation will be instrumental in addressing challenges posed by habitat loss, climate change, and genetic bottlenecks.

10.2.1 Whole-Genome Sequencing Approaches

Whole-genome sequencing (WGS) has revolutionized orchid biology by decoding the genetic blueprints underlying their ecological adaptations, stress resilience, and symbiotic relationships. Orchid genomes are notoriously challenging to assemble due to their large size (1–10+ gigabases), high repetitive DNA content, and heterozygosity. However, advancements in sequencing technologies and bioinformatics tools have enabled high-quality, chromosome-scale assemblies, unlocking insights into orchid evolution and conservation. Below, the detail sequenced orchid genomes, emerging tools, and case studies demonstrating their conservation applications is provided:

Phalaenopsis equestris, a cornerstone of the ornamental orchid industry, became the first orchid genome sequenced, offering foundational insights into epiphytic adaptations. Its 1.6 Gb genome ($2n = 38$), assembled via Illumina and Sanger sequencing, is highly repetitive (62% LTR retrotransposons) and revealed critical evolutionary innovations. Key findings include expanded MADS-box genes (*AP3*, *AGL6*) governing floral symmetry—a trait central to pollinator attraction—and genes associated with facultative crassulacean acid metabolism (CAM) pathways (*PEPC*, *PPDK*), potentially enabling drought resilience in arboreal habitats. The genome also uncovered retrotransposon-driven expansions, reshaping our understanding of orchid genome evolution and providing a roadmap for breeding stress-tolerant hybrids. These insights bridge fundamental biology and applied horticulture, guiding strategies to enhance resilience in commercial and endangered epiphytic orchids (Cai et al. 2015).

Dendrobium officinale (1.3 Gb, $2n = 38$), sequenced via PacBio and Illumina techologies, features a complex genome that provides insights into the biology of this important traditional Chinese medicinal orchid. This epiphytic orchid's genome reveals adaptations for survival in diverse ecological niches, including expanded resistance-related genes suggesting a powerful immune system. The genome contains evidence of whole-genome duplication (WGD) events shared with other orchids like Phalaenopsis, contributing to its ecological plasticity and adaptation to epiphytic lifestyles. Key genomic features include specialized gene clusters involved in polysaccharide biosynthesis, which are responsible for the production of medicinally important compounds that make this species highly valued in traditional Chinese medicine. These genomic insights directly inform cultivation strategies and conservation efforts crucial for sustaining wild populations threatened by overharvesting and habitat loss (Yan et al. 2015).

Apostasia shenzhenica, a basal orchid lineage critical for understanding early orchid evolution, features a compact genome (350 Mb, $2n = 54$) sequenced using PacBio CLR and Illumina. With only 45% repetitive content—primarily LTR retrotransposons—it retains ancestral traits such as functional photosynthetic genes (RBCL, LHC) and duplicated MYB transcription factors linked to early floral development. Like all orchids, its genome shows evidence of a whole-genome duplication (WGD) event that occurred early in orchid evolution, providing a blueprint for the evolutionary transition to mycorrhizal dependency and floral complexity. These findings clarify the genetic underpinnings of orchid diversification and adaptation (Zhang et al. 2017).

Gastrodia elata (1.06 Gb, $2n = 36$), a mycoheterotrophic orchid, lacks chlorophyll biosynthesis genes but harbors significant repetitive elements. Illumina paired-end sequencing uncovered enriched *SWEET* transporters for fungal nutrient scavenging, along with expanded gene families including glycoside hydrolases and urease that facilitate hyphal digestion. The genome reveals extensive gene loss, including most photosynthesis-related genes, while retaining genes essential for mycorrhizal symbiosis. This obligate mycoheterotrophic lifestyle requires sequential partnerships with two distinct fungal groups: Mycena species for seed germination and Armillaria species for tuber development and growt, emphasizing the need to conserve symbiotic fungal networks (Yuan et al. 2018).

Cymbidium sinense and C. goeringii (3.52 Gb, $2n = 40$), terrestrial and epiphytic orchids highly valued for their ornamental traits and cultural significance, showcase genomic adaptations for floral morphology and diverse ecological lifestyles. Both species were sequenced using advanced technologies including Illumina, PacBio, and Oxford Nanopore platforms. Their genomes are characterized by substantial repetitive element content and expanded gene families related to floral development, including MADS-box genes (e.g., AP3/PI-like, CsSVP/CsAP1/CsSOC1) that regulate floral patterning and seasonal flowering. Marked amplification of cytochrome- and photosystem-related genes was observed, consistent with their photosynthetic lifestyle and diverse habitat adaptations. Key genomic features include genes involved in anthocyanin and carotenoid biosynthesis pathways that drive the diverse flower colors prized in horticulture. These genomic resources highlight their ecological and economic importance, supporting marker-assisted breeding programs and conservation efforts to sustain these medicinally and ornamentally valuable species (Yang et al. 2021; Ai et al. 2021; Fan et al. 2023).

Vanilla planifolia (~2.4 Gb, $2n = 32$), the primary source of natural vanilla, boasts a genome sequenced using PacBio HiFi and Hi-C scaffolding, revealing critical insights into its ecological and economic value. With 59,128 genes and 67% repetitive DNA (predominantly Ty3/Gypsy retrotransposons), the genome maps the vanillin biosynthesis pathway (VpVAN, VpC4H) on chromosome 4 and identifies disease resistance loci (NBS-LRR) on chromosome 6. These discoveries are pivotal for enhancing cultivated varieties against fungal pathogens like *Fusarium oxysporum*, which threaten global vanilla production. Additionally, the genome highlights recent whole-genome duplication (WGD) events that amplified secondary metabolite production, offering a roadmap for sustainable farming and conservation of wild populations in tropical forests (Piet et al. 2022). This resource bridges genomic innovation with practical applications, safeguarding both agricultural productivity and biodiversity.

Platanthera zijinensis (2n = 42), a specialized orchid thriving in limestone karst ecosystems, possesses a large genome, with approximately 4.95 Gb and ~34,672 genes, sequenced via PacBio CLR technology. The P. zijinensis genome is marked by a ~71.6% repetitive DNA content, predominantly LTR retrotransposons, which underscores its evolutionary complexity. Key adaptations include terpene synthase (TPS) clusters responsible for producing pollinator-specific floral volatiles, critical for reproductive success. Such genomic insights elucidate mechanisms behind niche specialization in nutrient-poor, alkaline environments. By linking pollinator interaction networks and mycorrhizal associations, this research highlights the importance of symbiotic relationships between soil microbiota, pollinators, and orchids for conservation of karst habitats (Li et al. 2022). These findings bridge genomic data with ecological insights, informing conservation efforts for orchids in vulnerable karst ecosystems (Li et al. 2022).

Ophrys sphegodes (2n=36), the early spider orchid, recently achieved a high-quality chromosome-scale genome assembly (5.2 Gb) through PacBio HiFi and Hi-C technologies, marking a milestone in understanding sexually deceptive pollination. The genome reveals significant transposable element expansion, particularly LTR retrotransposons, contributing to its evolutionary complexity. Key adaptations include stearoyl-acyl carrier protein desaturases (SADs) and membrane-bound fatty acid desaturases (FADs), which drive the synthesis of pollinator-mimicking alkenes that mimic female bee pheromones to lure male pollinators. A highly differentiated region on chromosome 2 suggests genomic plasticity associated with pollinator-mediated evolution. This resource highlights the importance of maintaining Mediterranean pollinator networks and mycorrhizal associations critical for the orchid's survival. By identifying key genes and genomic regions, the genome informs ecological studies and supports conservation efforts for sexually deceptive orchids in fragmented Mediterranean habitats (Russo et al. 2024).

10.2.2 Population Genomics and Genetic Diversity Assessment

Population genomics and next-generation sequencing have proven invaluable for conservation of endangered orchids by revealing genetic diversity, evolutionary relationships, and adaptive potential. RAD-seq analysis of four orchid species demonstrated significant genetic diversity and population structure essential for conservation planning, with *Vanda tessellata* exhibiting the highest diversity among the studied species (Roy et al. 2017). In the rare terrestrial orchid *Habenaria linearifolia*, genome-wide SNP analysis revealed three distinct genetic clusters in South Korea, with populations exhibiting significant isolation-by-distance patterns and declining genetic diversity in smaller populations, highlighting urgent conservation needs (Lee et al. 2022). Adaptive sequence evolution driven by biotic stress has been demonstrated in *Dactylorhiza* orchids with distinct ecological optima, where genes involved in plant immune response showed signatures of positive selection (Balao et al. 2022). Human activities have significantly altered genetic patterns in *Phaius flavus* populations, with genomic analysis revealing reduced genetic diversity in areas under anthropogenic pressure (Liang et al. 2024). For cultivated orchids, high-density genetic maps and genome-wide association studies have identified QTLs associated with important traits in *Paphiopedilum* and *Phalaenopsis*, providing valuable resources for conservation breeding programs (Li and Zhu 2022; Hsu et al. 2022). Molecular markers such as SCoT polymorphisms have successfully characterized genetic variations in threatened medicinal orchids and associated them with important seed traits (Gholami et al. 2021), while SNP markers derived from disease resistance genes have effectively analyzed genetic diversity in *Phalaenopsis* orchids (Sukma et al. 2021). These genomic approaches collectively demonstrate the power of multiomics for establishing effective conservation strategies for endangered orchid species.

These population genomic studies provide critical diagnostic tools for assessing genetic erosion in endangered orchids and developing targeted conservation strategies that maintain evolutionary potential. By identifying genetically distinct populations, adaptive loci, and mechanisms of reproductive success, conservation practitioners can prioritize protection efforts and design more effective recovery programs for threatened orchid species.

10.2.3 Conservation Genetics and Phylogenetic Studies

Phylogenomic frameworks are indispensable for resolving evolutionary relationships and prioritizing conservation efforts for evolutionarily distinct lineages. Recent advances in orchid phylogenomics have significantly improved our understanding of evolutionary relationships within this diverse family. Plastome-based phylogenies incorporating 24 new orchid sequences have resolved long-standing taxonomic uncertainties and provided robust evolutionary frameworks essential for conservation prioritization (Kim et al. 2020). A plastome-based phylogeny of slipper orchids (Cypripedioideae) identified *Paphiopedilum* section *Parvisepalum* as the basal lineage within the genus, warranting urgent protection due to several critically endangered species and their unique evolutionary history (Guo et al. 2015a). In the slipper orchid genus *Cypripedium*, phylogenomic analysis using target enrichment and transcriptome data has uncovered rapid radiation and extensive hybridization events, highlighting the complex evolutionary history of this charismatic group and informing conservation strategies for evolutionarily distinct lineages (Lagou et al. 2024).

More comprehensive approaches combining hundreds of nuclear and plastid loci have yielded novel insights into orchid relationships, revealing previously undetected evolutionary patterns and divergence times crucial for biogeographical and macroevolutionary understanding (Pérez-Escobar et al. 2021; Serna-Sánchez et al. 2021). In Madagascar, molecular analyses of *Angraecum* orchids revealed multiple cryptic lineages occupying distinct ecological niches threatened by deforestation, leading to revised conservation assessments and targeted protection efforts (Micheneau et al. 2008). Similarly, floral transcriptome analysis of sexually deceptive and rewarding European orchids (*Ophrys* and *Gymnadenia*) has revealed genetic underpinnings of pollination strategies crucial for species boundaries and conservation (Piñeiro Fernández et al. 2019).

The ghost orchid (*Dendrophylax lindenii*) represents an important conservation case, with recent genetic studies confirming its limited genetic diversity and distinct evolutionary position, supporting enhanced legal protection of its habitat in Florida's swamp forests (Rodgers et al. 2014). Similarly, phylogenetic investigations of the endangered *Platanthera leucophaea* identified genetic threats from hybridization with related species in disturbed habitats, informing conservation strategies that include habitat buffer zones to minimize genetic introgression (Wallace 2003). Recent phylogenetic studies in *Cymbidium* orchids have provided new insights into systematics, diversification, and biogeography, identifying priority conservation targets based on evolutionary distinctiveness (Chen et al. 2024).

The application of evolutionary distinctiveness in orchid conservation provides a scientifically grounded approach to prioritize species and regions for protection. By integrating global frameworks, such as that proposed by Vitt et al. (2023), with regional studies like Pillon et al. (2006), conservation efforts can target species representing significant evolutionary history. This dual approach ensures the preservation of biodiversity within the Orchidaceae, addressing both global and regional challenges amid ongoing environmental pressures. In their study, historical biogeography analyses of *Dactylorhiza* orchids revealed ancient lineages at risk from climate change in Mediterranean refugia, prompting ex situ germplasm collections to preserve their unique evolutionary adaptations (Pillon et al. 2006). As orchid phylogenomics continues to evolve, the prospects of repurposing multi-tissue transcriptomes for phylogenetic analysis promise even greater resolution of evolutionary relationships and adaptive processes (Wong and Peakall 2022), providing essential frameworks for effective conservation planning in this threatened plant family (Freudenstein 2024). By integrating divergence dating, biogeographic models, and conservation assessments, these phylogenetic studies ensure that conservation strategies align with evolutionary trajectories, preserving not just species but evolutionary processes and potential.

10.2.4 Genome-Wide Association Studies (GWAS) in Orchid Traits

While genome-wide association studies (GWAS) have revolutionized trait mapping in many plant species, their application to orchid conservation remains limited. The challenges of applying GWAS to orchids include their large and complex genomes, long generation times, and difficulties in maintaining large mapping populations. Nevertheless, recent advances in orchid genomics are laying the groundwork for future association studies.

Presently, most genetic studies in orchids use targeted approaches rather than full GWAS. For example, Hsu et al. (2022) conducted high-density genetic mapping and GWAS in *Phalaenopsis* orchids, identifying loci linked to flower size, color intensity, and lip morphology. Similarly, Yang et al. (2023) performed GWAS on *Cymbidium sinense*, uncovering candidate genes (e.g., *MYB* transcription factors) associated with floral pigmentation and fragrance variation. These studies demonstrate the feasibility of GWAS in orchids despite their genomic complexity.

For endangered species like *Dendrobium officinale*, genomic resources such as chromosome-level assemblies (Niu et al. 2021) have enabled targeted analyses of stress-related genes. However, full GWAS remains rare due to insufficient population-scale data. Song et al. (2022a) highlighted the utility of high-throughput sequencing for functional genomics in orchids, emphasizing the need for larger datasets to unlock GWAS potential.

In orchids with specialized ecological relationships, such as sexually deceptive *Ophrys* species, Gibert et al. (2025) combined phenotypic divergence analyses with genomic data to identify early speciation signals, including floral scent-related loci. While not a traditional GWAS, this integrative approach bridges trait variation with genomic architecture. For mycoheterotrophic orchids like *Gastrodia elata*, genome sequencing (Yuan et al. 2018) revealed gene losses in photosynthesis pathways, but GWAS applications remain unexplored.

The potential of GWAS for orchid conservation is significant. As sequencing costs decrease, studies like Gawenda et al. (2012)—which used association mapping in *Phalaenopsis* to link molecular markers to flower size and shape—provide a roadmap for identifying adaptive alleles in wild populations. Future efforts could target traits like drought tolerance (*Dendrobium*) or pollinator specificity (*Ophrys*), informing strategies such as assisted gene flow or genotype prioritization in ex situ collections.

10.3 Transcriptomics in Understanding Orchid Biology

10.3.1 RNA-Seq Applications in Orchid Research

RNA-seq has become indispensable for orchid conservation, enabling gene discovery, stress response analysis, and symbiotic interaction studies in species lacking genomic resources. For instance, in *Dactylorhiza hatagirea*, an endangered alpine terrestrial orchid, de novo transcriptome assembly identified genes involved in resveratrol and trans-stilbenes biosynthesis, compounds with potential medicinal value (Dhiman et al. 2019). Similarly, in *Arundina graminifolia*, transcriptomic analysis revealed genes associated with flavonoid and bibenzyl biosynthesis pathways, highlighting the medicinal importance of this rare orchid species (Ahmad et al. 2022). These studies demonstrate RNA-seq's power to bridge genomic gaps and decode adaptive traits in non-model orchids.

Stress response mechanisms in orchids have been elucidated through transcriptome profiling. In *Dendrobium sinense*, an endemic orchid species in Hainan Island, RNA-seq uncovered genes involved in purine and phenylpropanoid metabolism pathways responding to drought stress (Zhang et al. 2021). Likewise, in *Cymbidium tracyanum*, transcriptome analysis revealed the nitrogen preference and regulatory pathways of nitrogen metabolism, providing insights into how this epiphytic orchid adapts to nutrient-limited environments (Dong et al. 2024). Such insights are vital for conserving species threatened by climate extremes and habitat degradation.

Symbiotic interactions, essential for orchid survival, have also been dissected using RNA-seq. In *Gastrodia elata*, transcriptome analysis of mycorrhizal interactions identified genes related to gastrodin biosynthesis, highlighting the importance of fungal symbiosis for this medicinal orchid's survival (Tsai et al. 2016). Similarly, transcriptomic analysis of *Serapias vomeracea* seeds during early interaction with the orchid mycorrhizal fungus *Tulasnella* sp. revealed important insights into the molecular mechanisms underlying this symbiotic relationship (De Rose et al. 2024). These findings can help optimize symbiotic propagation protocols for threatened species.

Floral diversity studies leverage RNA-seq to decode evolutionary innovations. In *Chiloglottis trapeziformis*, a sexually deceptive orchid, transcriptome analysis identified genes involved in the biosynthesis of its unique UV-B dependent floral volatile, chiloglottone 1, which is crucial for pollinator attraction (Wong et al. 2017). Meanwhile, *Habenaria radiata* transcriptomes revealed genes shaping its distinctive white bird-like flower morphology, offering insights into the genetic basis of its unique floral features (Takeda et al. 2025). Such work informs conservation of pollination systems and evolutionary adaptations.

Emerging technologies further expand RNA-seq applications. In *Vanilla planifolia*, pod transcriptome analysis elucidated genes central to vanillin biosynthesis, guiding sustainable cultivation to reduce wild harvesting (Rao et al. 2014). Additionally, transcriptomic studies in various orchid species have uncovered MADS-box genes regulating floral development and flowering, which are crucial for understanding reproductive biology and conservation (Lin et al. 2016; Huang et al. 2015).

By identifying stress-resilient genotypes, optimizing symbiosis, and decoding floral traits, RNA-seq equips conservationists with actionable insights. Integrating these approaches ensures adaptive potential is maintained in wild and cultivated populations, safeguarding orchids against habitat loss and climate change. Table 10.1 represents a summary of RNA-seq-based studies across diverse orchid taxa, including species, tissues sampled, sequencing platforms, key candidate genes, biological focus, metabolic pathways, and conservation applications. Data highlight the role of transcriptomics in unraveling adaptive traits, stress responses, and symbiotic mechanisms critical for safeguarding orchid biodiversity.

10.3.2 Gene Expression Studies in Stress Response

Under abiotic stress, orchids activate sophisticated genetic networks to mitigate damage. For instance, in the epiphytic orchid *Cymbidium sinense*, drought stress triggers the upregulation of antioxidant enzymes (e.g., superoxide dismutase) and osmoprotectant biosynthesis pathways, which collectively reduce oxidative damage and maintain cellular hydration. Concurrently, proteomic analyses revealed enhanced root morphological adaptations to improve water uptake efficiency under

Table 10.1 Genomic and transcriptomic studies of orchid species, highlighting key genes, pathways, and their relevance to conservation and breeding programs

Species (life type)	Tissue sampled	Sequencing platform	Number of transcripts/ unigenes/DEG	Key candidate genes identified	Study focus	Key pathways	Conservation or breeding relevance	References
Anoectochilus roxburghii (terrestrial)	Leaves	Illumina HiSeq 4000/NovaSeq	2740 DEGs from 85 Gb clean data	HSPs/HSFs, SOD, POD, CAT, CNGC1/2, WRKYs, BRI1, GID1C/GID2, SAUR32/50	Molecular regulation of tolerance to high temperatures	Plant-pathogen interaction, Plant hormone signal transduction, ROS scavenging pathways	Breeding heat-tolerant varieties and enhancing climate resilience	Zhang et al. (2024b)
Arundina graminifolia (terrestrial)	Flowers, leaves	Illumina NovaSeq 6000	49,300 transcripts	*F3'5'H, DFR, CHS, CHI, FNSII, CYP450s, UGTs*	Flavonoid/ bibenzyl biosynthesis	Phenylpropanoid, flavonoid, and bibenzyl biosynthesis	Medicinal orchid conservation	Ahmad et al. (2022)
Calanthe tsoongiana (terrestrial)	Rhizome, seeds, protocorms	Illumina HiSeq 2500	41,328 unigenes	CtrNAC, CtrWRKY, CtrLEA, CtrPIP	Seed germination and protocorm development	Hormone signaling, transcription regulation, stress response	Molecular basis for *in vitro* propagation and restoration of endangered montane orchids	Jiang et al. (2021)
Cattleya trianae (epiphytic)	Leaves, pseudobulbs, meristems, flower buds	Illumina NovaSeq 6000	~50,000 transcripts	CatrFT1B, CatrFT1C, CatrFT2A (FT-like), CatrTFL1 (TFL1-like)	Floral development and evolution	Reproductive transition regulation	*ex situ* conservation and control breeding	Ospina-Zapata et al. (2020)
Cymbidium ensifolium (terrestrial)	Green and red leaf tissues	Illumina HiSeq	12,217 DEGs	CHS, F3H, DFR, LDOX, UFGT, PSY, PDS, CRTISO, MYB, bHLH, WRKY	Leaf color variation	Pigment biosynthesis, hormonal signaling	Insights for ornamental breeding	Xiong et al. (2025)
Cymbidium goeringii (terrestrial)	Flower in developmental stages	Illumina HiSeq	85,868 unigenes	AACT, HMGS, HMGR, MVK, TPS, FDPS	Floral scent biosynthesis	Carotenoids, anthocyanins, and floral scent biosynthesis	Breeding fragrant cultivars, pollinator conservation and restoration	Ramya et al. (2019)

(continued)

Table 10.1 (continued)

Species (life type)	Tissue sampled	Sequencing platform	Number of transcripts/ unigenes/DEG	Key candidate genes identified	Study focus	Key pathways	Conservation or breeding relevance	References
Dendrobium catenatum (epiphytic)	Leaf	Illumina HiSeq	28,910 unigenes	Resistance-related genes, polysaccharide synthase genes	Genome sequencing and adaptive evolution	Polysaccharide biosynthesis, immune response pathways	Ecological adaptation and medicinal value	Zhang et al. (2016)
Dendrobium nobile (epiphytic)	Stem	PacBio SMRT, Illumina HiSeq 6000	80,135 unigenes	ALDH2B7, ADC, EPSPS-1, SHKA, DHAPS-1, GES, ACS1, SAHH, ACS2, CHLP, LOX2	Differential metabolite production under different epiphytic patterns	Primary and secondary metabolism	Cultivation optimization	Li et al. (2022)
Epipactis helleborine (terrestrial)	Mycorrhizalroots	Illumina HiSeq 1500	61,007 contigs	SWEET, GPAT (RAM2-like), subtilisin, CHS	Mycoheterotrophy, mycorrhizal symbiosis	Sugar transport and transport, flavonoid biosynthesis, phytohormone regulation, mycorrhizal symbiosis	Conservation of mycoheterotrophic orchids reliant on specific fungal partners	Suetsugu et al. (2017)
Gastrodia elata (mycoheterotrophic)	Tuber	PacBio SMRT	25,606 genes	NRT1 transporters	Fungal symbiosis/ heterotrophy	Nutrient uptake/ transport	Adaptation to low-light environments	Yuan et al. (2018)
Phaius flavus (terrestrial)	Flower parts	Illumina HiSeq 2500	65,264 transcripts, 17,134 DEGs	AS, PAI, IGPS, TSA, CYP450, UGT, GLU	Cold tress response	Indigo biosynthesis pathway, Tryptophan metabolism	Dptation to cold stress	Zhang et al. (2022)
Vanilla pompona (epiphytic)	Capsule	Illumina Novaseq	15,896 DEGs	CPLP, PAL, CYP, OMT, CSE, β-Glu, 4CL, UGTs, HCTVpVAN, VpMYB	Vanillin biosynthesis	Phenylpropanoid metabolism, starch & sucrose metabolism, lignin biosynthesis	Sustainable vanilla cultivation	Gastelbondo et al. (2025)
Species (life type)	**Tissue sampled**	**Sequencing platform**	**Number of transcripts/ unigenes**	**Key candidate genes identified**	**Study focus**	**Key pathways**	**Conservation relevance**	**References**

Phalaenopsis aphrodite (epiphytic)	Vegetative and reproductive	Illumina/ Solexa, Roche 454	42,590 annotated contigs	*AP3, AGL6, SEP1*	Floral development	MADS-box regulation	Breeding for ornamental traits	Su et al. (2013)
Erycina pusilla (epiphytic)	Fruit valves	Illumina Illumina NovaSeq 6000	83,137 transcripts; 5,785 DEGs	*EpSEP3, EpAG2, EpRPL, EpKOR1, EpMADS24/Bsister, EpMAH1*	Evolution of dehiscence zones and gene expression in orchid fruits	MADS-box regulation, lipid biosynthesis, fruit valve and DZ development	Understanding fruit development for model orchid evolution and adaptation	Pramanik et al. (2023)
Malaxis acuminata (terrestrial)	Tuber (pseudobulb), leaf	Illumina HiSeq 2500	48,921 unigenes	*PAL, C4H, 4CL*	Secondary metabolite biosynthesis	Phenylpropanoid pathways	Medicinal value conservation	Bhattacharyya et al. (2022)
Dactylorhiza hatagirea (terrestrial)	Tuber	Illumina HiSeq	46,785 unigenes	*STS, RS*	Resveratrol/ stilbene biosynthesis	Antioxidant pathways	Alpine habitat resilience	Dhiman et al. (2019)
Chiloglottis trapeziformis (terrestrial)	Labellum	Illumina HiSeq	41,200 transcripts	*TPS, LOX*	Floral volatile biosynthesis	Chiloglottone production	Pollinator deception conservation	Wong et al. (2017)
Orchis italica (terrestrial)	Inflorescence	Illumina HiSeq	38,984 unigenes	*DIV, RAD*	Floral symmetry	DIVARICATA/ RADIALIS pathways	Mediterranean habitat adaptation	De Paolo et al. (2014)
Vanda Mimi Palmer (epiphytic)	Labellum	Illumina HiSeq	35,600 transcripts	*TPS, PAL*	Floral fragrance	Terpenoid biosynthesis	Fragrance trait preservation	Toh et al. (2017)
Cypripedium macranthos (terrestrial)	Flowers	PacBio Sequel	68,342 transcripts	*AGL6, AP1*	Floral morphogenesis	MADS-box regulation	Habitat-specific pollination	Kambara et al. (2023)
Serapias vomeracea (terrestrial)	Seeds	Illumina NovaSeq	27,450 transcripts	*SWEET, PT4*	Fungal symbiosis	Nutrient transport	Mycorrhizal dependency in restoration	De Rose et al. (2024)
Limodorum abortivum (terrestrial)	Roots	Illumina HiSeq	33,120 transcripts	*RAM2, STR*	Mycorrhizal interaction	Lipid transfer signaling	Mediterranean forest conservation	Valadares et al. (2021)

(continued)

Table 10.1 (continued)

Species (life type)	Tissue sampled	Sequencing platform	Number of transcripts/unigenes/DEG	Key candidate genes identified	Study focus	Key pathways	Conservation or breeding relevance	References
Paphiopedilum armeniacum (terrestrial)	Seed	Illumina HiSeq	29,870 transcripts	*CAD, CCR*	Lignin biosynthesis	Non-methylated lignin pathways	Seed germination protocols	Fang et al. (2020)
Cymbidium tortisepalum (terrestrial)	Bisexual/ unisexual flowers	Illumina HiSeq	44,200 transcripts	*AG, AP3*	Sex determination	MADS-box regulation	Genetic diversity preservation	Ma et al. (2023)
Dendrobium sinense (epiphytic)	Leaves	Illumina NovaSeq	52,100 transcripts	*P5CS, NCED*	Drought response	Proline/ABA biosynthesis	Hainan Island endemic conservation	Zhang et al. (2021)
Phalaenopsis equestris (epiphytic)	Floral organ, leaf, root, meritem	Illumina HiSeq 4000/ NextSeq500	21,676 unigenes	*CAM-related genes*	Crassulacean acid metabolism	Carbon fixation	Epiphytic adaptation	Klepikova et al. (2021)
Vanilla planifolia (epiphytic)	Fruit, pod	Illumina HiSeq	50,321 transcripts	*VpVAN, VpMYB*	Vanillin biosynthesis	Phenylpropanoid metabolism	Sustainable cultivation	Rao et al. (2014)
Dendrobium officinale (epiphytic)	Whole plant	Illumina HiSeq	28,000 unigenes	*GA-GID1-DELLA*	Seed germination	Hormonal regulation	Symbiotic propagation	Meng et al. (2016)
Habenaria radiata (terrestrial)	Flowers	PacBio/Illumina	45,000 unigenes	*HrCHS, HrDFR*	Floral morphology	Flavonoid biosynthesis	Habitat fragmentation resilience	Takeda et al. (2025)
Gastrodia elata (mycoheterotrophic)	Tuber	PacBio SMRT	25,606 transcripts	*NRT1*	Fungal symbiosis	Nutrient transport	Low-light adaptation	Tsai et al. (2016)

arid conditions (Li et al. 2018). Beyond these ornamental genera, terrestrial orchids like *Bletilla ochracea* exhibit stress-responsive biosynthesis of secondary metabolites. For example, Yang et al. (2012) identified antimicrobial phenanthrenes and stilbenoids in *Bletilla ochracea* tubers, which are upregulated under biotic stress to enhance pathogen resistance. These findings suggest that secondary metabolite pathways in terrestrial orchids are dynamically regulated under environmental challenges, potentially serving as adaptive mechanisms for survival in ground-dwelling habitats. Such mechanisms highlight the metabolic and physiological plasticity of orchids in adapting to environmental challenges, and these studies highlight the interplay between conserved stress pathways and habitat-specific adaptations, such as the integration of crassulacean acid metabolism (CAM) with drought-responsive gene regulation.

Biotic stressors, including pathogens and herbivores, elicit distinct molecular defenses across diverse orchid lineages. The mycoheterotrophic orchid *Gastrodia elata* responds to environmental challenges by modulating expression of genes in the phenylpropanoid pathway, which synthesize bioactive compounds with medicinal properties (Yuan et al. 2018). In temperate terrestrial orchids like *Epipactis albensis*, proteomic studies have identified distinct defense mechanisms compared to tropical epiphytes. For instance, Jakubska-Busse et al. (2021) demonstrated that *Epipactis* species exhibit differential expression of pathogenesis-related (PR) proteins and phenylpropanoid pathway enzymes under biotic stress, contrasting with the terpenoid-dominated defenses observed in tropical orchids like *Phalaenopsis*. This divergence underscores habitat-specific evolutionary adaptations in pathogen resistance strategies across orchid lineages.

Orchids navigate complex symbiotic interactions through precise gene regulation. In terrestrial orchids like *Dendrobium officinale*, transcriptomic and proteomic analyses by Chen et al. (2017) identified genes (e.g., *DOF* transcription factors) and proteins critical for nutrient exchange during mycorrhizal symbiosis, enabling survival in nutrient-poor environments. Similarly, studies of fully mycoheterotrophic orchids such as *Gastrodia elata* revealed metabolic adaptations, including the suppression of photosynthetic gene networks and reliance on fungal carbon sources, as demonstrated by Zeng et al. (2018) through integrated metabolomic and proteomic profiling. These findings underscore the genetic and biochemical innovations underpinning orchid symbiosis and nutritional evolution. Among epiphytic orchids, Vanilla species represent particularly important subjects for genomic research due to their economic significance and unique hemiepiphytic growth habit. These plants utilize both terrestrial roots that anchor them within the soil and epiphytic roots that anchor them to tree trunks, representing a specialized climbing adaptation. Recent genomic advances, including chromosome-level genome assemblies of Vanilla planifolia, have provided unprecedented insights into the molecular basis of orchid biology and adaptation mechanisms in this commercially important epiphytic species (Piet et al. 2022).

Developmental processes in orchids are governed by intricate gene regulatory networks. Floral development studies in deceptive orchids like *Ophrys* have identified genes responsible for producing compounds that mimic insect pheromones, a remarkable adaptation for pollinator attraction (Sedeek et al. 2016). In *Apostasia* and other basal orchid lineages, expression studies have revealed ancestral developmental pathways that predate the evolution of the specialized labellum found in more derived orchids (Zhang et al. 2017). The economically important *Vanilla planifolia* shows distinctive expression patterns related to vanillin biosynthesis during fruit development, insights that are crucial for improving this valuable crop (Gallage et al. 2018).

Gene expression studies directly inform conservation practice across the orchid family. Transcriptomic profiling of endangered species like *Cypripedium macranthos* has provided valuable genomic resources through de novo assembly of multiple transcriptome datasets, enabling better understanding of this threatened genus's biology and potentially informing conservation strategies (Kambara et al. 2023). In neotropical orchids, studies of stress adaptation mechanisms have informed restoration efforts in degraded ecosystems. For instance, Herrera et al. (2018) investigated the symbiotic adaptation of *Bipinnula fimbriata* to heavy metal pollution, identifying mycorrhizal interactions and antioxidant enzyme upregulation as key mechanisms for survival in contaminated soils. These findings directly guide strategies for rehabilitating mining-impacted habitats. Emerging analytical approaches, such as co-expression network analyses (Toubiana et al. 2013), are resolving genetic hierarchies across orchid lineages, enabling systems-level insights into stress responses and developmental pathways. This integrative framework bridges gaps between model and non-model orchids, empowering conservationists to prioritize resilient genotypes for restoration.

10.3.3 Developmental Biology Insights

Understanding the molecular basis of orchid development—from seed germination to flowering—is critical for optimizing propagation protocols and conserving species with complex life cycles. Orchids exhibit unique developmental traits, such as symbiotic germination (Valadares et al. 2014), prolonged juvenile phases (Phillips et al. 2020), and intricate floral morphogenesis (Chen et al. 2021), which are governed by tightly regulated genetic and epigenetic networks.

Seed Germination and Protocorm Development Orchid seeds are minute and lack endosperm, relying entirely on mycorrhizal fungi for carbon and nutrient supply during germination in nature. While the molecular mechanisms underlying symbiotic germination are not fully characterized, studies such as De Rose et al. (2023) on orchid mycorrhizal protocorms identified patterns of plant and fungal gene expression coupled with stable isotope labeling, providing novel information on sulfur uptake and metabolism. Additionally, Jiang et al. (2021) revealed the mechanisms involved in seed germination and protocorm development of *Calanthe tsoongiana* through transcriptome sequencing and differential gene expression analysis, highlighting nutrient transport and hormone signaling genes critical for this developmental process. In *Dendrobium officinale*, Chen et al. (2017) revealed auxin- and cytokinin-related gene networks activated during symbiotic germination, with parallels observed in asymbiotic systems where exogenous hormone applications (e.g., TDZ) enhance protocorm formation. These findings have directly informed optimized media formulations for challenging species, such as *Cattleya tigrina*, where proteomic insights into hormone regulation improved asymbiotic protocols (De Conti et al. 2020).

The transition from protocorm to seedling involves complex developmental reprogramming. Studies in *Erycina pusilla* have identified key regulators like LEAFY COTYLEDON1 (LEC1) that promote embryonic development and somatic embryogenesis, critical processes for orchid propagation (Chen et al. 2019a, b). WUSCHEL-related homeobox (WOX) genes, known to maintain stem cell populations in model plants, have been identified in several orchid species with putative roles in reproductive development (Ramkumar et al. 2020). Understanding these molecular mechanisms has practical applications for improving germination rates in threatened species like *Orchis mascula* and species within *Cypripedium* genus where germination remains challenging even under laboratory conditions (Anghelescu et al. 2023).

Floral Morphogenesis and Pollination Orchid flowers are renowned for their structural complexity and pollinator-specific adaptations. Extensive research has focused on MADS-box genes that determine floral organ identity. In *Phalaenopsis equestris*, transcriptomic studies have identified expanded gene families of MADS-box genes, including DEFICIENS (DEF), GLOBOSA (GLO), and AGAMOUS-LIKE 6 (AGL6), which likely contribute to the highly specialized floral morphology of orchids (Cai et al. 2015). The characteristic labellum (lip) development involves complex interactions between these genes, with differential expression of DEF-like genes observed between the lip and other petals (Mondragón-Palomino and Theißen 2011).

Floral scent production, critical for pollinator attraction in many orchids, involves terpene synthase (TPS) genes and various biosynthetic pathways. In sexually deceptive orchids like *Ophrys*, transcriptome analysis has revealed active expression of genes involved in fatty acid biosynthesis and modification, producing compounds that mimic insect pheromones (Sedeek et al. 2013). The temporal regulation of these genes often coincides with pollinator activity periods, representing a remarkable example of molecular adaptation for reproductive success.

Vegetative Development and Stress Resilience Orchids exhibit remarkable plasticity in vegetative structures, such as pseudobulbs and velamen roots, to buffer environmental stress. Pseudobulbs function as water and nutrient storage organs, with transcriptomic studies in *Dendrobium catenatum* revealing upregulation of genes involved in carbohydrate metabolism and stress response during pseudobulb development (Zhang et al. 2016). These include genes related to raffinose biosynthesis and dehydration response, consistent with the role of pseudobulbs in drought tolerance.

Velamen roots, characteristic of epiphytic orchids, provide essential adaptations for water and nutrient capture in arboreal habitats. Research has demonstrated that charged ions are retained in the velamen probably due to positive and negative charges in the cell walls, while uncharged compounds are lost to the external medium, and that nutrient uptake follows biphasic kinetics with a highly efficient, active transport system at low external concentrations (Zotz and Winkler 2013). Additionally, studies have shown that the velamen's main function may lie in protecting the underlying cells against damaging UV rays (Chomicki et al. 2015). Research on *Cymbidium* roots has demonstrated differential expression of phosphate transporters and mycorrhizal-responsive genes under varying nutrient conditions, highlighting adaptations to nutrient-poor environments (Zhao et al. 2014).

10.3.4 Small RNA Analysis in Orchid Conservation

Small RNAs (sRNAs), including microRNAs (miRNAs) and small interfering RNAs (siRNAs), are emerging as important regulators of gene expression, stress adaptation, and development in orchids. These noncoding RNAs typically function by silencing target mRNAs through complementary base pairing or by directing DNA methylation, providing an additional layer of gene regulation beyond transcriptional control.

Early characterization of orchid miRNAs has been accomplished primarily through small RNA sequencing approaches. In *Phalaenopsis aphrodite*, conserved miRNA families including miR156, miR166, and miR399 have been identified, with expression patterns suggesting roles in vegetative development and stress response (An et al. 2011). Similarly, analysis of *Dendrobium officinale* small RNAs revealed both conserved and novel miRNAs, with many predicted to target transcription factors involved in development and secondary metabolism (Meng et al. 2016).

Stress-responsive microRNAs (miRNAs) have become a focal point in orchid research for their role in regulating gene expression under environmental stresses. In *Dendrobium sinense*, an endemic epiphytic orchid, drought-responsive miRNAs have been identified, including miR398, which targets copper superoxide dismutase genes, and miR408, which regulates genes involved in phenylpropanoid metabolism, both contributing to stress tolerance (Zhang et al. 2021). Through integrated transcriptomic and metabolomic analyses, the study revealed 622 differentially expressed genes and 391 differential metabolites under drought stress, with miR398 and miR408 playing critical roles in enhancing antioxidant defense and metabolic adjustments. These findings underscore the potential of miRNA-mediated regulation to bolster drought resilience in orchids, providing valuable insights for conservation and breeding strategies. Cold stress in *Cymbidium goeringii* induces significant physiological and molecular changes, with increased expression of cold-responsive genes, such as those encoding antioxidant enzymes and CBF-like transcription factors, which enhance tolerance to low temperatures (Fan et al. 2022). The study revealed that under 0°C stress, superoxide dismutase and peroxidase activities increased, alongside upregulation of genes involved in reactive oxygen species scavenging and cold acclimation pathways. These findings suggest potential applications for identifying stress-resilient genotypes in conservation programs.

Small RNAs play critical roles in orchid reproductive development, particularly in regulating floral organ specification. In Phalaenopsis, studies have identified differential expression of microRNAs (miRNAs) across floral organs, with miR172 predicted to target APETALA2-like transcription factors involved in determining floral organ identity (Huang et al. 2016). Through genome sequencing of the Phalaenopsis 'KHM190' cultivar, 188 miRNA families were identified, revealing their involvement in complex regulatory networks that govern floral morphology. The intricate floral structures characteristic of orchids, such as the zygomorphic perianth and specialized labellum, may partly arise from small RNA-mediated regulation of key developmental genes, offering insights for conservation and breeding programs.

The symbiotic relationships between orchids and mycorrhizal fungi involve sophisticated molecular regulation, including small RNA-mediated mechanisms that facilitate cross-kingdom communication. While research in orchids remains limited, studies in other plant-fungal systems demonstrate that small RNAs play pivotal roles in symbiotic establishment by modulating gene expression in both partners (Šečić et al. 2021). In orchids, transcriptomic and small RNA analyses have begun to unravel these dynamics. For example, in *Dendrobium officinale*, organ-specific transcriptome mapping revealed distinct regulatory networks governing symbiotic interactions, particularly in roots and protocorms (Meng et al. 2016). Further, microRNA profiling in *Phalaenopsis aphrodite* identified symbiosis-related miRNAs potentially involved in fungal communication (Chao et al. 2014), while studies in *Bletilla striata* highlighted miRNA-mediated regulation of seed germination pathways under asymbiotic conditions (Wang et al. 2021).

Notably, the endophytic fungus *Piriformospora indica* induces growth-promoting miRNAs in *Oncidium* orchid roots, suggesting a conserved role for small RNAs in sustaining mutualistic partnerships (Ye et al. 2014). These findings align with recent work showing that fungal sRNAs reprogram host gene expression during early symbiosis (Šečić et al. 2021), a mechanism likely critical for orchid seed germination, which depends entirely on mycorrhizal fungi for nutrient supply. This emerging field underscores the importance of small RNA regulation in orchid–fungal interactions, with direct implications for conservation strategies aimed at enhancing symbiotic germination in endangered species.

Practical applications of small RNA research for orchid conservation are still emerging. Potential approaches include using miRNA expression profiles as biomarkers for stress resilience or developmental competence, which could inform selection of materials for ex situ conservation. The possibility of engineering stress tolerance through manipulation of key miRNAs and their targets represents another frontier, though significant technical challenges remain for non-model plant species like orchids.

Despite promising advances, significant knowledge gaps persist in orchid small RNA biology. Most studies have focused on horticultural species, with limited information available for threatened wild species. The functional validation of miRNA–target relationships remains challenging in orchids due to transformation difficulties and long generation times. Future research integrating small RNA profiles with transcriptomic, genomic, and epigenomic data will likely provide more comprehensive understanding of orchid development and adaptation, informing conservation strategies for these ecologically and culturally important plants.

10.4 Orchid Proteomics: Insights into Adaptation and Conservation

10.4.1 Protein Profiling Techniques

Protein profiling techniques are indispensable for unraveling the molecular mechanisms that enable orchids to adapt to environmental stressors, engage in symbiotic relationships, and complete intricate developmental processes. These methodologies provide critical insights into protein expression, localization, and posttranslational modifications (PTMs), forming the backbone of orchid proteomic research.

Two-dimensional gel electrophoresis (2D-GE) remains one of the most widely used techniques for separating proteins based on their isoelectric point and molecular weight, resolving isoforms and PTMs. For instance, Li et al. (2018) used 2D-GE to identify differentially expressed proteins during drought stress in *Cymbidium sinense* and *C. tracyanum*, revealing distinct adaptive strategies between the two species. Similarly, Lai et al. (2013) employed 2D-GE to analyze the proteomic responses of *Phalaenopsis amabilis* to viral infections, identifying proteins involved in defense mechanisms.

Advancements in liquid chromatography–tandem mass spectrometry (LC-MS/MS) have revolutionized proteomics by enabling high-throughput identification and quantification of thousands of proteins. Song et al. (2022b) utilized label-free quantitative proteomics to investigate the impacts of salt stress on *Dendrobium huoshanense*, identifying key proteins involved in photosynthesis, carbohydrate metabolism, and stress responses. Similarly, Lu et al. (2020) used proteomic profiling to uncover the role of cytosolic superoxide dismutase (BsSOD1) in plant defense mechanisms in the medicinal orchid *Bletilla striata*.

The development of isobaric tags for relative and absolute quantitation (iTRAQ) has significantly enhanced quantitative proteomic analysis in orchids. Chen et al. (2017) employed iTRAQ combined with RNA-seq to investigate the regulation mechanisms of symbiotic germination in *Dendrobium officinale* seeds, revealing proteins critical for fungal recognition and seed development. Similarly, Chen et al. (2020) used the iTRAQ/MRM approach to analyze proteomic changes during flower bud development in *Phalaenopsis amabilis* under low-temperature induction, identifying proteins involved in flowering time regulation.

Posttranslational modifications (PTMs) have gained attention in orchid proteomic research. Feng et al. (2017) conducted a comprehensive succinyl-proteome profiling of *Dendrobium officinale*, revealing the involvement of protein succinylation in glycolysis and other metabolic pathways, providing insights into the molecular mechanisms underlying the medicinal properties of this traditional Chinese herb.

Orchid genome sequencing projects have significantly contributed to proteomic research by providing reference databases for protein identification. These genomic resources, coupled with advanced proteomic technologies, have enhanced our understanding of orchid biology and conservation. As highlighted by Chen and Harmon (2006) and Baginsky (2009), advances in plant proteomics continue to offer new opportunities for orchid research, particularly for non-model species (Champagne and Boutry 2013). Table 10.2 summarizes proteomic studies across Orchidaceae species highlighting key biological processes, molecular mechanisms, and their conservation or biotechnological applications. The table summarizes proteomic techniques, findings related to symbiotic interactions, stress responses, floral development, and other critical biological functions, with implications for orchid conservation and cultivation.

Applications in Conservation Proteomics has informed habitat restoration and cultivation strategies for orchids. For example, Valadares et al. (2020) characterized proteins involved in mycorrhizal symbiosis in *Oeceoclades maculata*, revealing metabolic changes and reduced defense responses in mycorrhizal roots. This knowledge provides insights into the molecular dialogue between orchids and fungi, with implications for optimizing symbiotic germination protocols for rare orchid species.

In terrestrial orchids, proteomic analyses have revealed important adaptations to specific environmental conditions. Li et al. (2018) identified proteins associated with drought adaptation in *Cymbidium sinense* and *C. tracyanum*, showing distinct physiological and proteomic responses between species, information critical for conservation planning. Similarly, Herrera et al. (2018) used proteomic approaches to investigate adaptation mechanisms in mycorrhizal *Bipinnula fimbriata* plantlets in heavy metal-polluted ecosystems, revealing tolerance mechanisms that could inform conservation strategies.

Symbiotic relationships are particularly important for orchid conservation. Zeng et al. (2018) revealed proteins associated with symbiotic germination of *Gastrodia elata* through proteomic analysis, identifying key enzymes involved in nutrient metabolism during mycorrhizal associations. Additionally, Xu et al. (2015) used proteomic analysis to understand how *Mycena dendrobii* promotes transplantation survival and growth of tissue culture seedlings of *Dendrobium officinale*, providing valuable insights for conservation programs.

Despite these advances, technical hurdles persist. Low-abundance proteins, such as transcription factors, are often obscured by high-abundance molecules like rubisco. As highlighted by van Wijk (2001) and Chen and Harmon (2006),

Table 10.2 Proteomic studies performed on the species within Orchidaceae family highlighting key biological processes, findings, and their implications for conservation and biotechnology

Orchid species	Proteomic technique	Biological process/ condition	Key findings	Conservation/biotechnological implications	References
Gastrodia elata	Proteomic analysis	Symbiotic germination	Identified proteins associated with symbiotic germination, including those involved in nutrient metabolism	Potential for improving in vitro propagation protocols for rare mycoheterotrophic orchids	Zeng et al. (2018)
Dendrobium officinale	Proteomic analysis	Seedling survival and growth	Revealed mechanisms by which *Mycena dendrobii* promotes transplantation survival	Direct application in improving acclimatization success rates for conservation programs	Xu et al. (2015)
Dendrobium chrysanthum	Comparative proteomics	Pollination response	Identified proteins differentially expressed during pollination	Insights for breeding programs and artificial pollination of endangered species	Wang et al. (2017)
Oeceoclades maculata	Proteomic and transcriptomic analyses	Mycorrhizal associations	Detected metabolic changes and reduced defense responses in mycorrhizal roots	Understanding natural symbiotic relationships for habitat restoration efforts	Valadares et al. (2020)
Vanilla planifolia	Proteomic analysis	Callus development	Characterized proteins involved in callus development	Improving micropropagation efficiency for this economically important orchid	Tan et al. (2013)
Dendrobium huoshanense	Label-free quantitative proteomics	Salt stress	Unraveled impacts of salt stress on protein expression	Selection of salt-tolerant genotypes for conservation in degraded habitats	Song et al. (2022b)
Sexually deceptive orchids	Transcriptome and proteome analysis	Pollinator attraction	Revealed candidate genes for pollinator attraction mechanisms	Understanding pollinator–orchid relationships crucial for in situ conservation	Sedeek et al. (2013)
Vanilla planifolia	Proteomic analysis	Shoot differentiation	Identified proteomic responses during early-stage shoot differentiation	Optimization of commercial propagation protocols for vanilla production	Palama et al. (2010)

Bletilla striata	Proteomic profiling	Plant defense	Discovered cytosolic superoxide dismutase BsSOD1 in defense	Potential development of disease-resistant varieties of this medicinal orchid	Lu et al. (2020)
Cymbidium sinense and *C. tracyanum*	Comparative physiological and proteomic analyses	Drought adaptation	Revealed different adaptive strategies between species	Selection of drought-resistant genotypes for climate change adaptation	Li et al. (2018)
Cymbidium ensifolium	Comparative proteomic analysis	Floral development	Identified differentially expressed proteins in floral organs	Understanding evolutionary and developmental aspects of orchid flowers	Li et al. (2014)
Phalaenopsis amabilis	Proteomics-based analysis	Viral infection response	Characterized proteomic responses to viral infections	Development of early detection methods for viral diseases in nurseries	Lai et al. (2013)
Epipactis spp.	Proteomics-based identification	Bacteria–orchid interactions	Identified orchid-associated bacteria	Potential use of beneficial bacteria in conservation and restoration	Jakubska-Busse et al. (2021)
Bipinnula fimbriata	Proteomic analysis	Heavy metal tolerance	Uncovered adaptation mechanisms to heavy metals	Phytoremediation applications and conservation in contaminated soils	Herrera et al. (2018)
Dendrobium officinale	Succinyl-proteome profiling	Posttranslational modifications	Revealed protein succinylation in metabolic pathways	Understanding molecular basis of medicinal properties for quality control	Feng et al. (2017)
Cattleya tigrina	Proteomic analysis	Protocorm-like body development	Identified proteins in protocorm-like body development	Improving asymbiotic germination protocols for endangered species	De Conti et al. (2020)
Cymbidium goeringii	Floral organ-specific proteome profiling	Floral organ development	Revealed proteins related to floral organ development	Insights for flower quality improvement in ornamental breeding	Chen et al. (2021)

(continued)

Table 10.2 (continued)

Orchid species	Proteomic technique	Biological process/ condition	Key findings	Conservation/biotechnological implications	References
Phalaenopsis 'Red Dragon'	Proteomic analysis	Petal senescence	Characterized changes during petal senescence	Extension of flower longevity for commercial orchid industry	Chen et al. (2018)
Phalaenopsis amabilis	iTRAQ/MRM approach	Flower bud development	Analyzed changes under low-temperature induction	Control of flowering time for commercial orchid production	Chen et al. (2020)
Dendrobium officinale	iTRAQ and RNA-Seq analyses	Symbiotic seed germination	Provided insights into symbiotic germination mechanisms	Improving propagation success rates for conservation programs	Chen et al. (2017)
Phalaenopsis	Protein analysis	Tissue browning	Identified proteins related to browning in leaf culture	Reducing tissue browning during micropropagation and tissue culture	Chen et al. (2012)

challenges in plant proteomics include protein extraction difficulties, dynamic range limitations, and functional annotation gaps, particularly for non-model orchid species lacking sequenced genomes.

Integrative approaches combining proteomics with genomics, transcriptomics, and metabolomics promise a holistic understanding of orchid biology. For instance, Chen et al. (2017) combined iTRAQ and RNA-seq analyses to study symbiotic germination in *Dendrobium officinale* seeds, identifying key proteins involved in nutrient exchange with fungal partners. Similarly, Sedeek et al. (2013) integrated transcriptome and proteome data to reveal candidate genes for pollinator attraction in sexually deceptive orchids. Such multiomics approaches, as reviewed by Mumtaz et al. (2017), will refine conservation strategies, from selecting resilient genotypes to optimizing cultivation conditions for endangered species.

10.4.2 Stress-Responsive Proteins

Orchids have evolved a sophisticated array of molecular and physiological mechanisms to cope with both biotic and abiotic stresses, making them remarkably resilient despite their often delicate appearance (Zhang et al. 2021). At the cellular level, heat shock proteins (HSPs) serve as crucial molecular chaperones, with HSP70 and HSP90 playing particularly important roles in maintaining protein stability under thermal stress (Chang et al. 2022). These chaperones are complemented by late embryogenesis abundant (LEA) proteins, which protect cellular structures during dehydration stress by replacing water molecules and maintaining membrane integrity (Chen et al. 2019a, b). The antioxidant defense system forms another critical component of stress response, featuring enzymes like superoxide dismutase (SOD), catalase (CAT), and peroxidases (POX) that work in concert to neutralize reactive oxygen species (ROS) generated under various stress conditions (Huehne et al. 2020).

Hormonal signaling pathways integrate these stress responses, with abscisic acid (ABA) playing a central role in drought adaptation by regulating stomatal closure and triggering the expression of stress-responsive genes (Zheng et al. 2025). The ABA signaling cascade interacts with other hormonal pathways, including those mediated by jasmonic acid (JA) and salicylic acid (SA), which orchestrate defenses against biotic threats such as pathogens and herbivores (Fu et al. 2012). Orchids employ specialized pathogenesis-related (PR) proteins, including chitinases and β-1,3-glucanases, to combat microbial invaders (Tsai et al. 2016), while secondary metabolites like bibenzyls and stilbenoids provide additional chemical defenses (Ahmad et al. 2022). Nutrient stress responses involve the upregulation of specific transporters for nitrogen (NRTs) and phosphorus (PHTs), as well as metabolic adjustments to cope with deficiencies (Dong et al. 2024).

These adaptive mechanisms are not isolated but function as interconnected networks, allowing orchids to fine-tune their responses to complex environmental challenges. For instance, the activation of antioxidant systems often coincides with the induction of HSPs, creating synergistic protection against multiple stressors (Song et al. 2022b). Similarly, hormonal cross-talk enables orchids to prioritize responses based on the nature and severity of the stress encountered (Ren et al. 2020). From an ecological perspective, these adaptations have enabled orchids to colonize diverse habitats, from tropical rainforests to alpine regions (Dhiman et al. 2019). From a conservation standpoint, understanding these stress response mechanisms provides valuable insights for protecting threatened species, optimizing cultivation practices, and developing strategies for habitat restoration (Sembi et al. 2019). The molecular components of these stress responses also offer potential biomarkers for assessing plant health and predicting resilience to environmental change, making them valuable tools for both ecological studies and horticultural applications (Ahmad et al. 2023).

10.4.3 Protein–Protein Interactions

Protein–protein interactions (PPIs) serve as fundamental regulatory mechanisms in orchids, governing diverse biological processes including floral development, stress responses, and symbiotic relationships. These molecular interactions form intricate networks that coordinate cellular functions and environmental adaptation in these specialized plants.

Several experimental approaches have been employed to characterize PPIs in orchids. Yeast two-hybrid (Y2H) systems have been particularly valuable, as demonstrated by Tsai et al. (2008) who mapped interactions between B-class MADS-box proteins during tepal development in *Phalaenopsis*. Similarly, Y2H screening in *Dendrobium* revealed key interactors of floral development regulators (Abdullakasim and Handa 2012). More recently, Yu et al. (2023) conducted comprehensive PPI network analyses across four orchid species, providing evolutionary insights into floral morphology diversification.

Co-immunoprecipitation (Co-IP) coupled with mass spectrometry has enabled the identification of functional protein complexes. Co-immunoprecipitation (Co-IP) coupled with mass spectrometry facilitates the identification of protein complexes critical to orchid development. Xu et al. (2024) employed Co-IP to confirm interactions among PeNAC67, PeKAN2, and PeMADS3 transcription factors in Phalaenopsis equestris, regulating lip development by modulating the perianth code's L complex. Similarly, Hua et al. (2024) used Co-IP to validate the OgPIF4–OgBZR1 interaction in Oncidium 'Gower Ramsey', promoting early flowering through brassinosteroid signaling. Advanced techniques like proximity-dependent biotinylation (BioID) are now being applied, as shown by Lin et al. (2015) who mapped virus–host protein interactions in orchid–pathogen systems.

Bimolecular fluorescence complementation (BiFC) has emerged as a powerful tool for visualizing dynamic PPIs in living cells. While orchid applications remain limited, recent methodological advances (Ren et al. 2024) suggest great potential for studying symbiotic interactions and stress responses. Computational approaches are increasingly important, with Zhang et al. (2024a) successfully predicting INDETERMINATE DOMAIN protein networks through integrative bioinformatics.

Conservation Applications Understanding protein–protein interactions (PPIs) has direct conservation implications. In *Gastrodia elata*, a mycoheterotrophic medicinal orchid, Zeng et al. (2018) identified critical host–fungus protein interactions during symbiotic germination with Mycena fungi, revealing 599 differentially accumulated proteins involved in energy metabolism, plant defense, and molecular signaling. These findings inform propagation protocols by elucidating the molecular mechanisms underlying nutrient transfer, aiding efforts to cultivate and conserve this vulnerable species. Similarly, López-Chávez et al. (2016) characterized mycorrhizal interaction networks in *Oncidium*, identifying 34 differentially expressed proteins, including NADH-dehydrogenase and cytochrome c oxidase, which enhance energy metabolism during symbiosis with Thanatephorus sp. RG26. Morphometric analysis revealed improved protocorm development and root hair formation, indicating enhanced nutrient uptake. These molecular targets can optimize fungal compatibility in propagation protocols, supporting conservation of Oncidium species. Similarly, Valadares et al. (2020) found 94 proteins in *Oeceoclades maculata* mycorrhizal roots, promoting nutrient uptake and informing habitat management. In *Dendrobium catenatum*, Chen et al. (2023) identified 1466 differentially expressed proteins under cold stress, highlighting PPI networks that enhance antioxidant defense and guide genotype selection for resilience. Yuan et al. (2020) revealed drought-responsive proteins in *Dendrobium huoshanense*, supporting conservation through stress-tolerant genotype selection. These insights strengthen propagation and habitat strategies for vulnerable orchids.

10.5 Metabolomics in Orchid Research

Metabolomics, the comprehensive study of small-molecule metabolites, has emerged as a transformative tool in orchid research. Orchids are renowned for their extraordinary chemical diversity, producing over 1200 bioactive compounds, including alkaloids, phenolics, terpenoids, and polysaccharides. These metabolites not only underpin their ecological resilience and medicinal value but also present unique challenges for conservation, particularly for overharvested medicinal species like *Dendrobium officinale* and *Gastrodia elata*. Metabolomics bridges these challenges by identifying biomarkers for sustainable cultivation, optimizing metabolite production in vitro, and monitoring stress responses to prioritize resilient genotypes. Below, we delve into the methodologies, applications, and implications of metabolomics in orchid research, supported by case studies and references.

10.5.1 Metabolic Profiling Methods

Metabolic profiling in orchids employs advanced analytical techniques to characterize both primary and secondary metabolites. Liquid chromatography–mass spectrometry (LC-MS) is widely utilized for secondary metabolite analysis due to its sensitivity and versatility. For example, ultrahigh-performance LC-MS (UHPLC-MS) identified 45 phenolic glycosides with antioxidant activity in

Dendrobium catenatum flowers, including quercetin and kaempferol derivatives (Zhang et al. 2019). Hyphenated techniques such as HPLC–ESI–HRMS/MS combined with HPLC–DAD–MS–SPE–UV–NMR were critical for identifying phenanthrene derivatives in *Aerides rosea*, including aeridin and coelonin, which are essential for phytochemical standardization (Cakova et al. 2015). Nuclear magnetic resonance (NMR) spectroscopy plays a key role in structural elucidation, as demonstrated in the resolution of cytotoxic phenanthrenequinone enantiomers from *Pleione bulbocodioides*, which exhibit distinct bioactivity profiles (Shao et al. 2019). Emerging technologies like UPLC-QTOF-MS have enabled comparative metabolomic studies of *Dendrobium* species, revealing species-specific alkaloids and flavonoids (Zhang et al. 2023). These methods collectively enhance the precision and depth of metabolite characterization in orchids.

10.5.2 Secondary Metabolite Analysis

Orchids produce a diverse array of secondary metabolites with significant pharmacological and ecological roles. Alkaloids such as dendrocrepidamine, isolated from *Dendrobium crepidatum*, demonstrate anti-inflammatory properties, highlighting their therapeutic potential (Ding et al. 2021). Bibenzyl derivatives from *Bletilla striata*, including militarin, exhibit antibacterial activity, validated through bioassay-guided isolation (Jiang et al. 2019). Phenolic compounds from *Dendrobium formosum* show dual inhibitory effects on α-glucosidase and pancreatic lipase, suggesting applications in managing metabolic disorders (Inthongkaew et al. 2017). Novel biphenanthrenes from *Aerides multiflora*, such as aeridin A, further underscore the chemical diversity of orchids, with demonstrated α-glucosidase inhibition (Thant et al. 2021). Polysaccharides from *Dendrobium officinale* activate dendritic cells via TLR2/4 pathways, supporting their immunomodulatory use (Lai et al. 2015), while polysaccharides from *Anoectochilus roxburghii* exhibit hepatoprotective and antioxidant effects (Zeng et al. 2016). Additionally, phytosterols like dendrosterone from *Dendrobium ochreatum* show in silico anticancer activity, bridging traditional knowledge with modern drug discovery (Banerjee and Karati 2022).

10.5.3 Stress-Induced Metabolic Changes

Orchids dynamically adjust their metabolic pathways in response to environmental stressors. Under drought conditions, *Cymbidium sinense* accumulates osmolytes such as proline and upregulates antioxidant flavonoids, enhancing cellular stability (Li et al. 2018). Similarly, *Dendrobium huoshanense* increases phenolic biosynthesis under salt stress, as revealed by label-free proteomic analysis (Song et al. 2022b). Heavy metal tolerance in mycorrhizal *Bipinnula fimbriata* involves symbiotic interactions that enhance root biomass and organic acid exudation (e.g., citrate, succinate, malate), facilitating metal accumulation and stress mitigation (Herrera et al. 2018). Although direct studies on pathogen responses are limited, phenolic glycosides in *Dendrobium catenatum* and polysaccharides in *Anoectochilus roxburghii* are known to enhance resilience to biotic stress, suggesting broad-spectrum defense mechanisms (Zhang et al. 2019; Zeng et al. 2016). These adaptive strategies highlight the metabolic plasticity of orchids in challenging environments.

10.5.4 Chemotaxonomic Applications

Metabolomics serves as a powerful tool for resolving taxonomic ambiguities and guiding conservation efforts. Chemotaxonomic markers such as bibenzyls and phenanthrenes distinguish closely related species like *Dendrobium draconis* and *D. falconeri*, addressing taxonomic uncertainties (Srittularak et al. 2009, 2011). Anthocyanins in *Dracula* orchids, including draculin, provide biogeographic insights and aid in species identification (Fossen and Øvstedal 2003). Geographic authentication of *Dendrobium officinale* relies on region-specific polysaccharide methylation patterns detected via NMR, which combat adulteration in commercial herbal products (Zhang et al. 2019). Unique indole glycosides like calanthoside from *Calanthe* species further link chemical profiles to habitat-specific adaptations, informing targeted conservation strategies (Yoshikawa et al. 1998). These applications underscore the dual role of metabolomics in advancing scientific understanding and ecological stewardship of orchid biodiversity.

10.6 Integration of Multiomics Data

The integration of multiomics data—spanning genomics, transcriptomics, proteomics, and metabolomics—provides a holistic framework to unravel the molecular intricacies of orchid biology. This integrative approach is pivotal for understanding stress resilience, symbiotic interactions, and adaptive traits, thereby informing targeted conservation strategies (Ching et al. 2024; Wang et al. 2022). Below, we explore the methodologies, tools, and applications driving multiomics integration in orchid research.

10.6.1 Data Integration Strategies

Integrating heterogeneous omics datasets requires robust computational frameworks to identify correlations and functional linkages. Correlation-based approaches connect molecular layers, such as linking gene expression with metabolite accumulation. For instance, in *Dendrobium sinense*, transcriptomic and metabolomic integration revealed drought-responsive genes coordinating with osmoprotectants like proline and trehalose, enhancing drought resilience (Zhang et al. 2021). Network-based methods, such as weighted gene co-expression network analysis (WGCNA), map interactions between genes, proteins, and metabolites. In *Cymbidium sinense*, WGCNA-based multiomics integration identified hub genes regulating floral development, revealing the crucial role of gibberellin–abscisic acid signaling pathways (Ahmad et al. 2024).

Pathway enrichment analysis overlays omics data onto metabolic pathways to pinpoint functional modules. A study of *Bletilla striata* integrated proteomic and metabolomic data to highlight key metabolic pathways involved in organ-specific development and secondary metabolite production (Ma et al. 2021). Advanced techniques like canonical correlation analysis (CCA) and partial least squares (PLS) are increasingly used for multidimensional data fusion. For example, multiomics analysis in *Gastrodia menghaiensis* revealed insights into orchid mycorrhizal interactions, uncovering regulators of this crucial symbiotic relationship (Jiang et al. 2022).

10.6.2 Systems Biology Approaches

Systems biology leverages multiomics data to model biological processes as interconnected networks. Genome-scale metabolic models (GEMs) predict metabolic flux distributions under varying conditions. A multiomics study of *Dendrobium catenatum* revealed the regulation mechanisms underlying freezing tolerance, identifying key pathways and genes that could guide the selection of cold-tolerant genotypes (Fu et al. 2024). Dynamic modeling captures temporal changes in molecular networks, as demonstrated in studies tracking the development and metabolite production in medicinal orchids.

Constraint-based reconstruction and analysis (COBRA) optimizes metabolic pathways for biotechnological applications. For example, integrated omics approaches in *Vanilla* species have been instrumental in understanding vanillin biosynthesis pathways, potentially leading to improved production methods. Multiomics studies in *Dendrobium officinale*, a traditional Chinese medicinal herb, have significantly advanced our understanding of its bioactive compounds and their biosynthetic pathways (Wang et al. 2022). These models enable predictive insights into orchid responses to environmental stressors, facilitating precision conservation.

10.6.3 Machine Learning Applications

Machine learning (ML) algorithms decode complex multiomics datasets to predict traits and classify samples. Supervised learning models, such as random forests and support vector machines (SVMs), classify stress-resilient genotypes. Recent advances in deep learning have been applied to orchid conservation, with automated assessment of conservation status using convolutional neural networks showing promising results (Zizka et al. 2021; Estopinan et al. 2024). These approaches can help prioritize conservation efforts by identifying threatened species and populations more efficiently.

Unsupervised learning methods like principal component analysis (PCA) reveal hidden patterns in multiomics data. Such techniques have been used to distinguish orchid species based on metabolic profiles and genetic markers, refining taxonomic classification (Du et al. 2024). Deep learning models, such as convolutional neural networks (CNNs), have been successfully employed for orchid classification and identification (Andono et al. 2021; Apriyanti et al. 2023), potentially aiding in monitoring rare species in the wild.

ML applications also extend to greenhouse management and cultivation, with self-supervised learning-based systems for monitoring orchid growth in precision agriculture settings (Chen et al. 2022). These technologies can help optimize cultivation conditions for endangered species in ex situ conservation programs.

10.6.4 Bioinformatics Tools and Databases

A suite of bioinformatics tools and databases supports multiomics data integration in orchid research. OrchidBase is a comprehensive resource for orchid genomics and transcriptomics, hosting data for species like *Phalaenopsis equestris* and *Dendrobium catenatum* (Chen et al. 2025; Fu et al. 2011). Similarly, Orchidstra provides integrated orchid functional genomics data that facilitates multiomics research (Su et al. 2013; Chao et al. 2017).

KEGG and PlantCyc provide curated metabolic pathways for pathway enrichment analysis. Cytoscape is widely used for network visualization and analysis of gene–metabolite networks in orchids (Ma et al. 2020). MetaboAnalyst and XCMS are popular tools for metabolomic data processing and analysis in studies identifying stress-responsive metabolites in orchids like *Dendrobium catenatum* and *Cymbidium tracyanum* (Tu et al. 2025).

Emerging bioinformatics approaches for orchid research include in silico analysis of key developmental genes, such as *Phalaenopsis* Orchid Homeobox1 (POH1) (Rasjid et al. 2023), and genome-wide analysis of gene families involved in stress responses and secondary metabolism (Zhao et al. 2023; Vishwakarma et al. 2023). These resources empower researchers to explore orchid biology at unprecedented resolution, supporting both basic research and conservation efforts.

10.7 Future Perspectives and Challenges

The integration of multiomics approaches in orchid conservation has opened new avenues for understanding and preserving these ecologically and culturally significant plants. However, several challenges remain, and emerging technologies offer exciting opportunities to address these issues. Below, we explore the future perspectives and challenges in orchid conservation, focusing on emerging technologies, data management, policy integration, and future research directions.

10.7.1 Emerging Technologies

Emerging technologies are poised to revolutionize orchid conservation by providing unprecedented insights into their biology and ecology. Single-cell omics enables the resolution of cell-type-specific responses, such as gene expression and metabolite profiles in orchid roots, leaves, and flowers. This approach is particularly valuable for understanding specialized cells involved in symbiotic interactions with mycorrhizal fungi, a critical aspect of orchid ecology and conservation (De Rose et al. 2024).

Spatial transcriptomics maps gene expression within tissues, offering insights into localized responses to environmental stressors. Recent advances in CRISPR-Cas9 genome editing offer potential for precise modifications to enhance stress resilience in endangered orchids (Hossain et al. 2013). While the application of gene editing in conservation remains controversial, it could provide valuable tools for understanding gene function and potentially enhancing ex situ conservation efforts.

Environmental DNA (eDNA) monitoring allows noninvasive tracking of orchid populations and their mycorrhizal partners. This technique is particularly valuable for rare and endangered species that may be difficult to locate through traditional surveys. Machine learning approaches combined with remote sensing data can help map orchid habitat suitability and prioritize areas for conservation (Estopinan et al. 2024).

Advanced bioinformatics tools specifically designed for orchid genomics, such as the latest versions of OrchidBase and Orchidstra, will continue to play crucial roles in organizing and analyzing the growing volume of multiomics data from these plants (Chen et al. 2025; Chao et al. 2017). The integration of these databases with conservation assessment tools could significantly enhance our ability to identify and protect threatened orchid species and populations.

10.7.2 Future Research Directions

Future research should focus on addressing key knowledge gaps and leveraging emerging technologies. Functional genomics is needed to validate the roles of candidate genes identified through omics studies. For example, studies on MYB transcription factors in *Phalaenopsis* and *Dendrobium* have confirmed their roles in regulating water-soluble polysaccharide biosynthesis and other important traits linked to adaptation and conservation value (He et al. 2019; Kaur et al. 2024). Multiomics integration will provide a systems-level understanding of orchid biology. Recent work combining transcriptomics and metabolomics in *Cymbidium tracyanum* has revealed gene–metabolite networks underlying floral scent biosynthesis, a key adaptation for pollinator attraction (Tu et al. 2025).

Climate adaptation studies are critical for predicting orchid responses to global change. Machine learning approaches integrated with multiomics data show promise for predicting habitat suitability for endangered orchids under future climate scenarios, potentially guiding assisted migration efforts (Estopinan et al. 2024; Zizka et al. 2021). Symbiosis research should explore the molecular mechanisms of orchid–mycorrhizal interactions. Recent proteomic and metabolomic studies of *Serapias vomeracea* roots colonized by *Tulasnella* fungi have identified key molecular components essential for symbiotic success (De Rose et al. 2024).

In this regard, advances in biotechnology offer exciting opportunities for conservation of endangered orchids. Studies on gene families involved in stress responses, such as GDSL-type esterase/lipase genes in *Dendrobium catenatum* (Zhan et al. 2022) and PAL genes across multiple orchid species (Vishwakarma et al. 2023), provide potential targets for enhancing resilience in vulnerable populations. Furthermore, the application of advanced breeding techniques guided by multiomics data could help maintain genetic diversity while reinforcing adaptive traits in conservation programs (Balilashaki et al. 2023).

10.8 Conclusion

The integration of multiomics approaches has revolutionized our understanding of orchid biology and opened new avenues for conservation of these ecologically and culturally significant plants. By combining genomics, transcriptomics, proteomics, and metabolomics with advanced computational methods, researchers can now unravel the complex molecular mechanisms underlying orchid adaptation, development, and stress responses (Ching et al. 2024; Mondal et al. 2022). These approaches are particularly valuable for understanding secondary metabolite production in medicinal orchids, which play an important role in traditional medicine and have significant potential for pharmaceutical applications. Various species of *Dendrobium*, *Bletilla*, *Gastrodia*, and other medicinal orchids produce valuable bioactive compounds whose biosynthetic pathways are now being elucidated through multiomics studies (Yuan et al. 2024; Wang et al. 2022).

Biotechnological methods including tissue culture, cell suspension cultures, and metabolic engineering have proven effective for producing and preserving these valuable compounds (Pandita et al. 2021; Hossain et al. 2013). Elicitation strategies, using various biotic and abiotic factors to enhance secondary metabolite production, offer promising approaches for both research and conservation purposes (Aehsas et al. 2024). The application of these techniques, guided by insights from multiomics data, provides opportunities to preserve the genetic diversity and metabolic potential of endangered orchid species.

Looking ahead, the continued development of integrated databases, advanced analytical tools, and novel biotechnological approaches will further enhance our ability to conserve orchids and their valuable traits. By bridging fundamental research with practical conservation strategies, multiomics approaches offer a powerful framework for preserving orchid diversity in the face of growing environmental challenges (Naresh et al. 2024; Ghai et al. 2022; Chao et al. 2021).

Acknowledgments The corresponding author thanks Vice Chancellor for Research of University of Kurdistan for support and help, Dr. Philipp M. Schlueter for scientific advice, and Ms. Zahra Almasi for collecting and preparing the references.

Competing Interests The authors have no relevant financial or nonfinancial interests to disclose.

References

Abdullakasim S, Handa T (2012) Genetic transformation and analysis of protein-protein interaction of Class B MADS-box genes from *Dendrobium moniliforme*. In: Current frontiers and perspectives in cell biology. InTech, Rijeka, Croatia, pp 163–178

Aehsas S, Banerjee S, Sahoo JP, Pradhan M, Mahapatra M, Nayak P (2024) Integration of omics and high-throughput technologies for enhancing plant secondary metabolites: a comprehensive overview. In: Biotechnology, multiple omics, and precision breeding in medicinal plants, Springer Nature, Cham, Switzerland, pp 248–257

Ahmad S, Gao J, Wei Y, Lu C, Zhu G, Yang F (2022) The transcriptome profiling of flavonoids and bibenzyls reveals medicinal importance of rare orchid *Arundina graminifolia*. Front Plant Sci 13:923000

Ahmad S, Wang F, Ahmad A, Ercisli S, Chen J, Guan Y, Peng D (2023) Suggesting a prospect for melatonin-mediated orchid flowering under extreme conditions. Sci Hortic 321:112340

Ahmad S, Lu C, Gao J, Wei Y, Xie Q, Jin J, Yang F (2024) Integrated proteomic, transcriptomic, and metabolomic profiling reveals that the gibberellin–abscisic acid hub runs flower development in the Chinese orchid *Cymbidium sinense*. Hortic Res 11(5):uhae073

Ahmadzadeh K, Mozafari AA, Vafaee Y, Khorshidi J, Khoddamzadeh AA (2025) Divergent in vitro responses to asymbiotic seed germination media in *Ophrys reinholdii* subsp. *straussii* and *Comperia comperiana*: the dual impact of species and media. In Vitro Cell Dev Biol Plant 61(2):427–442

Ai Y, Li Z, Sun WH, Chen J, Zhang D, Ma L, Zhang QH, Chen MK, Zheng QD, Liu JF, Jiang YT (2021) The *Cymbidium goeringii* genome provides insight into organ development and adaptive evolution in orchids. Ornam Plant Res 1:10

An F-M, Hsiao S-R, Chan M-T (2011) Sequencing-based approaches reveal low ambient temperature-responsive and tissue-specific microRNAs in Phalaenopsis orchid. PLoS One 6:e18937

Andono PN, Rachmawanto EH, Herman NS, Kondo K (2021) Orchid types classification using supervised learning algorithm based on feature and color extraction. Bull Elec Eng Inform 10(5):2530–2538

Anghelescu NE, Vafaee Y, Ahmadzadeh K, Chen JT (2023) Asymbiotic seed germination in terrestrial orchids: problems, progress, and prospects. In: Advances in orchid biology, biotechnology and omics. Springer, Singapore

Apriyanti DH, Spreeuwers LJ, Lucas PJ (2023) Deep neural networks for explainable feature extraction in orchid identification. Appl Intell 53(21):26270–26285

Baginsky S (2009) Plant proteomics: concepts, applications, and novel strategies for data interpretation. Mass Spectromet Rev 28(1):93–120

Balao F, Trucchi E, Wolfe T, Hao BH, Lorenzo MT, Baar J, Sedman L, Kosiol C, Amman F, Chase MW, Hedrén M, Paun O (2022) Adaptive sequence evolution is driven by biotic stress in a pair of orchid species (*Dactylorhiza*) with distinct ecological optima. Mol Ecol 31(9):2612–2626

Balilashaki K, Dehghanian Z, Gougerdchi V, Kavusi E, Feizi F, Tang X, Hossain MM (2023) Progress and prospect of orchid breeding: An overview. In: Advances in orchid biology, biotechnology and omics, Springer Nature, Singapore, pp 261–283

Banerjee J, Karati D (2022) In-silico pharmacology against cancer of a novel phytosterol Dendrosterone isolated from plant *Dendrobium ochreatum*. Pharmacol Res Mod Chin Med 4:100149

Bhattacharyya P, Sharma T, Yadav A, Lalthafamkimi L, Ritu, Swarnkar M K, Joshi R, Shankar R, Kumar S (2022) De novo transcriptome based insights into secondary metabolite biosynthesis in *Malaxis acuminata* (Jeevak)–A therapeutically important orchid. Front Plant Sci 13:954467

Cai J, Liu X, Vanneste K, Proost S, Tsai WC, Liu KW, Chen LJ, He Y, Xu Q, Bian C, Zheng Z, Sun F, Liu W, Hsiao YY, Pan ZJ, Hsu CC, Yang YP, Hsu YC, Chuang YC, Dievart A, Dufayard JF, Xu X, Wang JY, Wang J, Xiao XJ, Zhao XM, Du R, Zhang GQ, Wang M, Su YY, Xie GC, Liu GH, Li LQ, Huang LQ, Luo YB, Chen HH, Van de Peer Y, Liu ZJ (2015) The genome sequence of the orchid *Phalaenopsis equestris*. Nat Genet 47(1):65–72

Cakova V, Urbain A, Antheaume C, Rimlinger N, Wehrung P, Bonté F, Lobstein A (2015) Identification of phenanthrene derivatives in *Aerides rosea* (Orchidaceae) using the combined systems HPLC–ESI–HRMS/MS and HPLC–DAD–MS–SPE–UV–NMR. Phytochem Anal 26:34–39

Champagne A, Boutry M (2013) Proteomics of nonmodel plant species. Proteomics 13(3-4):663–673

Chang L, Tzean Y, Hsin KT, Lin CY, Wang CN, Yeh HH (2022) Stress associated proteins coordinate the activation of comprehensive antiviral immunity in *Phalaenopsis* orchids. New Phytol 233:145–155

Chao YT, Su CL, Jean WH, Chen WC, Chang YCA, Shih MC (2014) Identification and characterization of the microRNA transcriptome of a moth orchid *Phalaenopsis aphrodite*. Plant Mol Biol 84:529–548

Chao YT, Chen WC, Ho HY, Shih MC (2017) Orchidstra 2.0—a transcriptomics resource for the orchid family. Plant Cell Physiol 58(1):e9

Chao YT, Chen WC, Ho HY, Shih MC (2021) Genomics, Transcriptomics and miRNA family resources for Phalaenopsis aphrodite and the orchid family. In: The Orchid Genome. Springer, Cham, pp 117–131

Chase MW, Cameron KM, Freudenstein JV, Pridgeon AM, Salazar G, van den Berg C, Schuiteman A (2015) An updated classification of Orchidaceae. Bot J Linn Soc 177(2):151–174

Chen S, Harmon AC (2006) Advances in plant proteomics. Proteomics 6(20):5504–5516

Chen G, Chen D, Wang T, Xu C, Li L (2012) Analysis of the proteins related to browning in leaf culture of *Phalaenopsis*. Sci Hortic 141:17–22

Chen J, Liu SS, Kohler A, Yan B, Luo HM, Chen XM et al (2017) iTRAQ and RNA-Seq analyses provide new insights into regulation mechanism of symbiotic germination of *Dendrobium officinale* seeds (Orchidaceae). J Proteome Res 16(6):2174–2187

Chen C, Zeng L, Ye Q (2018) Proteomic and biochemical changes during senescence of Phalaenopsis 'Red Dragon' petals. Int J Mol Sci 19(5):1317

Chen J Y, Lee Y I, Chen Y C, Tsay H S, Yeh C H (2019) Increased expression of 9-cis-epoxycarotenoid dioxygenase, PtNCED1, associated with inhibited seed germination in a terrestrial orchid, *Phaius tankervilleae*. Front Plant Sci 10:1043

Chen J, Tang Y, Kohler A, Lebreton A, Xing Y, Zhou D, Guo S (2019a) Comparative transcriptomics analysis of the symbiotic germination of *D. officinale* (Orchidaceae) with emphasis on plant cell wall modification and cell wall-degrading enzymes. Front. Plant Sci 13:880600

Chen JC, Tong CG, Lin HY, Fang SC (2019b) Phalaenopsis LEAFY COTYLEDON1-induced somatic embryonic structures are morphologically distinct from protocorms. Front Plant Sci 10:1594

Chen C, Zeng L, Zhao H, Ye Q (2020) Proteomic analysis of the early development of the *Phalaenopsis amabilis* flower bud under low temperature induction using the iTRAQ/MRM approach. Molecules 25(5):1244

Chen Y, Xu Z, Shen Q, Sun C (2021) Floral organ-specific proteome profiling of the floral ornamental orchid (*Cymbidium goeringii*) reveals candidate proteins related to floral organ development. Bot Stud 62:1–12

Chen LB, Huang GZ, Huang XR, Wang WC (2022) A self-supervised learning-based intelligent greenhouse orchid growth inspection system for precision agriculture. IEEE Sensors J 22(24):24567–24577

Chen Y, Hu G, Zhang J, Ma C, Luo D, Zhang X, Fan J, Li K, Shen X (2023) Transcriptome and physiological analyses uncover the molecular mechanism of the response to low-temperature stress in *Dendrobium catenatum*. Front Plant Sci 14:1260728

Chen HY, Zhang ZR, Yao X, Ya JD, Jin XH, Wang L, Yu WB (2024) Plastid phylogenomics provides new insights into the systematics, diversification, and biogeography of *Cymbidium* (Orchidaceae). Plant Diver 46(4):448–461

Chen YY, Sun Y, Li CI, Lin ST, Zheng HC, Zhang ZB, Lee BR, Hsieh CL, Hsiao YY, Chow CN, Yang CW (2025) OrchidBase 6.0: increasing the number of *Cymbidium* (Orchidaceae) genomes and new bioinformatic tools for orchid genome analysis. BMC Plant Biol 25(1):1

Ching KS, Li CX, Xi L, Syamsumir DF, Ling MN (2024) Rebuilding of orchid germplasm conservation via molecular multi-omics approaches. J Sust Sci Manage 19(6):169–199

Chomicki G, Coiro M, Renner SS (2015) The velamen protects photosynthetic orchid roots against UV-B damage, and a large dated phylogeny implies multiple gains and losses of this function during the Cenozoic. New Phytol 205:1330–1341

Cribb P, Hermans J (2007) The conservation of Madagascar's orchids. A model for an integrated conservation project. Lankesteriana 7:255–261

De Paolo S, Salvemini M, Gaudio L, Aceto S (2014) De novo transcriptome assembly from inflorescence of *Orchis italica*: analysis of coding and non-coding transcripts. PLOS ONE 9:e102155

De Conti D, Corredor-Prado JP, Bueno JCF, Cangahuala-Inocente GC, Guerra MP, Pescador R (2020) Differentially expressed proteins in the development of protocorm-like bodies of *Cattleya tigrina* A. Richard. Acta Physiol Plant 42:1–14

De Rose S, Kuga Y, Sillo F, Fochi V, Sakamoto N, Calevo J, Perotto S, Balestrini R (2023) Plant and fungal gene expression coupled with stable isotope labeling provide novel information on sulfur uptake and metabolism in orchid mycorrhizal protocorms. Plant J 116(2):416–431

De Rose S, Sillo F, Ghirardo A, Perotto S, Schnitzler JP, Balestrini R (2024) Integration of fungal transcriptomics and metabolomics provides insights into the early interaction between the ORM fungus *Tulasnella* sp. and the orchid *Serapias vomeracea* seeds. IMA Fungus 15:1

de Vries J, Curtis BA, Gould SB, Archibald JM (2018) Embryophyte stress signaling evolved in the algal progenitors of land plants. Proc Natl Acad Sci USA 115(15):E3471

Dhiman N, Sharma NK, Thapa P, Sharma I, Swarnkar MK, Chawla A, Bhattacharya A (2019) De novo transcriptome provides insights into the growth behaviour and resveratrol and trans-stilbenes biosynthesis in *Dactylorhiza hatagirea*—An endangered alpine terrestrial orchid of western Himalaya. Sci Rep 9:13133

Ding XQ, Zou YQ, Liu J, Wang XC, Hu Y, Liu X, Zhang CF (2021) Dendrocrepidamine, a novel octahydroindolizine alkaloid from the roots of *Dendrobium crepidatum*. J Asian Nat Prod Res 23:1085–1092

Dong XM, Zhang W, Hu H, Gao TY, Wang XQ, Shi Q, Zhang SB (2024) Physiological and transcriptome analysis reveal the nitrogen preference and regulatory pathways of nitrogen metabolism in an epiphytic orchid, *Cymbidium tracyanum*. Environ Exp Bot 219:105618

Du Z, Jin Y, Yang X, Xia K, Chen Z (2024) Multi-omics analyses and botanical perfumer hypothesis provide insights into the formation and maintenance of aromatic characteristics of *Dendrobium loddigesii* flowers. Plant Physiol Biochem 214:108891

Estopinan J, Servajean M, Bonnet P, Joly A, Munoz F (2024) Mapping global orchid assemblages with deep learning provides novel conservation insights. Ecol Inform 81:102627

Fan Q, Zhang W, Ye Q, Su Y, Zhang X, Chen J (2022) Physiological and molecular responses of *Cymbidium goeringii* to low temperature stress. Horticulturae 8:1060

Fan W, He ZS, Zhe M, Feng JQ, Zhang L, Huang Y, Liu F, Huang JL, Ya JD, Zhang SB, Yang JB (2023) Highquality *Cymbidium mannii* genome and multifaceted regulation of crassulacean acid metabolism in epiphytes. Plant Commun 4:100619

Fang L, Xu X, Li J, Zheng F, Li M, Yan J, Li Y, Zhang X, Li L, Ma G, Zhang A, Lv F, Wu K, Zeng S (2020) Transcriptome analysis provides insights into the non-methylated lignin synthesis in Paphiopedilum armeniacum seed. BMC Genomics 21:524

Fatahi M, Vafaee Y, Nazari F, Tahir NA (2022a) *In vitro* asymbiotic seed germination, protocorm formation, and plantlet development of *Orchis simia* Lam.: a threatened terrestrial orchid species. S Afr J Bot 151:156–165

Fatahi M, Vafaee Y, Tahir NAR, Khorshidi J (2022b) Large-scale *in vitro* multiplication and phytochemical analysis of *Himantoglossum affine* (Boiss.) Schltr.: An endangered Euro-Mediterranean terrestrial orchid. Diversity 14(12):1137

Fatahi M, Anghelescu NE, Vafaee Y, Khoddamzadeh A (2023) Micropropagation of *Dactylorhiza umbrosa* (Kar. & Kir.) Nevski through asymbiotic seed germination and somatic embryogenesis: a promising tool for conservation of rare terrestrial orchids. S Afr J Bot 159:492–506

Fay MF (2018) Orchid conservation: how can we meet the challenges in the twenty-first century? Bot Stud 59:1–9

Fay MF, Bone R, Cook P, Kahandawala I, Greensmith J, Harris S, Pedersen HÆ, Marken J, Ségurel L (2018) Genetic diversity in *Cypripedium calceolus* (Orchidaceae) with a focus on North-Western Europe, as revealed by plastid DNA length polymorphisms. Ann Bot 122(1):109–119

Feng S, Jiao K, Guo H, Jiang M, Hao J, Wang H et al (2017) Succinyl-proteome profiling of *Dendrobium officinale*, an important traditional Chinese orchid herb, revealed involvement of succinylation in the glycolysis pathway. BMC Genomics 18:1–12

Fossen T, Øvstedal DO (2003) Anthocyanins from flowers of the orchids *Dracula chimaera* and *D. cordobae*. Phytochemistry 63:783–787

Freudenstein JV (2024) Orchid phylogenetics and evolution: history, current status and prospects. Ann Bot 135:805

Fu CH, Chen YW, Hsiao YY, Pan ZJ, Liu ZJ, Huang YM, Tsai WC, Chen HH (2011) OrchidBase: a collection of sequences of the transcriptome derived from orchids. Plant Cell Physiol 52(2):238–243

Fu SF, Tsai TM, Chen YR, Liu CP, Haiso LJ, Syue LH, Huang HJ (2012) Characterization of the early response of the orchid, *Phalaenopsis amabilis*, to *Erwinia chrysanthemi* infection using expression profiling. Physiol Plant 145:406–425

Fu K, Chen Y, Wang Y, Chen X, Gong Q, Shi Y, Han Z (2024) Multi-omics reveal the regulation mechanism in response to freezing tolerance in *Dendrobium catenatum*. Plant Growth Reg 104(1):535–547

Gallage NJ, Hansen EH, Kannangara R, Olsen CE, Motawia MS, Jørgensen K, Holme I, Hebelstrup K, Grisoni M, Møller BL (2018) Vanillin formation from ferulic acid in *Vanilla planifolia* is catalysed by a single enzyme. Nat Commun 9(1):3510

Gastelbondo M, Micheal V, Wang Y, Chambers A, Wu X (2025) Comparative transcriptome profiling of vanilla (*Vanilla planifolia)* capsule development provides insights of vanillin biosynthesis. BMC Plant Biol 25(1):1–6

Gawenda I, Schröder-Lorenz A, Debener T (2012) Markers for ornamental traits in *Phalaenopsis* orchids: population structure, linkage disequilibrium and association mapping. Mol Breed 30:305–316

Geppert C, Perazza G, Wilson RJ, Bertolli A, Prosser F, Melchiori G, Marini L (2020) Consistent population declines but idiosyncratic range shifts in Alpine orchids under global change. Nat Commun 11:5835

Ghai D, Kaur A, Kahlon PS, Pawar SV, Sembi JK (2022) A walk through the maze of secondary metabolism in orchids: a transcriptomic approach. Front Plant Sci 13:837563

Gholami S, Vafaee Y, Nazari F, Ghorbani A (2021) Exploring genetic variations in threatened medicinal orchids using start codon targeted (SCoT) polymorphism and marker-association with seed morphometric traits. Physiol Mol Biol Plants 27:769–785

Ghorbani A, Gravendeel B, Naghibi F, de Boer H (2014) Wild orchid tuber collection in Iran: a wake-up call for conservation. Biodivers Conserv 23(11):2749–2760

Gibert A, Schatz B, Buscail R, Nguyen D, Baguette M, Barthes N, Bertrand JA (2025) Floral phenotypic divergence and genomic insights in an *Ophrys* orchid: unraveling early speciation processes. New Phytol 245(2):849–868

Givnish TJ, Spalink D, Ames M, Lyon SP, Hunter SJ, Zuluaga A, Iles WJD, Clements MA, Arroyo MTK, Leebens-Mack J, Endara L, Kriebel R, Neubig KM, Whitten WM, Williams NH, Cameron KM (2015) Orchid phylogenomics and multiple drivers of their extraordinary diversification. Proc R Soc B 282(1814):20151553

Guo YY, Luo YB, Liu ZJ, Wang XQ (2015a) Evolution and biogeography of the slipper orchids: Eocene vicariance of the conduplicate genera in the old and new world tropics. PLoS One 10(11):e38788

Guo YY, Luo YB, Liu ZJ, Wang XQ (2015b) Reticulate evolution and sea-level fluctuations together drove species diversification of slipper orchids (*Paphiopedilum*) in South-East Asia. Mol Ecol 24(12):2838–2855

Guo JL, Cao WJ, Li ZM, Zhang YH, Volis S (2019) Conservation implications of population genetic structure in a threatened orchid *Cypripedium tibeticum*. Plant Divers 41:13–18

He C, Teixeira da Silva JA, Wang H, Si C, Zhang M, Zhang X, Li M, Tan J, Duan J (2019) Mining MYB transcription factors from the genomes of orchids (*Phalaenopsis* and *Dendrobium*) and characterization of an orchid R2R3-MYB gene involved in water-soluble polysaccharide biosynthesis. Sci Rep 9(1):13818

Herrera H, Valadares R, Oliveira G, Fuentes A, Almonacid L, Do Nascimento SV et al (2018) Adaptation and tolerance mechanisms developed by mycorrhizal *Bipinnula fimbriata* plantlets (Orchidaceae) in a heavy metal-polluted ecosystem. Mycorrhiza 28(7):651–663

Hickerson NMN, Samuel MA (2024) CRISPR-Cas9 technology in plant genomics Research: a review and tutorial for implementation with considerations for orchid genetic studies. In: Yeung ECT, Lee YI (eds) Orchid propagation. Springer protocols handbooks. Humana, New York

Hinsley A, de Boer HJ, Fay MF, Gale SW, Gardiner LM, Gunasekara RS, Phelps J (2018) A review of the trade in orchids and its implications for conservation. Bot J Linn Soc 186(4):435–455

Hoban S, Archer FI, Bertola LD, Bragg JG, Breed MF, Bruford MW, Coleman MA, Ekblom R, Funk WC, Grueber CE, Hand BK, Jaffé R, Jensen EL, Johnson J, Kershaw F, Liggins L, MacDonald AJ, Mergeay J, Miller JM, Muller-Karger FE, O'Brien D, Paz-Vinas I, Potter KM, Razgour O, Vernesi C, Hunter ME (2022) Genetic diversity targets and indicators in the CBD post-2020 Global Biodiversity Framework must be improved. Biol Conserv 248:108654

Hossain MM, Kant R, Van PT, Winarto B, Zeng S, Teixeira da Silva JA (2013) The application of biotechnology to orchids. Crit Rev Plant Sci 32(2):69–139

Hsiao YY, Chen YW, Huang SC, Pan ZJ, Fu CH, Chen WH, Tsai WC, Chen HH (2011) Gene discovery using next-generation pyrosequencing to develop ESTs for *Phalaenopsis* orchids. BMC Genomics 12:1–3

Hsu CC, Chen SY, Chiu SY, Lai CY, Lai PH, Shehzad T, Chen HH (2022) High-density genetic map and genome-wide association studies of aesthetic traits in *Phalaenopsis* orchids. Sci Rep 12(1):3346

Hua MDS, Rajendran SK, Yeh KW, Yeh CH (2024) Brassinosteroid signaling synergizing with AsA/GSH pathway promotes flowering through PIF4/bzr1 interaction under prolonged high ambient temperature stress in Oncidium orchid. SSRN 4514064. https://ssrn.com/abstract=4514064

Huang JZ, Lin CP, Cheng TC, Chang BCH, Cheng SY, Chen YW, Lee CY, Chin SW, Chen FC (2015) A de novo floral transcriptome reveals clues into *Phalaenopsis* orchid flower development. PLoS One 10(5):e0123474

Huang JZ, Lin CP, Cheng TC, Huang YW, Tsai YJ, Cheng SY, Chen FC (2016) The genome and transcriptome of *Phalaenopsis* yield insights into floral organ development and flowering regulation. PeerJ 4:e2017

Huehne PS, Bhinija K, Srisomsap C, Chokchaichamnankit D, Weeraphan C, Svasti J, Mongkolsuk S (2020) Detection of superoxide dismutase (Cu-Zn) isoenzymes in leaves and pseudobulbs of *Bulbophyllum morphologlorum* Kraenzl orchid by comparative proteomic analysis. Biochem Biophys Rep 22:100762

Inthongkaew P, Chatsumpun N, Supasuteekul C, Kitisripanya T, Putalun W, Likhitwitayawuid K, Sritularak B (2017) α-Glucosidase and pancreatic lipase inhibitory activities and glucose uptake stimulatory effect of phenolic compounds from *Dendrobium formosum*. Rev Bras 27:480–487

Jakubska-Busse A, Kędziora A, Cieniuch G, Korzeniowska-Kowal A, Bugla-Płoskońska G (2021) Proteomics-based identification of orchid-associated bacteria colonizing the *Epipactis albensis*, *E. helleborine*, and *E. purpurata* (Orchidaceae, Neottieae). Saudi J Biol Sci 28(7):4029–4038

Jiang S, Wan K, Lou HY, Yi P, Zhang N, Zhou M, Pan WD (2019) Antibacterial bibenzyl derivatives from the tubers of *Bletilla striata*. Phytochemistry 162:216–223

Jiang Y, Tian M, Wang C, Zhang Y (2021) Transcriptome sequencing and differential gene expression analysis reveal the mechanisms involved in seed germination and protocorm development of *Calanthe tsoongiana*. Gene 772:145355

Jiang Y, Hu X, Yuan Y, Guo X, Chase MW, Ge S, Jin X (2022) The *Gastrodia menghaiensis* (Orchidaceae) genome provides new insights of orchid mycorrhizal interactions. BMC Plant Biol 22(1):179

Jie X, Feng Y, Jiahao F, Ganggul L, Jiani Y, Zhongyu X et al (2023) Comprehensive chemical profiling of two *Dendrobium* species and identification of anti-hepatoma active constituents from *Dendrobium chrysotoxum* by network pharmacology. BMC Complem Med Ther 23(1):217

Kaki A, Vafaee Y, Khadivi A (2020) Genetic variation of *Anacamptis coriophora*, *Dactylorhiza umbrosa*, *Himantoglossum affine*, *Orchis mascula*, and *Ophrys schulzei* in the western parts of Iran. Ind Crop Prod 156:112854

Kambara K, Fujino K, Shimura H (2023) Construction of a de novo assembly pipeline using multiple transcriptome data sets from *Cypripedium macranthos* (Orchidaceae). PLoS One 18(6):e0286804

Kanehisa M (2016) KEGG bioinformatics resource for plant genomics and metabolomics. In: Plant bioinformatics: methods and protocols. Humana Press, New York, pp 55–70

Kaur A, Yadav VG, Pawar SV, Sembi JK (2024) Insights to phenylalanine ammonia lyase (PAL) and secondary metabolism in orchids: An in silico approach. Biochem Genet 62(1):413–435

Kim YK, Jo S, Cheon SH, Joo MJ, Hong JR, Kwak M, Kim KJ (2020) Plastome evolution and phylogeny of Orchidaceae, with 24 new sequences. Front Plant Sci 11:22

Kim S, Park J, Lee H, Kim Y (2022) Drought-responsive transcription factors in *Calanthe triplicata* identified by RNA-seq. Plant Cell Rep 41:1453–1465

Klepikova AV, Kasianov AS, Ezhova MA, Penin AA, Logacheva MD (2021) Transcriptome atlas of *Phalaenopsis equestris*. PeerJ 10;9:e12600

Kolanowska M (2021) The future of a montane orchid species and the impact of climate change on the distribution of its pollinators and magnet species. Glob Ecol Conserv 32:e01939

Kui L, Chen H, Zhang W, He S, Xiong Z, Zhang Y et al (2017) Building a genetic manipulation tool box for orchid biology: identification of constitutive promoters and application of CRISPR/Cas9 in the orchid, *Dendrobium officinale*. Front Plant Sci 7:2036

Kull T, Selgis U, Peciña MV, Metsare M, Ilves A, Tali K, Sepp K, Kull K, Shefferson RP (2016) Factors influencing IUCN threat levels to orchids across Europe on the basis of national red lists. Ecol Evol 6(17):6245–6265

Lagou LJ, Kadereit G, Morales-Briones DF (2024) Phylogenomic analysis of target enrichment and transcriptome data uncovers rapid radiation and extensive hybridization in the slipper orchid genus Cypripedium. Ann Bot 134(7):1229–1250

Lai T, Deng Y, Zhang P, Chen Z, Hu F, Zhang Q, Shi N (2013) Proteomics-based analysis of Phalaenopsis amabilis in response toward Cymbidium mosaic virus and/or Odontoglossum ringspot virus infection. Am J Plant Sci 4(09):1853

Lai CY, Yang LC, Lin WC (2015) Type II arabinogalactan from *Anoectochilus formosanus* induced dendritic cell maturation through TLR2 and TLR4. Phytomedicine 22:1207–1214

Lee SR, Choi TY, Jung SY (2022) Genetic diversity on a rare terrestrial orchid, Habenaria linearifolia in South Korea: implications for conservation offered by Genome-Wide Single Nucleotide Polymorphisms. Front Plant Sci 13:772621

Li DM, Zhu GF (2022) High-density genetic linkage map construction and QTLs Identification Associated with four leaf-related traits in lady's slipper orchids (*Paphiopedilum concolor* × *Paphiopedilum hirsutissimum*). Horticulturae 8(9):842

Li X, Xu W, Chowdhury MR, Jin F (2014) Comparative proteomic analysis of labellum and inner lateral petals in Cymbidium ensifolium flowers. Int J Mol Sci 15(11):19877–19897

Li JW, Chen XD, Hu XY, Ma L, Zhang SB (2018) Comparative physiological and proteomic analyses reveal different adaptive strategies by *Cymbidium sinense* and *C. tracyanum* to drought. Planta 247(1):69–97

Li C, Dong N, Zhao Y, Wu S, Liu Z, Zhai J (2021) A review for the breeding of orchids: current achievements and prospects. Hortic Plant J 7(5):380–392

Li MH, Liu KW, Li Z, Lu HC, Ye QL, Zhang D, Wang JY, Li YF, Zhong ZM, Liu X, Yu X (2022) Genomes of leafy and leafless *Platanthera* orchids illuminate the evolution of mycoheterotrophy. Nat Plants 8(4):373–388

Li Q, Liu C, Huang C, Wang M, Long T, Liu J, Shi J, Shi J, Li L, He Y, Xu DL (2022) Transcriptome and metabonomics analysis revealed the molecular mechanism of differential metabolite production of *Dendrobium nobile* under different epiphytic patterns. Front Plant Sci 13:868472

Liang C, Li J, Li S, Zhang H, Zheng J, Miao J, Zhu X, He X, Zhai J (2024) Human activity changed the genetic pattern of the orchid *Phaius flavus* population. Diversity 16(11):685

Lin PC, Hu WC, Lee SC, Chen YL, Lee CY, Chen YR, Chang YC (2015) Application of an integrated omics approach for identifying host proteins that interact with Odontoglossum ringspot virus capsid protein. Mol Plant-Microbe Interact 28:711–726

Lin CS, Hsu CT, Liao DC, Chang WJ, Chou ML, Huang YT, Shih MC (2016) Transcriptome-wide analysis of the MADS-box gene family in the orchid *Erycina pusilla*. Plant Biotechnol J 14:284–298

Liu H, Jacquemyn H, He X, Chen W, Huang Y, Yu S, Lu Y, Zhang Y (2021) The impact of human pressure and climate change on the habitat availability and protection of *Cypripedium* (Orchidaceae) in Northeast China. Plan Theory 10(1):1–15

López-Chávez MY, Guillén-Navarro K, Bertolini V, Encarnación S, Hernández-Ortiz M, Sánchez-Moreno I, Damon A (2016) Proteomic and morphometric study of the in vitro interaction between *Oncidium sphacelatum* Lindl. and Thanatephorus sp. Mycorrhiza 26:353–365

Lu BW, An FX, Cao LJ, Yang YJ, Liu PM, Wang X, Liu J (2020) Proteomic profiling uncovered the cytosolic superoxide dismutase BsSOD1 associated with plant defence in the herbal orchid *Bletilla striata*. Funct Plant Biol 47(10):937–944

Ma X, Meng Y, Wang P, Tang Z, Wang H, Xie T (2020) Bioinformatics-assisted, integrated omics studies on medicinal plants. Brief Bioinform 21(6):1857–1874

Ma X, Tang K, Tang Z, Dong A, Meng Y, Wang P (2021) Organ-specific, integrated omics data-based study on the metabolic pathways of the medicinal plant *Bletilla striata* (Orchidaceae). BMC Plant Biol 21:1–15

Ma X, Ju S, Lin H, Huang H, Huang J, Peng D, Ming R, Lan S, Liu Z-J (2023) Sex-related gene network revealed by transcriptome differentiation of bisexual and unisexual flowers of orchid *Cymbidium tortisepalum*. Int J Mol Sci 24:16627

Meng Y, Yu D, Xue J, Lu J, Feng S, Shen C, Wang H (2016) A transcriptome-wide, organ-specific regulatory map of *Dendrobium officinale*, an important traditional Chinese orchid herb. Sci Rep 6:18864

Micheneau C, Carlsward BS, Fay MF, Bytebier B, Pailler T, Chase MW (2008) Phylogenetics and biogeography of Mascarene angraecoid orchids (Vandeae, Orchidaceae). Mol Phylogenet Evol 46(3):908–922

Mondal S, Song H, Zhang L, Cao Y (2022) Multi-omics and computational biology in horticultural plants: from genotype to phenotype. Front Plant Sci 13:1073266

Mondragón-Palomino M, Theißen G (2011) Why are orchids so diverse? Reduction of evolutionary constraints by paralogues of class B floral homeotic genes. Genome Biol 10(7):1–11

Mumtaz MW, Hamid AA, Akhtar MT, Anwar F, Rashid U, Al-Zuaidy MH (2017) An overview of recent developments in metabolomics and proteomics–phytotherapic research perspectives. Front Life Sci 10(1):1–37

Naresh R, Tomar P, Singh RK (2024) State of the art of Omics Technologies in ornamental plant research. In: Ornamental horticulture: latest cultivation practices and breeding technologies, Springer Nature, Singapore, pp 175–191

Niu Z, Zhu F, Fan Y, Li C, Zhang B, Zhu S, Ding X (2021) The chromosome-level reference genome assembly for *Dendrobium officinale* and its utility of functional genomics research and molecular breeding study. Acta Pharm Sin B 11(7):2080–2092

Ospina-Zapata DA, Madrigal Y, Alzate JF, Pabón-Mora N (2020) Evolution and expression of reproductive transition regulatory genes FT/TFL1 with emphasis in selected Neotropical orchids. Front Plant Sci 11:469

Palama TL, Menard P, Fock I, Choi YH, Bourdon E, Govinden-Soulange J, Bahut M, Payet B, Verpoorte R, Kodja H (2010) Shoot differentiation from protocorm callus cultures of *Vanilla planifolia* (Orchidaceae): proteomic and metabolic responses at early stage. BMC Plant Biol 10:82

Pandita D, Pandita A, Wani SH, Abdelmohsen SA, Alyousef HA, Abdelbacki AM, Elansary HO (2021) Crosstalk of multi-omics platforms with plants of therapeutic importance. Cells 10(6):1296

Pérez-Escobar OA, Dodsworth S, Bogarín D, Bellot S, Balbuena JA, Schley RJ, Drew BT, Chomicki G, Fambrini M, Sánchez-Vilas J, Cástor Oyola-Leal PA, Poulsen AD, Rosselló JA, Sayou C, Tizard J, Cascante-Marín A, Culham A, Baker WJ (2021) Hundreds of nuclear and plastid loci yield novel insights into orchid relationships. Am J Bot 108(7):1166–1180

Pérez-Escobar OA, Bogarín D, Przelomska NA, Ackerman JD, Balbuena JA, Bellot S, Bühlmann RP, Cabrera B, Cano JA, Charitonidou M, Chomicki G (2024) The origin and speciation of orchids. New Phytol 242(2):700–716

Phelps J, Biggs D, Webb EL (2016) Tools and terms for understanding illegal wildlife trade. Front Ecol Environ 14:479–489. https://doi.org/10.1002/fee.1325

Phillips RD, Reiter N, Peakall R (2020) Orchid conservation: from theory to practice. Ann Bot 126(3):345–362

Piet Q, Droc G, Marande W, Sarah G, Bocs S, Klopp C, Martin G, Sarah M, Poucheret P, Mournet P, Rouard M, Cibrián-Jaramillo A, Bellec A, Engelen S, Ortega-Abboud E, Risterucci AM, Omore A, Grapin A, Baurens FC, Rodier-Goud M, Guignon V, Lanaud C, Breton C, Summo M, De Lamotte F, Rivallan R, Leclercq J, Argout X (2022) A chromosome-level, haplotype-phased *Vanilla planifolia* genome highlights the challenge of partial endoreplication for accurate whole-genome assembly. Plant Commun 3:100330

Pillon Y, Fay MF, Shipunov AB, Chase MW (2006) Species diversity versus phylogenetic diversity: a practical study in the taxonomically difficult genus *Dactylorhiza* (Orchidaceae). Biol Conserv 129(1):4–13

Piñeiro Fernández L, Byers KJ, Cai J, Sedeek KE, Kellenberger RT, Russo A, Schlüter PM (2019) A phylogenomic analysis of the floral transcriptomes of sexually deceptive and rewarding European orchids, Ophrys and Gymnadenia. Front Plant Sci 10:1553

Pramanik D, Becker A, Roessner C, Rupp O, Bogarín D, Pérez-Escobar OA, Dirks-Mulder A, Droppert K, Kocyan A, Smets E, Grande B (2023) Evolution and development of fruits of *Erycina pusilla* and other orchid species. PLoS ONE 18(10):e0286846

Ramkumar TR, Kanchan M, Sembi JK (2020) Genome wide characterization of WUSCHEL-related homeobox (WOX) gene family in *Apostasia shenzhenica*, a primeval orchid. Plant Sci Today 4(7):164–171

Ramya M, Park PH, Chuang YC, Kwon OK, An HR, Park PM, Baek YS, Kang BC, Tsai WC, Chen HH (2019) RNA sequencing analysis of *Cymbidium goeringii* identifies floral scent biosynthesis related genes. BMC Plant Biol 19:337

Rao X, Krom N, Tang Y, Widiez T, Havkin-Frenkel D, Belanger FC, Chen F (2014) A deep transcriptomic analysis of pod development in the vanilla orchid (*Vanilla planifolia*). BMC Genomics 15:964

Rasjid N, Kurniawan FY, Putri SU, Linggabuwana A, Prasojo IS, Semiarti E (2023) In Silico analysis of Phalaenopsis Orchid Homeobox1 (POH1) functional gene for shoot development in Phalaenopsis orchid. J Trop Biodiver Biotechnol 8(3):83934

Ren R, Wei Y, Ahmad S, Jin J, Gao J, Lu C, Yang F (2020) Identification and characterization of NPR1 and PR1 homologs in Cymbidium orchids in response to multiple hormones, salinity and viral stresses. Int J Mol Sci 21:1977

Ren H, Ou Q, Pu Q, Lou Y, Yang X, Han Y, Liu S (2024) Comprehensive review on bimolecular fluorescence complementation and its application in deciphering protein-protein interactions in cell signaling pathways. Biomol Ther 14:859

Ren Z, Ji X, Jiao Z, Luo Y, Zhang GQ, Tao S, Lei Z, Zhang J, Wang Y, Liu ZJ, Wei G (2020) Functional analysis of a novel C-glycosyltransferase in the orchid *Dendrobium catenatum*. Hortic Res 1:7

Robbirt KM, Roberts DL, Hutchings MJ, Davy AJ (2014) Potential disruption of pollination in a sexually deceptive orchid by climatic change. Curr Biol 24(23):2845–2849

Rodgers L, Pernas T, Hill SD (2014) Mapping invasive plant distributions in the Florida Everglades using the digital aerial sketch mapping technique. IPSM 7(2):360–374

Roy SC, Moitra K, De Sarker D (2017) Assessment of genetic diversity among four orchids based on ddRAD sequencing data for conservation purposes. Physiol Mol Biol Plants 23(1):169–183

Russo A, Alessandrini M, El Baidouri M, Frei D, Galise TR, Gaidusch L, Oertel HF, Garcia Morales SE, Potente G, Tian Q, Smetanin D, Bertrand JAM, Onstein RE, Panaud O, Frey JE, Cozzolino S, Wicker T, Xu S, Grossniklaus U (2024) Genome of the early spider-orchid *Ophrys sphegodes* provides insights into sexual deception and pollinator adaptation. Nat Commun 15:6308

Schelkunov MI, Shtratnikova VY, Nuraliev MS, Selosse MA, Penin AA, Logacheva MD (2015) Exploring the limits for reduction of plastid genomes: a case study of the mycoheterotrophic orchids *Epipogium aphyllum* and *Epipogium roseum*. Genome Biol Evol 7(4):1179–1191

Šečić E, Zanini S, Wibberg D, Jelonek L, Busche T, Kalinowski J, Nasfi S, Thielmann J, Imani J, Steinbrenner J, Kogel KH (2021) A novel plant-fungal association reveals fundamental sRNA and gene expression reprogramming at the onset of symbiosis. BMC Biol 19:171

Sedeek KE, Qi W, Schauer MA, Gupta AK, Poveda L, Xu S, Schlüter PM (2013) Transcriptome and proteome data reveal candidate genes for pollinator attraction in sexually deceptive orchids. PLoS One 8(5):e64621

Sedeek KEM, Qi W, Schauer MA, Gupta AK, Poveda L, Xu S, Liu F, Courdavault V, Andrade-Navarro MA, Guimarães RA, Hampel M, Mithöfer A, Svatoš A, Becker A, Cozzolino S, Schiestl FP, Grossniklaus U (2016) Amino acid change in an orchid desaturase enables mimicry of the pollinator's sex pheromone. Curr Biol 26:1505–1511

Selosse MA, Schneider-Maunoury L, Martos F (2017) Time to re-think fungal ecology? Fungal ecological niches are often prejudged. New Phytol 213(3):968–972

Serna-Sánchez MA, Pérez-Escobar OA, Bogarín D, Torres-Jimenez MF, Alvarez-Yela AC, Arcila-Galvis JE, Hall CF, de Barros F, Pinheiro F, Dodsworth S, Chase MW, Antonelli A, Arias T (2021) Plastid phylogenomics resolves ambiguous relationships within the orchid family and provides a solid timeframe for biogeography and macroevolution. Sci Rep 11(1):6858

Sezik E (2022) Destruction and conservation of Turkish orchids. In: Biodiversity: biomolecular aspects of biodiversity and innovative utilization. Springer US, Boston, pp 391–400

Shao SY, Wang C, Han SW, Sun MH, Li S (2019) Phenanthrenequinone enantiomers with cytotoxic activities from the tubers of *Pleione bulbocodioides*. Org Biomol Chem 17:567–572

Sembi JK, Ghai D, Verma J (2019) Molecular interventions to ameliorate environmental stresses in orchids. In: Molecular Approaches in Plant Biology and Environmental Challenges, 449–474. Springer, Singapore

Song C, Wang Y, Manzoor MA, Mao D, Wei P, Cao Y, Zhu F (2022a) In-depth analysis of genomes and functional genomics of orchids using cutting-edge high-throughput sequencing. Front Plant Sci 13:1018029

Song C, Zhang Y, Chen R, Zhu F, Wei P, Pan H, Chen C, Dai J (2022b) Label-free quantitative proteomics unravel the impacts of salt stress on *Dendrobium huoshanense*. Fron Plant Sci 12(13):874579

Srittularak B, Likhtiwitayawuid K (2009) New bisbibenzyls from *Dendrobium falconeri*. Helv Chim Acta 92:740–744

Srittularak B, Anuwat M, Likhtiwitayawuid K (2011) A new phenanthrenequinone from *Dendrobium draconis*. J Asian Nat Prod Res 13:251–255

Su CL, Chen WC, Lee AY, Chen CY, Chang YC, Chao YT, Shih MC (2013) A modified ABCDE model of flowering in orchids based on gene expression profiling studies of the moth orchid *Phalaenopsis aphrodite*. PLoS One 8(11):e80462

Su CL, Chao YT, Yen SH, Chen CY, Chen WC, Chang YC, Shih MC (2013) Orchidstra: an integrated orchid functional genomics database. Plant Cell Physiol 54(2):e11

Suetsugu K, Yamato M, Miura C, Yamaguchi K, Takahashi K, Ida Y, Shigenobu S, Kaminaka H (2017) Comparison of green and albino individuals of the partially mycoheterotrophic orchid *Epipactis helleborine* on molecular identities of mycorrhizal fungi, nutritional modes and gene expression in mycorrhizal roots. Mol Ecol 26(6):1652–1669

Sukma D, Elina J, Raynalta E, Aisyah S, Aziz SA, Chan MT (2021) Analysis of the genetic diversity of Phalaenopsis orchids with single nucleotide polymorphisms and snap markers derived from the Pto GENE. SABRAO J Breed Genet 53(4):620

Swarts ND, Dixon KW (2009) Terrestrial orchid conservation in the age of extinction. Ann Bot 104(3):543–556

Takeda S, Nishikawa Y, Tachibana T, Higaki T, Sakamoto T, Kimura S (2025) Morphological and transcriptome analysis of the near-threatened orchid *Habenaria radiata* with petals shaped like a flying white bird. Plan Theory 14:393

Tan BC, Chin CF, Liddell S, Alderson P (2013) Proteomic analysis of callus development in Vanilla planifolia Andrews. Plant Mol Biol Report 31:1220–1229

Tenenbaum D (2016) KEGGREST: client-side REST access to KEGG. R Package Version 1(1)

Thant MT, Sritularak B, Chatsumpun N, Mekboonsonglarp W, Punpreuk Y, Likhitwitayawuid K (2021) Three novel biphenanthrene derivatives and a new phenylpropanoid ester from *Aerides multiflora* and their α-glucosidase inhibitory activity. Plan Theory 10:385

Thomsen PF, Willerslev E (2015) Environmental DNA – An emerging tool in conservation for monitoring past and present biodiversity. Biol Conserv 183:4–18

Toh C, Mohd-Hairul AR, Ain NM, Namasivayam P, Go R, Abdullah NAP, Abdullah JO (2017) Floral micromorphology and transcriptome analyses of a fragrant Vandaceous Orchid, *Vanda Mimi Palmer*, for its fragrance production sites. BMC Res Notes 10:554

Tong CG, Wu FH, Yuan YH, Chen YR, Lin CS (2019) High-efficiency CRISPR/Cas-based editing of *Phalaenopsis* orchid MADS genes. Plant Biotechnol J 18(4):889–891

Toubiana D, Fernie AR, Nikoloski Z, Fait A (2013) Network analysis: tackling complex data to study plant metabolism. Trends Biotechnol 31(1):29–36

Trávníček P, Chumová Z, Záveská E, Hanzlíčková J, Kupková L, Kučera J, Gbúrová Štubňová E, Rejlová L, Mandáková T, Ponert J (2021) Integrative study of genotypic and phenotypic diversity in the Eurasian orchid genus *Neotinea*. Front Plant Sci 12:734240

Tsai WC, Pan ZJ, Hsiao YY, Jeng MF, Wu TF, Chen WH, Chen HH (2008) Interactions of B-class complex proteins involved in tepal development in Phalaenopsis orchid. Plant Cell Physiol 49:814–824

Tsai WC, Fu CH, Hsiao YY, Huang YM, Chen LJ, Wang M, Liu ZJ, Chen HH (2013) OrchidBase 2.0: comprehensive collection of Orchidaceae floral transcriptomes. Plant Cell Physiol 54(2):e7

Tsai CC, Wu KM, Chiang TY, Huang CY, Chou CH, Li SJ, Chiang YC (2016) Comparative transcriptome analysis of *Gastrodia elata* (Orchidaceae) in response to fungus symbiosis to identify gastrodin biosynthesis-related genes. BMC Genomics 17:212

Tu M, Liu N, He ZS, Dong XM, Gao TY, Zhu A, Zhang SB (2025) Integrative omics reveals mechanisms of biosynthesis and regulation of floral scent in *Cymbidium tracyanum*. Plant Biotech J 23(6): 2162-2181

Vafaee Y, Mohammadi G, Nazari F, Fatahi M, Kaki A, Gholami S, Ghorbani A, Khadivi A (2021) Phenotypic characterization and seed-micromorphology diversity of the threatened terrestrial orchids: implications for conservation. S Afr J Bot 137:386–398

Valadares RBS, Perotto S, Santos EC, Lambais MR (2014) Proteome changes in *Oncidium sphacelatum* (Orchidaceae) at different trophic stages of symbiotic germination. Mycorrhiza 24(5):349–360

Valadares RB, Perotto S, Lucheta AR, Santos EC, Oliveira RM, Lambais MR (2020) Proteomic and transcriptomic analyses indicate metabolic changes and reduced defense responses in mycorrhizal roots of Oeceoclades maculata (Orchidaceae) collected in nature. J Fungi 6(3):148

Valadares RB, Marroni F, Sillo F, Oliveira RR, Balestrini R, Perotto S (2021) A transcriptomic approach provides insights on the mycorrhizal symbiosis of the Mediterranean orchid *Limodorum abortivum* in nature. Plan Theory 10:251

van Wijk KJ (2001) Challenges and prospects of plant proteomics. Plant Physiol 126(2):501–508

Vishwakarma SK, Singh N, Kumaria S (2023) Genome-wide identification and analysis of the PAL genes from the orchids *Apostasia shenzhenica*, *Dendrobium catenatum* and *Phalaenopsis equestris*. J Biomol Struct Dynam 41(4):1295–1308

Vitt P, Taylor A, Rakosy D, Kreft H, Meyer A, Weigelt P, Knight TM (2023) Global conservation prioritization for the Orchidaceae. Sci Rep 13(1):1–11

Wäldchen J, Mäder P (2018) Machine learning for image based species identification. Methods Ecol Evol 9(11):2216–2225

Wallace LE (2003) The cost of inbreeding in *Platanthera leucophaea* (Orchidaceae). Am J Bot 90:235–242

Wallace L (2006) Spatial genetic structure and frequency of interspecific hybridization in *Platanthera aquilonis* and *Platanthera dilatata* (Orchidaceae) occurring in sympatry. Am J Bot 93(7):1001–1009

Wan X, Zou LH, Zheng BQ, Wang Y (2019) Transcriptomic profiling for prolonged drought in *Dendrobium catenatum*. Sci Data 6:293

Wang W, Yu H, Li T, Li L, Zhang G, Liu Z, Huang T, Zhang Y (2017) Comparative proteomics analyses of pollination response in endangered orchid species *Dendrobium chrysanthum*. Int J Mol Sci 18(12):2496

Wang C, Tian M, Zhang Y (2021) Characterization of microRNAs involved in asymbiotic germination of *Bletilla striata* (Orchidaceae) seeds. Plant Physiol Biochem 167:163–173

Wang Y, Tong Y, Adejobi OI, Wang Y, Liu A (2022) Research advances in multi-omics on the traditional Chinese herb Dendrobium officinale. Front Plant Sci 12:808228

Wang Y, Wang H, Ye C, Wang Z, Ma C, Lin D, Jin X (2024) Progress in systematics and biogeography of Orchidaceae. Plant Divers 46(4):425–434

Wong DC, Peakall R (2022) Orchid phylotranscriptomics: the prospects of repurposing multi-tissue transcriptomes for phylogenetic analysis and beyond. Front Plant Sci 13:910362

Wong DC, Amarasinghe R, Rodriguez-Delgado C, Eyles R, Pichersky E, Peakall R (2017) Tissue-specific floral transcriptome analysis of the sexually deceptive orchid *Chiloglottis trapeziformis* provides insights into the biosynthesis and regulation of its unique UV-B dependent floral volatile, chiloglottone 1. Front Plant Sci 8:1260

Xiong L, Ou Y, Li P, Tian Y, Guo L, Li J, Li Y, Su B, Peng D, Liu Z, Ai Y (2025) Comparative physiology and transcriptome analysis revealed the mechanisms underlying red leaf formation in *Cymbidium ensifolium*. Sci Hortic 340:113933

Xu XB, Ma XY, Lei HH, Song HM, Ying QC, Xu MJ, Wang HZ (2015) Proteomic analysis reveals the mechanisms of *Mycena dendrobii* promoting transplantation survival and growth of tissue culture seedlings of *Dendrobium officinale*. J Appl Microbiol 118(6):1444–1455

Xu Q, Yang Z, Jia Y, Wang R, Zhang Q, Gai R, Wu Y, Yang Q, He G, Wu JH, Ming F (2024) PeNAC67-PeKAN2-PeSCL23 and B-class MADS-box transcription factors synergistically regulate the specialization process from petal to lip in *Phalaenopsis equestris*. Mol Hortic 4:15. https://doi.org/10.1186/s43897-023-00079-8

Yan L, Wang X, Liu H, Tian Y, Lian J, Yang R, Hao S, Wang X, Yang S, Li Q, Lu Y, Zhang G, Li J, Wong GK, Wang J, Wang Q, Rao Z, Liu ZJ (2015) The genome of *Dendrobium officinale* illuminates the biology of the important traditional Chinese orchid herb. Mol Plant 8:922–934

Yang X, Tang C, Zhao P, Shu G, Mei Z (2012) Antimicrobial constituents from the tubers of *Bletilla ochracea*. Planta Med 78(06):606–610

Yang FX, Gao J, Wei YL, Ren R, Zhang GQ, Lu CQ, Jin JP, Ai Y, Wang YQ, Chen LJ, Ahmad S (2021) The genome of *Cymbidium sinense* revealed the evolution of orchid traits. Plant Biotechnol J 19:2501–2516

Yang F, Guo Y, Li J, Lu C, Wei Y, Gao J, Xie Q, Jin J, Zhu G (2023) Genome-wide association analysis identified molecular markers and candidate genes for flower traits in Chinese orchid (*Cymbidium sinense*). Hortic Res 10(11):uhad206

Ye W, Shen CH, Lin Y, Chen PJ, Xu X, Oelmüller R, Yeh KW, Lai Z (2014) Growth promotion-related miRNAs in *Oncidium* orchid roots colonized by the endophytic fungus *Piriformospora indica*. PLoS One 9:e84920

Yoshikawa M, Murakami T, Kishi A, Sakurama T, Matsuda H, Nomura M, Kubo M (1998) Novel indole S, O-bisdesmoside, calanthoside, the precursor glycoside of tryptanthrin, indirubin, and isatin, with increasing skin blood flow promoting effects, from two *Calanthe* species (Orchidaceae). Chem Pharm Bull 46:886–888

Yu CH, Wang Y, Sun B, Zhang R, Zhu Y (2023) Analysis of four orchid protein-protein interactions networks to elucidate the mechanism of floral morphology diversity formation. https://doi.org/10.21203/rs.3.rs-2859543/v1

Yuan Y, Jin X, Liu J, Zhao X, Zhou J, Wang X, Wang D, Lai C, Xu W, Huang J, Zha L, Liu D, Ma X, Wang L, Zhou M, Jiang Z, Meng H, Peng H, Xiao W (2018) The *Gastrodia elata* genome provides insights into plant adaptation to heterotrophy. Nat Plants 4:731–741

Yuan Y, Zhang J, Wang Y, Ma C, Hu G, Fan J, Shen X (2020) Transcriptomic and proteomic analyses reveal mechanisms of drought tolerance in *Dendrobium huoshanense*. Plant Sci 298:110592

Yuan Y, Zuo J, Wan X, Zhou R, Xing W, Liu S (2024) Multi-omics profiling reveal responses of three major Dendrobium species from different growth years to medicinal components. Front Plant Sci 15:1333989

Zeng B, Su M, Chen Q, Chang Q, Wang W, Li H (2016) Antioxidant and hepatoprotective activities of polysaccharides from *Anoectochilus roxburghii*. Carbohydr Polym 153:391–398

Zeng X, Li Y, Ling H, Chen J, Guo S (2018) Revealing proteins associated with symbiotic germination of *Gastrodia elata* by proteomic analysis. Bot Stud 59:1–7

Zhan X, Qi J, Zhou B, Mao B (2020) Metabolomic and transcriptomic analyses reveal the regulation of pigmentation in the purple variety of *Dendrobium officinale*. Sci Rep 10(1):17700

Zhan X, Qian Y, Mao B (2022) Identification of two GDSL-type esterase/lipase genes related to tissue-specific lipolysis in Dendrobium catenatum by multi-omics analysis. Life 12(10):1563

Zhang GQ, Xu Q, Bian C, Tsai WC, Yeh CM, Liu KW, Yoshida K, Zhang LS, Chang SB, Chen F, Shi Y, Su YY, Zhang YQ, Chen LJ, Yin Y, Lin M, Huang H, Deng H, Wang ZW, Zhu SL, Zhao X, Deng C, Niu SC, Huang J, Wang M, Liu GH, Yang HJ, Xiao XJ, Hsiao YY, Wu WL, Chen YY, Mitsuda N, Ohme-Takagi M, Luo YB, Van de Peer Y, Liu ZJ (2016) The *Dendrobium catenatum* Lindl. genome sequence provides insights into polysaccharide synthase, floral development and adaptive evolution. Sci Rep 6:19029

Zhang GQ, Liu KW, Li Z, Lohaus R, Hsiao YY, Niu SC, Wang JY, Lin YC, Xu Q, Chen LJ, Yoshida K, Fujiwara S, Wang ZW, Zhang YQ, Mitsuda N, Wang M, Liu GH, Pecoraro L, Huang HX, Xiao XJ, Lin M, Wu XY, Wu WL, Chen YY, Chang SB, Sakamoto S, Ohme-Takagi M, Yagi M, Zeng SJ, Shen CY, Yeh CM, Luo YB, Tsai WC, Van de Peer Y, Liu ZJ (2017) The Apostasia genome and the evolution of orchids. Nature 549(7672):379–383

Zhang X, Zhang S, Gao B, Qian Z, Liu J, Wu S, Si J (2019) Identification and quantitative analysis of phenolic glycosides with antioxidant activity in methanolic extract of *Dendrobium catenatum* flowers and selection of quality control herb-markers. Food Res Int 123:732–745

Zhang C, Chen J, Huang W, Song X, Niu J (2021) Transcriptomics and metabolomics reveal purine and phenylpropanoid metabolism response to drought stress in *Dendrobium sinense*, an endemic orchid species in Hainan Island. Front Genet 12:692702

Zhang L, Chen Y, Ma Y, Chen L, Ma Y, Jin R, Liang S, Duan J (2022) Integration of the metabolome and transcriptome reveals indigo biosynthesis in *Phaius flavus* flowers under freezing treatment. PeerJ 10:e13106

Zhang T, Yang X, Wang F, Liu P, Xie M, Lu C, Fan B (2023) Comparison of the metabolomics of different *Dendrobium* species by UPLC-QTOF-MS. Int J Mol Sci 24:17148

Zhang T, Yu X, Liu D, Zhu D, Yi Q (2024a) Genome-wide identification, expression pattern and interacting protein analysis of INDETERMINATE DOMAIN (IDD) gene family in *Phalaenopsis equestris*. PeerJ 12:e18073

Zhang L, Yang H, Zheng M, Zhou G, Yang Y, Liu S (2024b) Physiological and transcriptomic analyses reveal the regulatory mechanisms of *Anoectochilus roxburghii* in response to high-temperature stress. BMC Plant Biol 24(1):584

Zhao X, Zhang J, Chen C, Yang J, Zhu H, Liu M, Lv F (2014) Deep sequencing-based comparative transcriptional profiles of *Cymbidium hybridum* roots in response to mycorrhizal and non-mycorrhizal beneficial fungi. BMC Genomics 15:747

Zhao X, Zhang M, He X, Zheng Q, Huang Y, Li Y, Ahmad S, Liu D, Lan S, Liu Z (2023) Genome-wide identification and expression analysis of the SPL gene family in three orchids. Int J Mol Sci 24(12):10039

Zheng R, Zhao K, Chen J, Zhu X, Peng Y, Shen M, Zhou Y (2025) Genomic signatures of SnRK s highlighted conserved evolution within orchids and stress responses through ABA signaling in the *Cymbidium ensifolium*. BMC Plant Biol 25:277

Zizka A, Silvestro D, Vitt P, Knight TM (2021) Automated conservation assessment of the orchid family with deep learning. Conserv Biol 35(3):897–908

Zotz G, Winkler U (2013) Aerial roots of epiphytic orchids: the velamen radicum and its role in water and nutrient uptake. Oecologia 171:733–741

The manufacturer's authorised representative in the EU is Springer Nature Customer Service Centre GmbH, Europaplatz 3, 69115 Heidelberg, Germany. If you have any concerns regarding our products, please contact ProductSafety@springernature.com

Printed and bound by CPI Group (UK) Ltd, Croydon, CR0 4YY
01/10/2026
02225923-0015